RARE HALOS, MIRAGES, ANOMALOUS RAINBOWS
AND RELATED ELECTROMAGNETIC PHENOMENA

A CATALOG OF GEOPHYSICAL ANOMALIES

Compiled by:

William R. Corliss

Published and Distributed by

The Sourcebook Project   P.O. Box 107   Glen Arm, MD 21057

Library of Congress Catalog Number 84-50491

ISBN 0-915554-12-7

First printing: May 1984

Printed in the United States of America

TABLE OF CONTENTS

LIST OF PROJECT PUBLICATIONS

CATALOGS:

Lightning, Auroras, Nocturnal Lights, and Related Luminous Phenomena (category GL)

Tornados, Dark Days, Anomalous Precipitation, and Related Weather Phenomena (category GW)

Earthquakes, Tides, Unidentified Sounds, and Related Phenomena (categories GH, GQ, GS)

Rare Halos, Mirages, Anomalous Rainbows, and Related Electromagnetic Phenomena (category GE)

HANDBOOKS:

The Unfathomed Mind: A Handbook of Unusual Mental Phenomena

Incredible Life: A Handbook of Biological Mysteries

Unknown Earth: A Handbook of Geological Enigmas

Mysterious Universe: A Handbook of Astronomical Anomalies

Ancient Man: A Handbook of Puzzling Artifacts

Handbook of Unusual Natural Phenomena

SOURCEBOOKS:

Strange Phenomena, vols. G1 and G2

Strange Artifacts, vols. M1 and M2

Strange Universe, vols. A1 and A2

Strange Planet, vols. E1 and E2

Strange Life, vol. B1

Strange Minds, vol. P1

NEWSLETTERS:

Science Frontiers (current anomaly reports)

Anomaly Register (Catalog addenda)

For information on the availability, prices, and ordering procedures, write:

SOURCEBOOK PROJECT
P.O. Box 107
Glen Arm, MD 21057

PREFACE

After more than ten years of scouring the scientific and semiscientific literature for anomalies, my major conclusion is that this is an amazingly fruitful activity. In fact, organized science should have been doing the same searching for the past 200 years. It is simply astounding that a Catalog of Anomalies does not already exist to guide scientific thinking and research. It is at least as important to realize what is not known as it is to realize what is well-explained. Nevertheless, here begins the first publication of a Catalog of Anomalies; the product largely of one person's library research, carried forward without a single public or foundation dollar.

Under the aegis of the Sourcebook Project, I have already published 20 volumes, comprising well over 7,000 pages, of source material on scientific anomalies. (See page iv for a list of titles.) As of this moment these 20 volumes represent only about 25% of my collection of scientific anomalies. New material is being added at the rate of about 1,200 new articles and items per year, 300 of which are from the current literature. These rates could be easily multiplied several times over by spending more time in the libraries. After ten years only the scientific journals of the United States and England have received my serious attention. There remain the English-language journals of the rest of the world, those publications in other languages, university theses, government reports, bulletins of scientific research facilities, conference papers, and not the least, books and newspapers. The cataloging task is just beginning, for the anomalies in the world's scientific and semiscientific literature seem nearly infinite in number.

Given this rough assessment of the magnitude of the anomaly literature, one can understand why the planned Catalog of Anomalies will require at least 25 volumes of about the same size as the one you now hold. I visualize a shelf of 25 volumes, with master indexes, to be only the initial step in providing scientists with ready access to what is not, in my opinion, well-explained.

Will the Catalog of Anomalies revolutionize science? Probably not---at least not immediately. Quite often the initial reaction to the books of anomalies already published has been disbelief. The data must be in error; the data are mainly testimonial; the data are too old; the supposed anomaly was explained long ago. Germs of truth reside in all these complaints. But for every anomaly or example that can be legitimately demolished, ten more take its place. Nature is very anomalous or, equivalently, Nature is not yet well-understood by science. Much remains to be done.

William R. Corliss

P.O. Box 107
Glen Arm, MD 21057
March 1, 1984

"ROUND ABOUT THE ACCREDITED AND ORDERLY FACTS OF EVERY SCIENCE THERE EVER FLOATS A SORT OF DUST-CLOUD OF EXCEPTIONAL OBSERVATIONS, OF OCCURRENCES MINUTE AND IRREGULAR AND SELDOM MET WITH, WHICH IT ALWAYS PROVES MORE EASY TO IGNORE THAN TO ATTEND TO . . . . ANYONE WILL RENOVATE HIS SCIENCE WHO WILL STEADILY LOOK AFTER THE IRREGULAR PHENOMENA. AND WHEN THE SCIENCE IS RENEWED, ITS NEW FORMULAS OFTEN HAVE MORE OF THE VOICE OF THE EXCEPTIONS IN THEM THAN OF WHAT WERE SUPPOSED TO BE THE RULES." **William James**

HOW THE CATALOG IS ORGANIZED

Purpose of the Catalog

The Catalog of Anomalies is designed to collect and categorize all phenomena that cannot be explained readily by prevailing scientific theories. Following its definition, each recognized anomaly is rated in terms of: (1) its substantiating data; and (2) the challenge the anomaly poses to science. Next, all examples of the anomaly discovered so far are noted, some in detail. Finally, all examined references are listed. Thus, the Catalog is a descriptive guide as well as a compendium of examples and references. Scientific researchers therefore have a ready-made foundation for beginning further investigations into these intriguing phenomena. This is the basic purpose of the Catalog: the collection and consolidation of the unknown and poorly explained to facilitate future research and explanation.

General Plan of the Catalog

It was tempting to organize this Catalog alphabetically, making it an "encyclopedia of anomalies." But many of the phenomena have obscure names or, even worse, no names at all. Under these circumstances, access to the data base would be difficult. Therefore, a system of classification was designed based upon readily recognized classes of phenomena and the means by which the observer detects them. Subject matter is first divided into nine general classes of scientific endeavor, as illustrated in the diagram. Few persons would have difficulty classifying a phenomenon as biological, astronomical, etc. The second, third, and fourth levels of classification are also based upon generally recognized attributes, such as luminosity, sound, etc. The similarity of this method of categorization to those employed in natural history field guides is purely intentional. Like bird identification, phenomenon classification soon becomes second nature. In fact, almost all of the phenomena described in the Catalog are accessible to everyone with five normal senses and perhaps binoculars and telescope.

Most catalogs boast numbering systems and this one is no exception. Rather than employ a purely numerical system, the first three levels of classification are designated by letters. The triplets of letters chosen have some mnemonic value. Thus, a GEH anomaly is easily recognized as being in the geophysics class (G), involving electromagnetic radiation (E), and having halo attributes (H). The number added to the triplet of letters marks the fourth level of classification, so that GEH2 signifies noncircular halos, the second type of anomalous halos. Every anomaly type has such a unique alphanumeric code. All cross references and indexes are based on this system. Catalog additions and revisions are made easier with this scheme.

The Catalog codes may seem cumbersome at first but their mnemonic value to the compiler has been significant. Hopefully, they will help other users, too. The codes are simple, yet flexible enough to encompass the several thousand types of anomalies identified so far.

A glance through this volume of the Catalog will reveal that each example of a specific anomaly type bears an X-number, and each reference an R-number. GEH2-X3 therefore specifies the third example of noncircular halos; and GEH2-R4 the fourth reference to noncircular halos. Indexes and cross references can consequently be made more precise than page numbers.

How Data and Anomalies Are Evaluated

Each anomaly type is rated twice on four-level scales for data "validity" and "anomalousness," as defined below. These evaluations represent only the opinion of the compiler and are really only rough guides.

Data Evaluation Scale

1 Many high-quality observations. Almost certainly a real phenomenon.

2 Several good observations or one or two high quality observations. Probably real.

3 Only a few observations, some of doubtful quality. Phenomenon validity questionable.

4 Unacceptable, poor-quality data. Such phenomena are included only for purposes of comparison or amplification. Only two such items are registered in this volume.

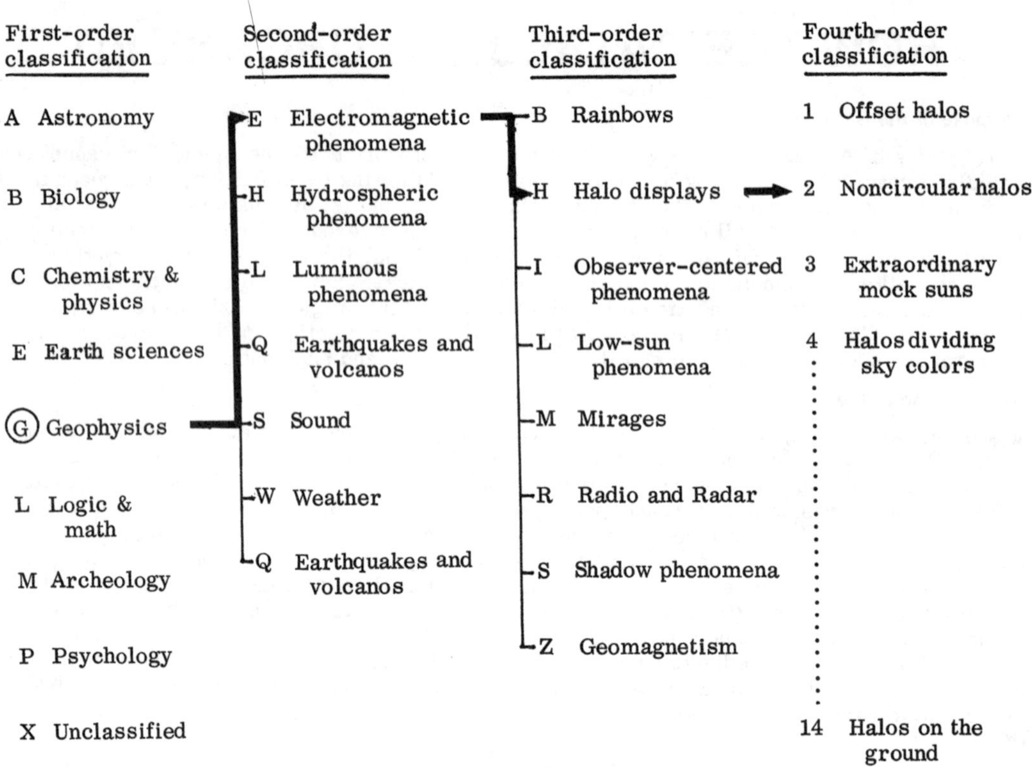

Catalog Coding Scheme

Anomaly Evaluation Scale

1 Anomaly cannot be explained by modifications of present laws. Revolutionary.

2 Can probably be explained through relatively minor modifications of present laws.

3 Can probably be explained using current theories. Primarily of curiosity value.

4 Well-explained. Included only for purposes of comparison and amplification.

Anomalies that rate "1" on both scales are very rare, with no entries at all in this volume of the Catalog, and only one in the first Catalog volume. Such anomalies, however, are the most important because of their potential for forcing scientific revolutions. As additional Catalog volumes are published, the relative proportion of "double-1s" will increase, especially in the fields of biology and psychology. The following matrix summarizes the ratings assigned in this volume.

Anomaly Ratings

Data Ratings	1	2	3	4
1-	0	8	14	6
2-	0	16	14	4
3-	1	15	10	0
4-	1	1	0	0

Total: 90 anomaly types

Anomaly Examples

Examples of anomaly types are designated by the letter X in the body of the Catalog. All examples discovered so far are listed. If the example is of the event type, time and place are specified if available. Some of the more significant geophysical events, such as the remarkable Haicheng earthquake of 1975, may appear in several references, all of which are proivded in full and keyed to the X-number of the example. Powerful earthquakes are represented by many examples, but only a single example of an electrostatic phenomenon associated with an earthquake has been found. When library research has made many examples of an anomaly type available, several of the most interesting and instructive are reported in some detail. Direct quotations from the eye-witness accounts are frequently employed to convey accurately the characteristics of the phenomenon to the reader. These selected examples are printed full-width in the Catalog columns. Examples of less import are indented to separate them visually.

The References and Sources

Each anomaly type and the examples of it are buttressed by all references that have been collected and examined. Since some references describe several examples, each reference includes the X-numbers of the examples mentioned. When a reference covers more than one type of anomaly, it is repeated in the bibliography following each anomaly type. Actually, there is little repetition of this sort in the Catalog.

Perusal of the Source Index will demonstrate that the great majority of references come from the scientific literature. Heavily represented in this volume of the Catalog are: Nature, Science, Monthly Weather Review, Meteorological Magazine, and the Quarterly Journal of the Royal Meteorological Society. Several less technical publications are also frequently mentioned: Marine Observer, Scientific American, and English Mechanic. The Marine Observer is a publication of the U.K. Meteorological Office. The English Mechanic was for many years an important English technical magazine, much like Scientific American. All of the serials mentioned so far are considered generally reliable, although one must always be wary. In addition to these often-referenced journals, there is a very wide selection of other publications. Some of these are more popular in character. In this context, it should be remembered that unusual phenomena do not seek out scientists, and the laymen who observe many anomalies have no knowledge of or access to scientific journals.

The time span covered by the sources ranges over 150 years. Many excellent reports of anomalies come from the latter half of the Nineteenth Century. Although many scientists frown on such old reports, the quality is often good, and they should not be discarded arbitrarily. Not only were the observers of a century ago competent, they were unbiased by knowledge of modern theories and lived in an environment unpolluted by modern vehicles, effluxes, and other contaminants. It should also be mentioned here that the modern meteorological literature tends heavily toward mathematical modelling and theory as opposed to eye-witness accounts of strange phenomena. Two journals that are exceptions to this rule are the Marine Observer and the U.K. Journal of Meteorology.

The Indexes

Each Catalog volume concludes with five separate indexes. At first glance this may seem to be too much of a good thing. But in the context of a science-wide catalog of anomalous events and unusual natural phenomena, each idex has its special utility.

The subject index is essential in any work of this type. It is placed last for easy access. The time-of-event and place-of-event indexes are analytical tools for the anomaly researcher. They help identify phenomena that are reported separately (perhaps in widely different journals) but are really different aspects of the same event. This integrating feature will become more apparent as additional Catalog volumes appear. To illustrate, the 1908 Tunguska Event produced luminous and barometric phenomena as well as atmospheric optical effects. It is possible that composite time-of-event and place-of-event indexes covering all scientific fields will reveal cause-and-effect relationships that have not been recognized before. It is the intent of the Catalog effort to generate a composite set of indexes when all four Catalog volumes in geophysics are published as well as a final master index covering all of science.

The source index shows immediately the dependence of this Catalog upon scientific literature rather than newspapers and other popular publications. Its real purpose, though, is the rapid checking to determine if a specific reference has or has not been caught already in the fishing nets of this Catalog project. The source index is doubly valuable because many footnotes and bibliographies in the scientific literature display sources only; that is, titles and authors are omitted entirely. The researcher also comes across many vague references to such-and-such an article by so-and-so back in 1950 in Nature. The exhaustive and rather ponderous source and first-author indexes can help pin down many references lacking specifics.

All five indexes use the catalog codes described above rather than page numbers. The codes are permanent whereas the page numbers will change as addenda and revised volumes are produced. The mnenonic value of the catalog codes is useful here, too, because the approximate nature of each index entry is readily apparent, while page numbers give only location.

Supporting Publications of the Sourcebook Project

The Catalog volumes are actually only distillations of huge quantities of source material. The Anomaly Data Research Center/Sourcebook Project has already published 16 volumes of source material, as detailed on page iv.

Catalog Addenda and Revisions

Over 1,000 new reports of anomalies are collected from current and older scientific journals each year. New anomaly types and additional examples of types already cataloged are accumulating rapidly. When new material assignable to a Catalog volume already in print is acquired, it appears first in the serial ANOMALY REGISTER, published frequently by the Anomaly Data Research Center. (Ordering information on page iv.) When sufficient new material has been assembled in a Catalog area, that volume of the Catalog will be revised and published with all additions and corrections incorporated.

Request for Additions and Corrections

The Anomaly Data Research Center welcomes reports of new anomalies and examples of unusual phenomena. If you discover a report of an anomaly not listed in the Catalog, send a xerox of the article to the Anomaly Data Research Center for evaluation. If the report is added to our data base, credit will be given to the submitter in the ANOMALY REGISTER and the revised edition of the Catalog. Reports from recognized scientific journals are preferred but everything is grist for the mill! The address of the Anomaly Data Research Center is: P.O. Box 107, Glen Arm, MD 21057.

INTRODUCTION TO ELECTROMAGNETIC PHENOMENA IN THE ATMOSPHERE

The atmosphere plays many curious optical tricks, from sideways mirages to all-red rainbows to radar "angels". The venerable laws of optical reflection, refraction, and color dispersion are adequate to explain the great majority of these frequently breathtaking displays of colors, shadows, halos, and distorted images. Undoubtedly, the same laws will ultimately explain most of the rarer phenomena described below, but the theoretical attempts made so far have not been completely satisfactory. How does one account for elliptical halos, kaleidoscopic suns, long-delayed radio echos, and all the trickery of mirages?

Actually, only a few scientists take much interest in atmospheric optical phenomena nowadays. They are for the most part content that these "minor" esoterica can be brought into the fold of the scientifically explained with little effort, that we already know all that is really important about light, color, and the propagation of radio waves in the atmosphere. This attitude is, of course, smug and dangerous. Some of the ephemeral phenomena of this book pinpoint areas of genuine scientific ignorance---perhaps not so much ignorance of the basic laws of optics as of the fine structure of our atmosphere and its interfaces with the earth's surface and outer space.

It is regrettable that so many of the irregular atmospheric phenomena transpire at sea, on mountaintops, in the polar reaches, and in other remote places. Only a favored few witnesses can marvel at these natural nuances of sunlight and moonlight. By necessity, this Catalog volume is mainly the testimony of mariners, explorers, and those who frequent the remote, open places of the world.

GEB RARE RAINBOWS AND ALLIED SPECTRAL PHENOMENA

Key to Phenomena

GEB0 Introduction

Everyone knows the rainbow and just about everyone also knows that it is produced by sunlight falling on a veil of raindrops. In more scientific terms, the sunlight (or moonlight) is dispersed into the spectral colors by refractions and reflections inside the transparent raindrops. Two internal reflections create the primary bow; three reflections produce the larger, rarer secondary bow with a reverse color sequence. Such commonplace rainbows are excluded here as unmysterious, but this still leaves offset rainbows, distorted rainbows, sandbows, moving rainbows, and many other intriguing color phenomena associated with naturally dispersed water droplets.

How do all these variations of the basic rainbow phenomenon arise? The classical rainbow situation may be complicated by several factors. Most important is the presence of an additional source of light, which usually is due to the reflection of sunlight from an unruffled body of water. Vertically displaced rainbows are just one result of reflected sunlight; one can imagine ripples and swell introducing additional distortion effects. A low sun is a second source of distortion, because its light has lost some of the normally present violets and blues due to scattering through the longer atmospheric path. The presence of thin, intervening sheets of rain also superimpose new colors and bows on the basic rainbow. Even with these known perturbations, many rainbow phenomena cannot be explained using conventional rainbow theory.

Since droplets of water are the source of spectral bows, we should look for colored bows on cloud tops (from aircraft or mountain tops), on fields covered with dew, and wherever nature spreads out layers of transparent, spherical droplets. In searching for rainbow phenomena, one must not be led astray by the closely related halos and glories described elsewhere in this Catalog (GEH). Although even an expert may sometimes be perplexed as to how to classify an enigmatic strip of colors in the sky. It might be a segment of rainbow or a portion of some rare form of halo.

GEB1 Unusual Multiple Rainbows

Description. Rainbows or segments thereof appearing in addition to the ordinary primary and secondary bows. These extra rainbows are displaced vertically upwards or downwards; their colors may be in the same order as those of the primary or reversed. Such additional bows should not be confused with supernumerary bows (GEB15), which are much narrower than the primary, show only a few colors, and appear just inside the primary. Extra bows that can be explained in terms of a second source of illumination---usually the sun reflected from a placid water surface---are eliminated in most instances as rare but explicable.

Data Evaluation. A handful of good observations with careful descriptions of color order, bow position, and optical environment. Rating: 2.

Anomaly Evaluation. If it is assumed that "reflection bows" have been eliminated (except X1), we have rainbows that do not yield to standard rainbow theory. The phenomena portrayed below, in fact, may encompass several different effects. It is impossible to say at this time whether they are the results of unrecognized optical phenomena and thus important anomalies or whether some second source of illumination has been overlooked. Rating: 2.

Possible Explanations. Extra bows may be created when the sun is reflected from water, ice, and even glass in manmade structures. The possibility also exists that two or more curtains of raindrops at different distances from the observer may create more than one bow along the same axis.

Similar and Related Phenomena. Supernumerary bows (GEB15); horizontally displaced or offset rainbows (GEB6); intersecting rainbows (GEB2).

Examples of Unusual Multiple Rainbows

X1. July 9, 1792. Alverstoke, England. "Some ninety years ago a fragmentary quarternary bow was seen. The observer was a Mr. Sturge; the place Alverstoke, on the seacoast of Hampshire; and the time evening, July 9, 1792, with the sun low in the horizon to the N. W. Portions of the primary and secondary were accompanied by fragments of tertiary and quarternary bows, as in the accompanying diagram. I have dotted the continuations of these fragments to give an idea of what would have been seen if the atmospheric conditions had been favourable. Where the secondary and tertiary bows overlap, there would have been, as in the case of Halley's observation, a portion of a white arch." (R1) The remarkable aspect of this item is the extreme rarity of the duplicate of the secondary bow, formed undoubtedly by a reflection of the sun. Note also that a vertically displaced reflection primary bow could intersect a normal secondary bow. See GEB2. (WRC)

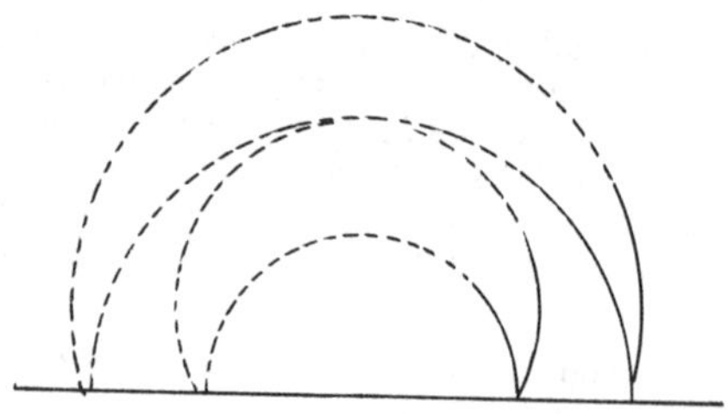

The two inner bows are the primary (inside) and reflected primary (outside). The outer bows are the secondary and reflected secondary. Only the solid lines were seen. (X1)

X2. 1873. England. "I send a sketch of two rainbows, eccentric to each other, even at the same time, a few evenings since, just before sunset. An explanation of the phenomenon is simple. A magnificent rainbow was being refracted from a dense shower, when a narrow (if I may use the term) shower came in front of the denser, showing a rainbow on itself, but not intercepting, only refracting, the sun's rays on and from the further shower." (R2) Most multiple rainbows are explained in terms of sunlight reflected from a nearby water surface; but here we have a new twist. (WRC)

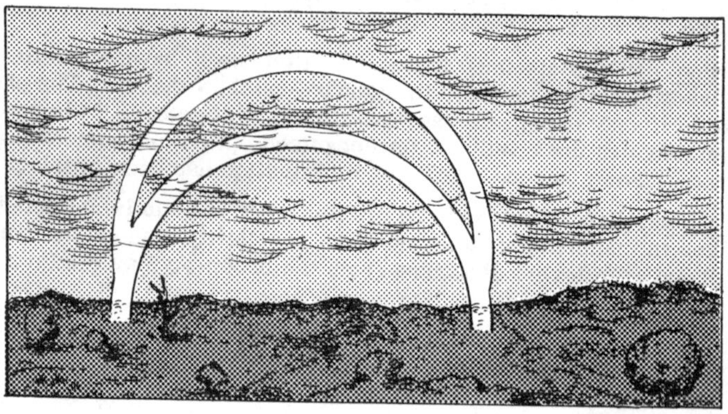

Multiple rainbows apparently created by two showers (X2)

X3. November 5, 1913. Umtali, Rhodesia. "Yesterday, November 5, at 5.40 p.m., a little previous to the sun disappearing behind the mountains, I noticed in the east six brilliant rainbows, extending for about one-quarter of the semicircle. Five were quite close together, and the sixth some little distance away as per accompanying rough sketch." (R3) These extra bows do not appear to be supernumerary because they are well-separated from the (assumed) primary bow on the right, they are equal to the primary in width, and they show all the colors. (WRC)

X4. March 25, 1955. North Atlantic Ocean. "Observed part of a rainbow, bearing between 078° and 110° (the sun bearing 255°), against a background of dark Cb cloud without anvil. The colours were very pronounced and were repeated in an inverted form. Altitude of sun 22 1/2°.....<u>Note</u>. This interesting observation is of an abnormal rainbow. The colour sequence of the upper half of the combined bow corresponds in colour sequence to that of a primary bow. The lower half could not be a supplementary rainbow since the colour of this would be in the same order as in the primary bow, viz. red on top. Furthermore, a supplementary bow is much narrower than the primary bow and seldom shows more than two colours, green and red." The note was added by the <u>Marine Observer</u> editor. No figure was published with the article. (WRC) (R4)

X5. November 26, 1961. Indian Ocean. "At

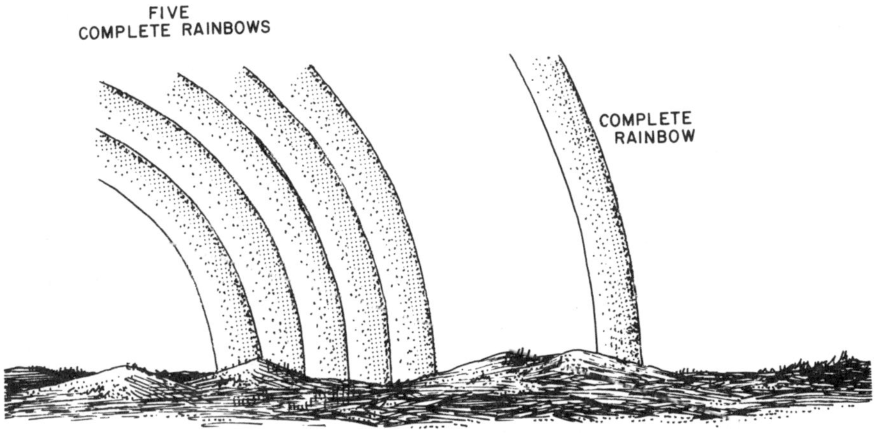

Multiple rainbows of enigmatic origin (X3)

0030 GMT a line squall was observed ahead of the vessel, moving approx. SE'S; a single rainbow was seen to form which very clearly showed the colours of the spectrum. Ten min. later as the vessel was passing through the line squall, then lying NE'N-SW'W, a triple rainbow was seen directly ahead. All three bows had the same order of colours ---red on the outside and violet inside. The colours of the primary (A) were very clear, almost vivid; this was the broadest of the three bows and it curved down almost to the sea surface, on each side of the forward part of the vessel. This bow was reflected on the sea surface, but the other two were not. At 0045 there was a moderate shower which lasted for 3 min., after which the bows disappeared. There was no wind and the sea was like a mirror, but with a moderate swell from 150°.....Note. m.v. City of Khartoum observed a most unusual rainbow phenomenon. Bow A in the diagram appears to be the primary bow. Bow B, covering more than a semi-circle, appears to have been caused by the reflection of the sun's light at the smooth sea surface before its being refracted by the raindrops. Bow C is similar to a secondary bow but with a reversed order of colours. We cannot explain this." (R5)

X6. June 24, 1976. Aspendale, Australia. Abstract. "The positions of the primary (P) and secondary (S) arcs with the horizon and the implied apparent center were consistent with the altitude and azimuth of the Sun; moreover, there appeared a third arc just above the horizon between the P and S arcs, having the same point of contact with the horizon as P. The observed phenome-

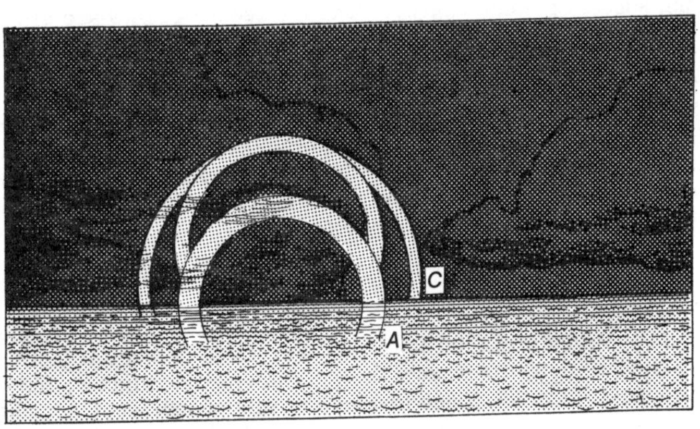

Triple rainbow (X5)

non is explained as being part of a primary bow resulting from the scattering by raindrops of sunlight reflected from the water surface between the Sun and the observer. Reflection at a wavy surface with low angle incidence occurs with maximum intensity at an angle of reflection equal to the angle of incidence and in the vertical plane containing the Sun, the observer, and the image Sun." (R6)

X7. June 6, 1979. Reading, England. "On 6 June 1979, at about 6 45 pm, I observed an unusual rainbow from the University athletics pavilion at Whiteknights, Reading. A heavy shower had just passed towards the south-east and a rainbow with primary and secondary bows was visible. The primary bow appeared to be double, being formed of two bows---one being displaced upwards by the width of a single bow from the other. Both bows had the same radius, this being shown by the two bows merging near the horizon. After a minute or two the uppermost bow faded leaving the ordinary single primary bow. Both bows were complete, with bright colours, and this event was not to be confused with a distinct supernumary bow." (R7)

References

R1. Ackroyd, William; "Singular Rainbow," Knowledge, 2:282, 1882. (X1)
R2. Mackaeg; "Rainbow," English Mechanic, 16:410, 1873. (X2)
R3. "Six Rainbows Seen at Once," Royal Meteorological Society, Quarterly Journal, 40:75, 1914. (X3)
R4. Harrison, P.P.O.; "Rainbow," Marine Observer, 26:15, 1956. (X4)
R5. O'Neill, F.C.; "Triple Rainbow," Marine Observer, 32:185, 1962. (X5)
R6. Garratt, J.R.; "Anomalous Rainbow," Weather, 35:12, 1980. (X6)
R7. Judd, S.; "Double Primary Rainbow," Weather, 34:445, 1979. (X7)

GEB2 Intersecting Rainbows

Description. Bows or segments of bows that intersect normal primary and/or secondary rainbows. The center of curvature of the intersecting arcs, which may be highly colored, is above the horizon; while that of the normal bows is below. The stage is thus set for intersection. Most often, only a single segment of the intersecting bow will rise from near the horizon toward the secondary bow, if present. Either or both of the normal primary and secondary bows may be cut by the arc. This phenomenon should not be confused with offset white arcs that intersect normal bows (GEB3, GEB6).

Data Evaluation. Intersecting rainbows have been well-observed on numerous occasions. Rating: 1.

Anomaly Evaluation. All of the observations collected so far can be explained by the presence of reflecting sheets of water nearby. The intersecting bows are simply reflection rainbows with centers of curvature higher than those of the normal bows. Rating: 4.

Possible Explanations. See above.

Similar and Related Phenomena. Multiple rainbows (GEB1); white arcs intersecting rainbows (GEB3, GEB6); halo phenomena consisting or intersecting arcs that are sometimes colored, but which are generally confined to the vicinity of the sun, whereas most rainbows are viewed with the sun at one's back. See rainbow prismatic pillars (GEB16).

Examples of Intersecting Rainbows

X1. August 10, 1821. Valley of Valders, Norway. Reflected primary intersected normal primary. (R3, R5) Since intersecting rainbows are common and easily explained, only one example (X10) will be detailed. (WRC)

X2. 1844. Clyde River, England. (R4)

X3. 1859. Byson, Illinois. (R1) Although termed a rainbow, the description of this phenomenon sounds more like a halo display with sun dogs. (WRC)

X4. 1874. St. Andrews, Scotland. A spectacular engraving accompanies this

report. (R8)
X5. June 1875. Sweden. (R2)
X6. 1876. Widness, England. (R6)
X7. May 13, 1886. Prairie du Chien, Wisconsin. (R6)
X8. July 14, 1911. Tezpur, India. (R7)
X9. May 10, 1919. Europe. (R9)

X10. August 12, 1930. Solway Firth, Scotland. "I observed an unusually brilliant primary rainbow together with a secondary with numerous spurious bows on the under side of the southorn-most prtion of the primary. The bows were not complete as there was an interrupting cloud layer at approximately 2,000 ft. level. The important part, however, of the observation was the appearance of an intersecting bow between the southern extremity of the primary and the secondary bow, the intersection at the primary being sharp at the extremity whilst the intersection and ending of the secondary bow was less sharp, inclining towards a fade out, the intersection here taking place near the cloud level. The colours of the intersecting bow were brilliant and of the same order as those of the primary. The whole phenomena on the southern portion of the rainbow resembled the letter N, but with the connecting stroke in the reverse direction, as shown in the sketch. Such an intersecting bow arises from the presence of two sources of parallel rays. In the present instance one set was that coming directly from the sun giving the primary and the secondary bows and the other set from the image of the sun in the bay to the westward of the point of observation. The image being as much below the horizon as the sun was above it, the centre of the intersecting bow was thus as much above the horizon as the centre of the primary was below it." (R10)

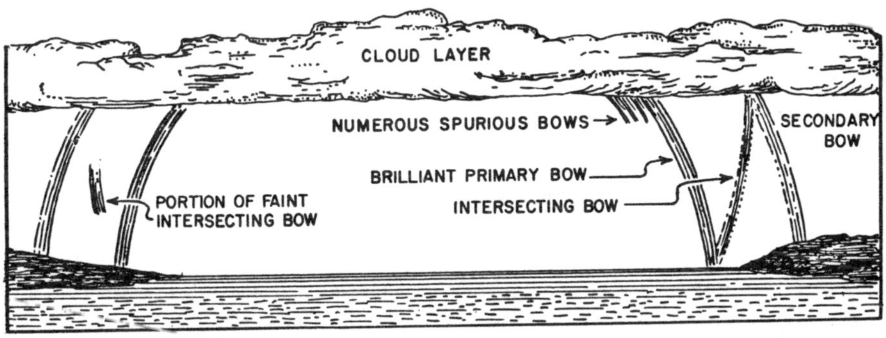

Intersecting bows caused by reflection from the water (X10)

X11. December 19, 1950. Jamaica. (R12)
X12. No date given. Herne Bay, England. (R11)
X13. No date given. Nya Kopparberg, Sweden. (R8)

References

R1. "Atmospheric Phenomenon," Scientific American, 14:157, 1859. (X3)
R2. English Mechanic, 22:297, 1875. (X5)
R3. Geelmuyden, H.; "Intersecting Rainbows," Observatory, 11:229, 1888. (X1)
R4. "Intersecting Rainbows," Observatory, 11:326, 1888. (X2)
R5. Geelmuyden, H.; "Intersecting Rainbows," Sidereal Messenger, 7:219, 1888. (X1)
R6. Hagen, John G.; "Intersecting Rainbows," Sidereal Messenger, 7:266, 1888. (X6, X7)
R7. Hucklebridge, T.; "A Remarkable Rainbow," English Mechanic, 94:68, 1911. (X8)
R8. Talman, C. Fitzhugh; "Little-Known Rainbows," Scientific American, 106:305, 1912. (X4, X13)
R9. Fryer, J. H.; "Unusual Rainbow," Symons's Meteorological Magazine, 54:88, 1919. (X9)
R10. Crichton, J.; "Intersecting Rainbows," Meteorological Magazine, 65:213, 1930. (X10)
R11. "A Triple Rainbow," Nature, 132:437. 1933. (X12)
R12. Myhura, H.; "Unusual Rainbow Phenomena," Weather, 6:219, 1951. (X11)

GEB3 Lunar Rainbows with Offset White Arcs and Bows

Description. Complete bows or segments thereof that appear in conjunction with normal lunar rainbows but offset from them in azimuth. Such anomalous bows may intersect the normal bow or be displaced as much as 180° in azimuth. They are normally white. All observed to date have been at sea, predominantly in the Indian Ocean.

Background. Lunar rainbows are seldom seen because the moon's light is so much weaker than the sun's. When they do become visible, they are generally pale and white. Secondary lunar rainbows are exceedingly rare, and the offset bows reported here even more so.

Data Evaluation. The handful of reports available are from ships at sea. Usually they are carefully recorded, complete with angular measurements and weather data. Rating: 1.

Anomaly Evaluation. Offset lunar rainbows are not "recognized" optical phenomena; that is no reasonable explanation exists in terms of current optical theory. Rating: 2.

Possible Explanations. Offset bows exactly 180° from the normal rainbow; that is, in the direction of the moon; could be tertiary bows, although such bows are hardly ever seen even in strong sunlight. Note that offset rainbows cannot be explained in terms of reflected light sources as vertically displaced bows are. Some unrecognized halo phenomena may be involved.

Similar and Related Phenomena. Similar offset bows sometimes appear with solar rainbows (GEB6); anomalous halo phenomena (GEH).

Examples of Lunar Rainbows with Offset White Arcs and Bows

X1. July 29, 1947. English Channel, North Atlantic Ocean. "The moon, three-quarters full, with an approximate altitude of 25°, was observed dead ahead. Astern of the ship and covering some 3/10 of the sky was a mass of white cumulus. Ahead, was about 2/10 stratocumulus, through a gap in which the moon shone. The remainder of the sky was clear. Immediately astern of the ship was observed a lunar rainbow of a startling bright milky hue and, just ahead, another, larger but not so vividly bright as the one astern. These bows appeared to be not much more than the ship's length apart (600 ft.) and both were completely visible throughout their entire arcs. There had been no rain or drizzle for the previous eight hours, the sky being mainly clear of cloud. Neither had there been any suspicion of fog or haze, although these had been experienced 12 hours earlier..... Note.---This observation of the second bow, that ahead of the ship and thus between the moon and the observer, is a very remarkable one, but in the absence of angular measurements it cannot be fully explained. Two points of interest arise: (i) a (tertiary) rainbow is known to be formed in the position stated but it is so faint that observations of it are extremely rare---furthermore, according to optical theory it should be of approximately the same size as the primary rainbow, whereas the bow observed was larger; (ii) if conditions were so exceptional that this tertiary bow was fairly bright, one would have expected that the ordinary secondary bow, seen above and outside the primary bow, would have been readily visible. It does not, however, appear to have been seen." (R1) Since the two bows are offset 180°, two separate phenomena may be involved; also, the tertiary bow is really extremely rare. (WRC)

X2. November 10, 1948. North Atlantic Ocean. "Two complete lunar rainbows were observed and one arc of a third bow. The lower complete bow showed the spectrum colors very distinctly, but the upper bow and remaining arc were white. Bearing NNE, height of the centre 28°.... Note. This is a very interesting observation. The secondary lunar rainbow is not very often seen, as it is usually too faint to be visible. The most remarkable part of the observation is the arc of the third bow, which is an example of an abnormal bow. Such bows are very rarely seen and cannot be explained on the basis of the ordinary optical theory of rainbows." (R2)

X3. January 1, 1953. Off the Seychelles, Indian Ocean. "After several hours of squally showers in the vicinity, and periods

Primary and secondary lunar rainbows with offset white arc (X2)

of slight rain at the vessel from thick Cu and Cb clouds, a lunar rainbow was observed with clear colours, violet, yellow and green. This formation remained pronounced for a few minutes only. A second arc which formed beside the full moon was white and higher in the sky than the first, although its arch was not so wide and tended to fade near sea level. Nevertheless it was plainly visible for 20 min..... Note. This is an interesting observation of a lunar rainbow showing clear colours. The white arc observed near the moon is very remarkable. It is not a rainbow nor is it any of the recognised halo phenomena. Occasional observations have been made of halos or arcs which are not centred on the sun or moon and which do not occupy the known positions of halo phenomena in other parts of the sky. Unfortunately the observation gives no indication as to whether there was any thin cirrostratus or other form of upper cloud in the region of the moon." (R3)

X4. June 16, 1956. Indian Ocean. "At 1855 G. M. T. a rainbow-like arc of pure white light appeared, bearing approximately 060°. Up to an altitude of 40° above the horizon it was quite strong and easily distinguished; above this it gradually lost its brilliance and finally disappeared from view at an altitude of 75°, bearing approximately 140°. The arc appeared to be behind thin Sc cloud rather than superimposed upon it, as, at its brightest point, viz, near the horizon, no cloud obscured the arc. To the westward the sky was clear and the moon bore approximately 260°. This phenomenon persisted until 1913 and then disappeared in a matter of seconds, the upper part of the arc instantaneously and the lower part gradually. Bright moonlight throughout." (R4) This mysterious observation is included with lunar rainbows because it occurred in the opposite direction of the moon. (WRC)

X5. August 31, 1963. Indian Ocean. "Di-

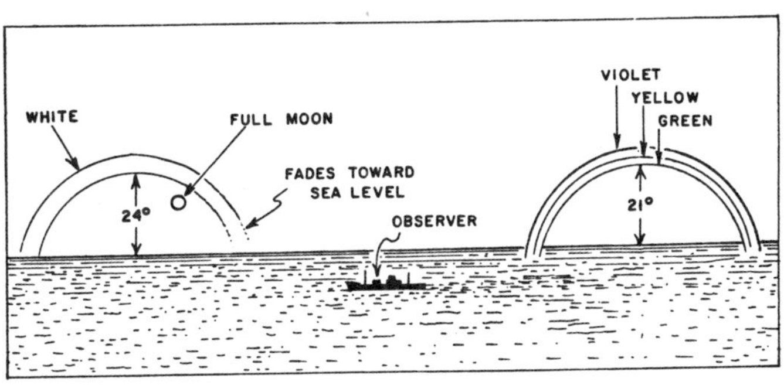

Lunar rainbow with anomalous white arc near moon (X3)

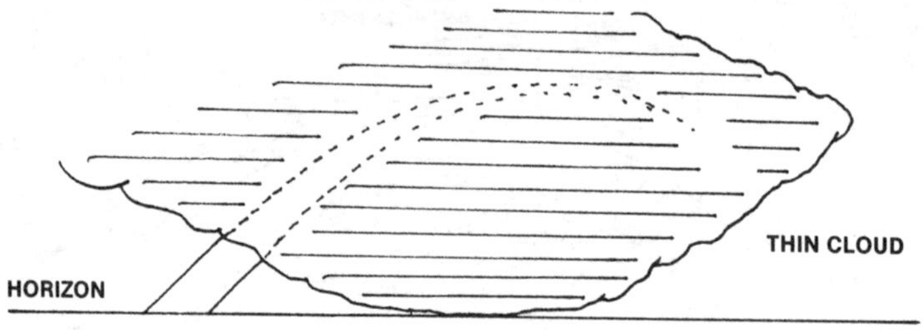

Arc of pure white light, Indian Ocean (X4)

rectly after a shower of rain had passed at 2315 GMT a bright lunar rainbow was seen, having an altitude of about 25°. A very pale bow which seemed to have a slightly larger radius, intersected the first bow as shown in the sketch. Both were of very short duration enabling a measurement of the altitude of the bright bow, only, to be made. Neither bow showed any colouring." (R5)

References

R1. Owen, G.; "Lunar Rainbow," Marine Observer, 18:145, 1948. (X1)
R2. Robertson, G. M.; "Lunar Rainbows," Marine Observer, 19:188, 1949. (X2)
R3. Greig, J. W.; "Lunar Rainbow and Arc," Marine Observer, 24:13, 1954. (X3)
R4. Wood, J. F.; "Unidentified Phenomenon," Marine Observer, 27:87, 1957. (X4)
R5. Collett, C. T.; "Intersecting Lunar Rainbows," Marine Observer, 34:181, 1964. (X5)

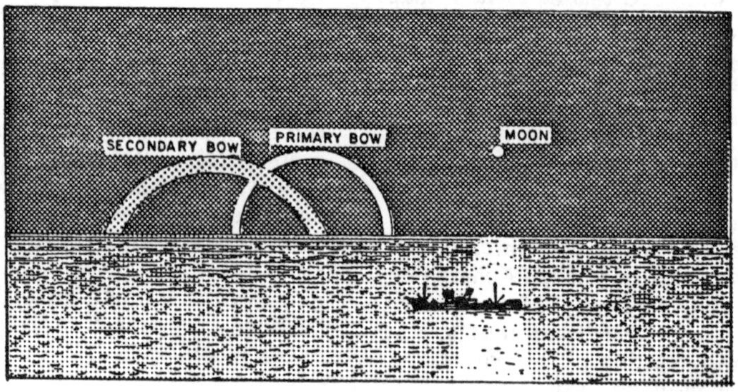

Lunar rainbow with offset bow. The moon, of course, was 180° from the primary bow. (X5)

GEB4 Red Rainbows: Rare But Explicable

Description. Rainbows that are red or pink, with possibly a touch of green. Since most colors are absent, red rainbows are much narrower than normal rainbows. Seen only at sunrise and sunset.

Data Evaluation. Although rare, red rainbows are easily explained, and the few observations recorded in the scientific literature are fully accepted by science. Rating: 1.

Anomaly Evaluation. Since red rainbows yield easily to optical theory, no anomaly exists. This entry is included because of the general interest in the phenomenon and its possible relevance to other rainbows occurring when the sun is low. Rating: 4.

Possible Explanations. When the sun is low in the sky, the shorter wavelengths in sunlight are scattered by the atmosphere, leaving the longer wavelengths---predominantly red---to form a rainbow.

Similar and Related Phenomena. Low-sun phenomena (GEL); purple rainbows (GEB14); distorted rainbows (GEB18).

Examples of Red Rainbows

X1. June 30, 1922. Hove, England. (R2)
X2. October 8, 1928. Harlingham, England. (R3)
X3. August 18, 1972. Polson, Montana, (R5)
X4. No date given. Fermoyle, England. (R1)

X5. No date given. Wellfleet, England. "There was a sharp shower for about 15 min just as the sun was sinking toward a cloudless western horizon. Looking toward the east from a point about 80 ft above Wellfleet Bay, I saw an unusually brilliant primary bow at its maximum altitude, with one end in the water and the other end on the shore. As I watched, the blue, green, yellow and orange portions were quickly wiped out, the entire operation taking place in not more than 1 sec. There remained a bow of a single color, red, only slightly less brilliant than before, and in width about a quarter of that of the original bow. I turned to the west and found that the sun had disappeared completely below the horizon, though the familiar red afterglow was still strong. No further change took place in the appearance of the rainbow for, perhaps, 30 sec, when it suddenly vanished. A bow of but a single color was visible when there was no sun in the sky!" (R4)

References

R1. English Mechanic, 66:317, 1897. (X4)
R2. Harries, H.; "A Red Rainbow," Meteorological Magazine, 57:246, 1922. (X1)
R3. Russell, Spencer; "A Pink Rainbow," Meteorological Magazine, 63:234, 1928. (X2)
R4. Palmer, Frederic; "Unusual Rainbows," American Journal of Physics, 13:203, 1945. (X5)
R5. Thompson, A. H.; "Water Bows: White Bows and Red Bows," Weather, 29:178, 1974. (X3)

GEB5 Moving Rainbows

Description. Rainbows that move laterally (in azimuth)with respect to a stationary observer, or seem relatively stationary to a moving observer, or in some way violate the fixed angular relationship between the sun, the observer, and the rainbow stipulated in rainbow theory.

Data Evaluation. A very few observations have been uncovered, most of which are of rather poor quality. Rating: 3.

Anomaly Evaluation. Any dynamic change in the sun-observer-rainbow angular relationship (except for that due to a moving sun) is anomalous. Fortunately, prevailing rainbow theory permits at least one simple explanation of small movements of bows. Rating: 3.

Possible Explanations. Lateral mirages (GEM3), although mildly anomalous in their own right, might well account for slight shifts in rainbow azimuths.

Similar and Related Phenomena. Lateral mirages (GEM3); jumping halos (GEH7).

Examples of Moving Rainbows

X1. January 24, 1851. England. "While riding on top of an omnibus from Plymouth to Redruth, I observed a phenomenon which I described as a stationary rainbow. Rain beginning to fall, an imperfect rainbow was produced; one half of the bow only was seen, the foot of the arch resting on a hill to which the omnibus was approaching, and passed over, to a considerable extent, a field on which some cattle were grazing. As the omnibus approached the field in question, the rainbow appeared to shift it position but not to recede from the observer; it approached the right hand, the observer facing it by about 50 yards. As we neared the field in which it was seen we came nearer to it, and it certainly did not in any degree appear in any way to recede, but was, with the exception alluded to, perfectly stationary, and it thus continued until we arrived even with the field, when we saw it at our side, immediately upon which it disappeared." (R1) One is not supposed to be able to approach the foot of a rainbow, since the arch is supposed to maintain a specific angular relationship with the observer and the sun. Thus, even though the rainbow mentioned above is called "stationary", it moved from where theory says it should have been! (WRC)

X2. December 16, 1952. Benbecula Airport, Outer Hebrides, United Kingdom. "The arcs observed were travelling slowly through the straits separating Benbecula from North Uist. The weather had been cloudy with moderate rain from 8 oktas of nimbostratus at 1,500 ft. with patches of fractostratus at 1,000 ft., the wind at the time being 180° 15 kt. At 1055 G.M.T. a break in the cloud became visible in the south-west, and at 1115 the wind veered to 280° 12-15 kt. indicating the passage of a trough. After this, the wind dropped slightly and the rain became intermittent, and during this period a faint but complete rainbow was observed at 1125 between Benbecula and North Uist. At 1135 an arc formed in the opposite direction to the original bow which had faded and now had only an arc composed of the western half of the bow, red being uppermost in both arcs. Subsequent developments are shown in Fig. 1, red being uppermost in all the arcs shown. In this diagram the lengths of the arcs and the angles between the bows are only approximations." (R3)

X3. No date given. India. "I had just taken some readings with a theodolite in the early morning when an incomplete rainbow appeared in the sky above the point which I had been observing. The sky was clear, with the exception of a few clouds in the neighborhood of the shower responsible for the bow. I elevated the telescope to look at it; as I did so the bow began to move rapidly out of the field of view so that I had to depress the instrument to follow it. The movement was about 3°. I could detect no change in colour sequence or in radius, though the final bow seemed brighter and was complete. The movement lasted less than thirty seconds." A mirage (refraction) effect was suggested as the cause of the motion. (R2)

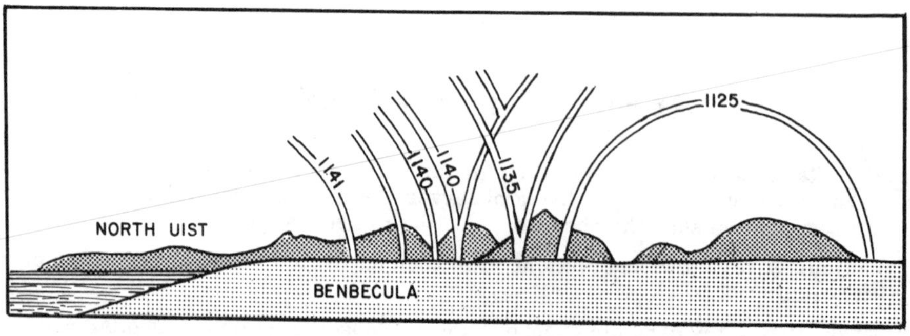

Moving rainbows in the Outer Hebrides. Numbers give times. (X2)

References

R1. Birt, W.R.; "Remarkable Rainbows," English Mechanic, 12:519, 1871. (X1)
R2. "Moving Rainbows," Nature, 115:548, 1925. (X3)
R3. Watson, L.E.; "Rainbow Phenomenon," Meteorological Magazine, 82:218, 1953. (X2)

GEB6 Solar Rainbows with Offset White Arcs

Description. Segments of offset white bows that intersect normal solar rainbows. All observations collected to date have been at sea.

Background. A similar phenomenon, possibly identical in origin, occurs with lunar rainbows (GEB3). In the lunar case, complete white bows offset by as much as 180° appear in conjunction with normal lunar rainbows.

Data Evaluation. A few carefully made observations have been found in the nautical literature. Rating: 2.

Anomaly Evaluation. As the notes appended to the observations recorded below emphasize, the white bows have no ready explanation using current optical theory. Rating: 2.

Possible Explanations. Abnormal halo phenomena (GEH).

Similar and Related Phenomena. White bows and arcs offset from lunar rainbows (GEB3); abnormal halo phenomena, such as offset halos (GEH).

Examples of Solar Rainbows with Offset White Arcs

X1. October 31, 1950. Off Florida coast, North Atlantic Ocean. "Off Mollasses Reef an arc of a rainbow was observed extending from right to left for approximately a quarter of a semi-circle. At the same time a white arc intersected the other as shown in the accompanying sketch. Note. This is an unusual and interesting observation and no explanation can be given of the cause of the white bow intersecting the primary rainbow." (R1)

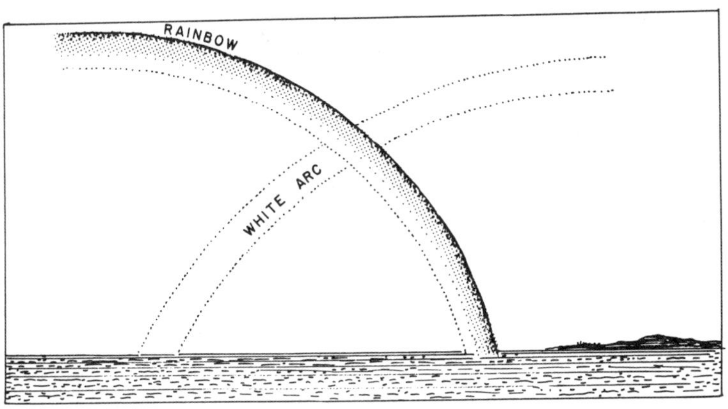

Solar rainbow with offset white arc off Florida (X1)

X2. October 22, 1956. South Atlantic Ocean. "At 0802 G. M. T. during generally cloudy conditions giving light showers of rain near the vessel, the end of a rainbow of large radius was seen on the horizon, the rest of the bow being hidden by low Sc (sketch 1). The bow was 4° wide and fairly bright. Reading from the inside, the colours were red, yellow, green, blue, yellow, white, pink and red. The sun was then clear of clouds. At 0805, the colour sequence from the inside was yellow, purple, blue, green, orange and red. By now the sun was shining through thin As. At 0806, during showers in the area, the bright, white ends of a second bow became visible inside the rainbow (sketch 2). The ends of these bands were 1° wide. At 0813 yet another fainter white bow, 2° wide, appeared outside the other. The bearings then were as shown in sketch 3. When the rain ceased at 0820, the original rainbow disappeared, as did the second white arc, leaving the first two faintly visible (sketch 4). At 0833 these also faded out..... Note. This is a very interesting observation. The colours of the main rainbow seem to be somewhat abnormal even if allowance is made for possible supernumerary bows inside and in contact with it. The white rainbows are abnormal phenomena which cannot be explained on the ordinary geometrical theory which accounts for the primary, secondary and supernumerary bows." (R2)

References

R1. Eckford, R. D. S.; "Rainbow and White Arc," Marine Observer, 21:215, 1951. (X1)

R2. Arthur, R. B.; "Rainbows," Marine Observer, 27:210, 1957. (X2)

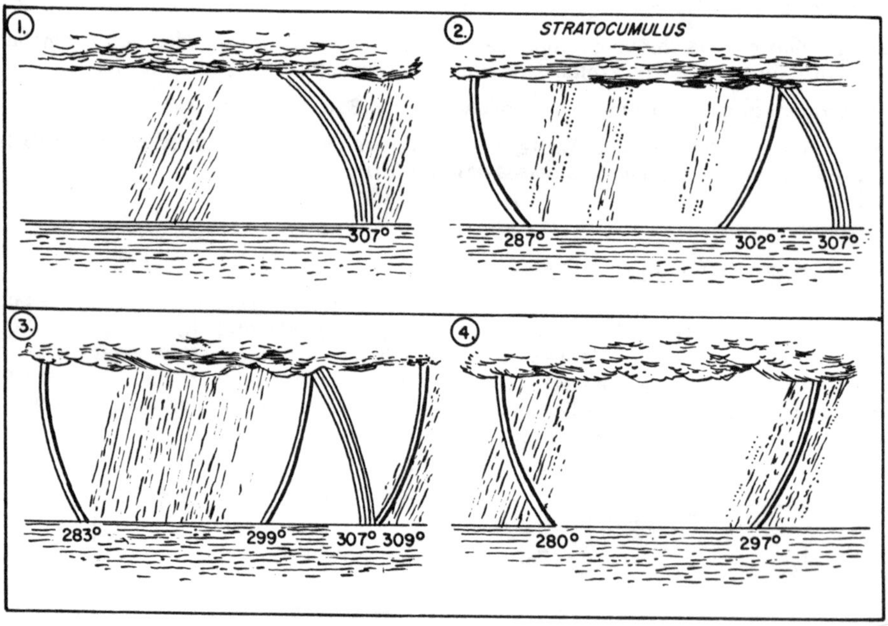

South Atlantic rainbows with anomalous white arcs (X2)

GEB7 Lunar Rainbows Transforming to Disks

Description. Primary lunar rainbows which broaden downward, changing into segments of disks. The disks are white.

Data Evaluation. One observation so far, but it seems to be of good quality. Rating: 3.

Anomaly Evaluation. Rainbow theory does not account for the expansion of a bow into a partial disk. Rating: 2.

Possible Explanations. Since lunar rainbows involve low light levels, it is possible that some sort of psychological contrast effect may be involved here.

Similar and Related Phenomena. Some abnormal halo phenomena incorporate disks (GEH); luminous patches on the horizon (GLA15).

Examples of Lunar Rainbows Transforming into Disks

X1. September 22, 1923. Lake Windermere, England. "At 7h, 30m, on September 22nd, passengers about to embark on a steamer at Bowness, on Lake Windermere, observed a lunar rainbow, the red and violet tints of which were faintly visible. Later, as we steamed southwards towards Lakeside, standing alone on deck, I saw another duller bow appear outside the first bow. The two bows now spanned from earth to earth. Then to my surprise, the inner and brighter bow broadened centripetally, losing its colours and extending downwards until it became a white half disc, like a gigantic half moon, still fringed by the outer bow. Lastly, I watched both outer bow and luminous half-disc fade away. At no time did I see any colour in the outer bow as it was too faint." (R1)

References

R1. Wallee, J. C.; "A Curious Lunar Rainbow," Meteorological Magazine, 58:232, 1923. (X1)

GEB8 Radial Streaks Crossing Rainbows

Description. Bright and dark radial streaks that intersect both primary and secondary rainbows, converging on the antisolar point, which is identical to the center of all normal rainbows. This point is below the horizon for normal rainbows. These streaks are often visible in the open sky outside the secondary and inside the primary. Reports are conflicting as to whether they cross the dark space between primary and secondary.

Data Evaluation. Although rarely reported in the literature, radial streaks are probably more common than the bibliography suggests. A careful examination of the rainbow often reveals them, especially when the sun is low. Experienced scientists have made many of the observations, and there seems no reason to doubt the phenomenon. Rating: 1.

Anomaly Evaluation. Dark and light streaks are often observed converging on the antisolar point in the absence of rainbows. Called antisolar rays or anticrepuscular rays, they seem identical to the radial streaks of rainbows. (See GEL2.) Antisolar rays intersecting rainbows pose three minor puzzles: (1) Why do some of the rays have a changing, cloudy, aurora-like appearance; (2) Why is there such divergent opinion as to whether the streaks cross or do not cross various parts of the rainbow display; and (3) What is the significance of the brightening at the intersection of the streaks with anomalous white arcs accompanying some rainbows (X6)?

Possible Explanations. Radial streaks crossing rainbows almost certainly derive from antisolar rays (themselves mildly enigmatic). The bright streaks, representing stronger regions of illumination may enhance those portions of the rainbow that they cross.

Similar and Related Phenomena. Antisolar rays (GEL2); crepuscular rays (GEL2); radial streaks in halo displays (GEH11).

Examples of Radial Streaks Crossing Rainbows

X1. August 13, 1857. Belfast Lough, Northern Ireland. "At about 6 P.M., after a thunderstorm---which was the second on that day---a rainbow was formed. The phenomenon to be described was seen about the left-hand part of the rainbow. The sky outside the rainbow was of that purple-black which accompanies thunder-storms. Inside it was lighter, and diversified by alternate dark and bright rays, like those of an aurora, which appeared to converge to a point a little above the centre of the rainbow, but below the horizon. These rays fluctuated like those of an aurora, but much less rapidly, their changes being scarcely quick enough to be visible. The dark rays extended through the rainbow, almost blotting it out where they crossed it. The rays were continued outside the rainbow at the period of their greatest intensity, but not all the time, and were never so distinct outside as inside. The dark rays were of the same purple black colour as the sky outside the rainbow; the light rays were like what we call "watery sunshine." This appearance lasted in all about three-quarters of an hour." (R1) In this example, the crepuscular ray explanation seems quite adequate. (WRC)

X2. August 23, 1866. Washingtonville, New York. A brilliant primary separated from the secondary by purplish band. "A third curious feature of this scene was the fact that the primary or lower arc was in many places crossed and obscured by what seemed to be clouds which had arranged themselves in the form of radii of the circle of which the arc was a part. Their form was not that of rays of the sun as seen through clouds at sunset, but rather that of the irregular streamers or patches of aurora borealis, except that they were of the color of the other clouds, and had not the motion of auroral flashes. Had these streaks occurred only at the crown of the arc or vertical to the earth, it might be thought they were the ordinary appearance of rain falling from a distant cloud; but, in fact, they were much more plentiful at the ends of the arc than at top, and were unmistakably arranged with reference to the center of the circle. They also continued as long as the arc was visible." (R2) The patchy, aurora-like character of the rays is unusual. (WRC)

X3. July 8, 1877. Summit of the Drachenfels, Switzerland. This and the following two observations were made by a scientist to demonstrate that careful study of rainbows would reveal many cases of radial streaks. (R3, R4)

X4. July 15, 1877. Lucerne, Switzerland. (R3)

X5. July 16, 1877. Sarnen, Switzerland. Each streak corresponded to a patch of sunlight on the hills behind the rainbow. (R3)

X6. July 14, 1958. Indian Ocean. "At 0215 G.M.T. a complete rainbow of normal brilliance was seen, reaching an elevation of 25°, accompanied by a portion of the secondary which also extended, on the left side of the primary, from the horizon to about 25° elevation. The secondary lasted for 2 min. Inside the primary, on the right hand side and parallel to it, a portion of a colourless bow, which persisted for 5 min, extended from the horizon---the spacing between the two bows being 7° of arc. A number of crepuscular rays crossed the primary and secondary, while another set crossed the primary and the colourless bow. A very interesting feature was the well-marked band of colouring that occurred in the place where the crepuscular rays and the colourless bow overlapped. The order of colouring was the same as in the primary---i.e. with red on the outside." (R6)

X7. General observations. Thompson (R3) has never seen radial streaks cross the dark space between the primary and secondary bows or the colored portions of bright bows. He has,however, seen bright radial streaks cross faint bows, with the points of intersection showing brighter colors. Minnaert (R5) on the other hand states that radial streaks appear in the diffuse area between primary and secondary bows. The streaks in X6 crossed both bows as well as the interbow space. Obviously, some differences of opinion exist.

References

R1. Murphy, J.J.; "Account of an Instance of Converging Rays Seen at Greenisland ....;" Report of the British Association, 1857, p. 35. (X1)

R2. Leute, J.R.; "A Meteorological Phenomenon," Scientific American, 14:200, 1866. (X2)

R3. Thompson, Silvanus P.; "On Certain Phenomena Accompanying Rainbows," Philosophical Magazine, 5:6:272, 1878. (X3-X5)

R4. Thompson, Silvanus P.; "On Certain Phenomena Accompanying Rainbows," Report of the British Association, 1878, p. 450. (X3)

R5. Minnaert, M.; The Nature of Light and Color in the Open Air, New York, p. 179. (X7)

R6. Harrison, P.P.O.; "Unusual Rainbow," Marine Observer, 29:114, 1959. (X6)

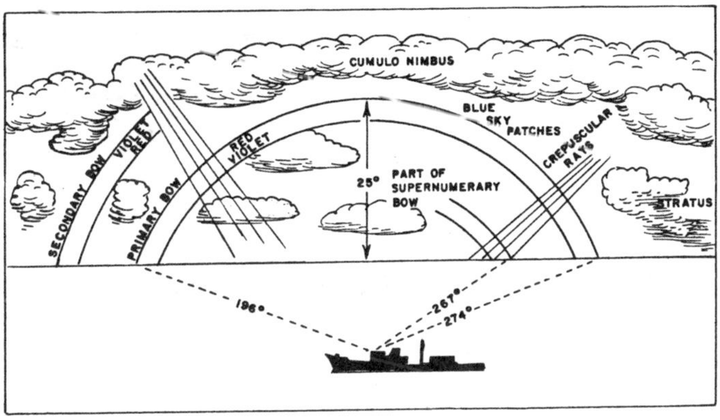

Radial streaks (crepuscular rays here) crossing rainbows (X6)

GEB9 Rainbows Perturbed by Thunder and Lightning

Description. The blurring of rainbow colors and definition during wave-like perturbations following thunder and lightning. Secondary rainbows seem more sensitive than primaries.

Data Evaluation. A single published observation. Rating: 3.

Anomaly Evaluation. Upon reflection, one is not surprised that sound waves and electric field perturbations by lightning discharges affect something as delicate as a rainbow. The issue is really which physical mechanism is responsible for the effect. Three have been proposed below. Rating: 3.

Possible Explanations. Sound waves force the combining of water droplets; the change in electric field strength reduces droplet surface tension, facilitating merger; the shock waves from thunder temporarily reduce the droplet population in the thin region where the rainbows are formed.

Similar and Related Phenomena. Dark bands crossing halos (GEH11); visible shock waves. (GES2).

Examples of Rainbows Perturbed by Thunder and Lightning

X1. Summer 1908. Finland. "A remarkable phenomenon was observed during a thunder shower in Finland in the summer of 1908 by V.J. Laine, who was making meteorological observations for the Finnish scientific society. The shower approached from the east and the thunder, which had been heard at intervals during a half hour, ceased before any rain fell at the place of observation. During approximately the same half hour the eastern horizon was completely spanned by a double rainbow, which likewise vanished before the rain came. Immediately after each peal of thunder the colors of both bows, and especially those of the secondary bow, became confused and indistinct, and the whole double rainbow appeared to vibrate rapidly. Laine explains this singular phenomenon in accordance with the Airy-Pernter theory of the rainbow, by assuming that the concussion of the air which produced the sound of thunder, also increased the diameter of the raindrops from less than 1/250 inch to between 1/50 and 1/25 inch." (R1, R2) Minnaert thinks the above explanation unlikely and proposes that the surface tension of the rain drops is affected by the lightning, allowing the drops to merge together more easily. (R3)

References

R1. Scientific American, 102:138, 1910. (X1)

R2. "Influence of Thunder on Size of Raindrops," American Journal of Science, 4:29:190, 1910. (X1)

R3. Minnaert, M.; "The Effect of Lightning on a Rainbow," The Nature of Light and Color in the Open Air, New York, 1954, p. 182. (X1)

GEB10 Anomalous Fogbows or White Rainbows

Description. Radial structures and small prismatic bows observed in fogbows that cannot be accounted for by rainbow theory.

Background. While not at all uncommon in foggy climes, fogbows are rather startling to the uninitiated, for they appear close at hand and rather "spooky." Fogbows have smaller angular diameters than normal rainbows, but the width of the band is much broader. Most fogbows or white rainbows are pure white, although sometimes some orange can be detected on the outside, blue inside. Supernumerary bows frequently appear with fogbows, but the order of colors is opposite that in the normal colored primary; that is,the red is inside rather than outside. If rainbow theory is applied to very tiny water droplets rather then the large drops typical of summer showers, the fogbows and its supernumeraries can be accounted for nicely. Since ordinary fogbows are not anomalous, the focus here is on a few idiosyncracies of fogbows that have been recorded.

Data Evaluation. Anomalous fogbows seem to be one-of-a-kind. The observations are old and provide little detail. Rating: 3.

Anomaly Evaluation. Rainbow theory, as modified for very small droplets, does not encompass the radial and prismatic structures presented below. Rating: 2.

Possible Explanations. None.

Similar and Related Phenomena. Radial streaks in normal colored rainbows (GEB8).

Examples of Anomalous Fogbows or White Rainbows

X0. January 7, 1928. North Atlantic Ocean. "A single halo was observed at intervals, opposite the moon, i.e., 180°. The angle between the horizon and upper limb varied in size from about 6° to 10°, the lower limb appearing to reach to side of the ship. Note. A fog bow is formed in the same manner as an ordinary rainbow by refraction and reflection of light from the sun or moon within drops of water. In the case of fog the phenomenon is produced by the very small drops of water suspended in the air. It can be proved that when the drops of water are very small, either in the form of fog or very fine rain, the resulting bow is white and not coloured as in the ordinary rainbow, because of the feebleness and overlapping of the colours." (R3)

X1. January 1808. Potomac River, Virginia. "Early in the morning a milk-white fog came on, so thick that the captain of the packet found it necessary to anchor, not knowing where he was. About half-past eleven he came up to Sir George, and remarked that they should have all clear soon, 'for the fog-eater was come.' The captain explained him-

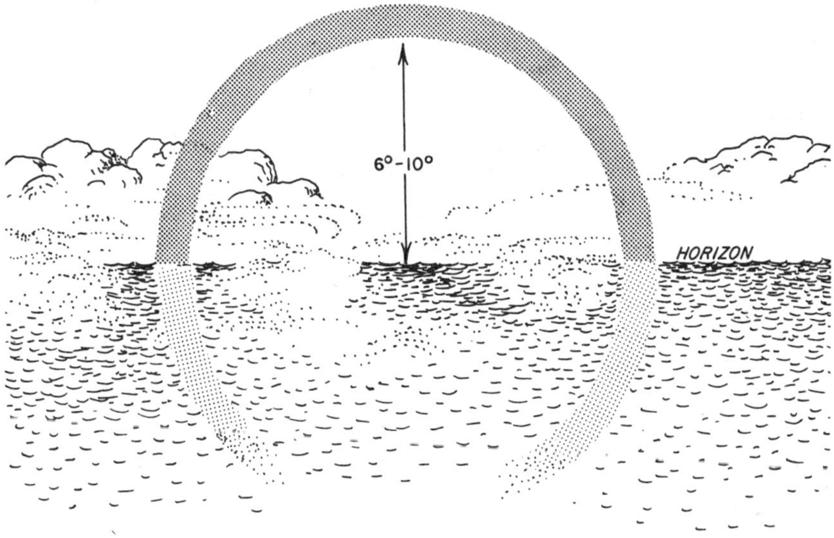

Typical fogbow. Note the small radius. (X0)

self by pointing to the head of the vessel, where there was visible a ring of thicker white fog than that in which they were enveloped, apparently about sixty feet in diameter, the belt of the ring appearing about two feet broad. Within this ring was another, two feet in diameter, suspended in its centre, and with prismatic colours. It lasted about 20' or 30', after which the fog cleared away." (R1) Normal fogbows do not have small prismatic bows at their centers. Supernumerary bows would have been much larger. (WRC)

X2. July 14, 1876. Kanturk, Ireland. Small needles or lines of grey or whitish light were observed crossing the arcs of the supernumerary bows inside a fogbow. (R2)

References

R1. Brewster, David; "On Fog-Rings Observed in America," Report of the British Association, 1845, p. 19. (X1)

R2. Burton, Philip; "An Account of Unusual Coloured Bows Observed in Fogs," Report of the British Association, 1884, p. 656. (X2)

R3. Gardner, G. F.; "Fog Bow," Marine Observer, 6:11, 1929. (X0)

GEB11 Anomalous Dewbows, Cloudbows, and Horizontal Rainbows

Description. Prismatic bows and strips of colors that appear very close to the observer and are usually spread out across a horizontal surface. The structures may have unexplained geometries and color sequences, and they may move relative to the viewer. Water surfaces and bedewed fields produce many of these unexpected spectral manifestations.

Background. Hyperbolic dewbows are well-understood (X0); so are elliptical bows seen from high altitudes (X8).

Data Evaluation. The observations of deviant dewbows and horizontal rainbows are uncommonly good and are often accompanied by carefully made drawings. The phenomenon, however, possesses many facets, the majority of which have been recorded only once. Rating: 2.

Anomaly Evaluation. Several attributes of these bows are not covered by standard rainbow theory: (1) their frequent nearness (within arm's reach); (2) their occasional motion relative to the observer; (3) their radical departures from the predicted hyperbola with its apex at the observer's feet; and (4) their occasional flanking dark and light bands. Rating: 2.

Possible Explanations. None.

Similar and Related Phenomena. Heiligenschein (GEI2); moving rainbows (GEB5); rainbows parallel to the horizon (GEB13); grossly distorted rainbows (GEB18); circumzenithal arcs which may adorn the sky with colored patches resembling rainbows (GEH).

Examples of Anomalous Dewbows and Horizontal Rainbows

X0. General observations. "The heather, on autumn mornings, is covered with millions of cobwebs, which, otherwise unnoticeable, are now sprinkled with dew-drops and lit up by the sunbeams. In that play of light we may see a rainbow standing out on the ground in front of us, not as a circle, but in the shape of a wide-open hyperbola. The explanation is simple: the light reaches our eye from all directions, forming an angle of 42° with the axis through the sun and our eye. The cone so formed intersects the surface of the ground in a hyperbolic curve as long as the sun is low. It would in the course of the day become an ellipse, although this has very seldom been seen..... The dew-bow has also been observed in the following circumstances: (a) On a pond covered with duckweed; on a lawn. (b) On a pond with an oily surface on which the dew-drops can lie without mixing with the water; a surface of this kind may be formed, for instance, by the sooty smoke from factories. In one case

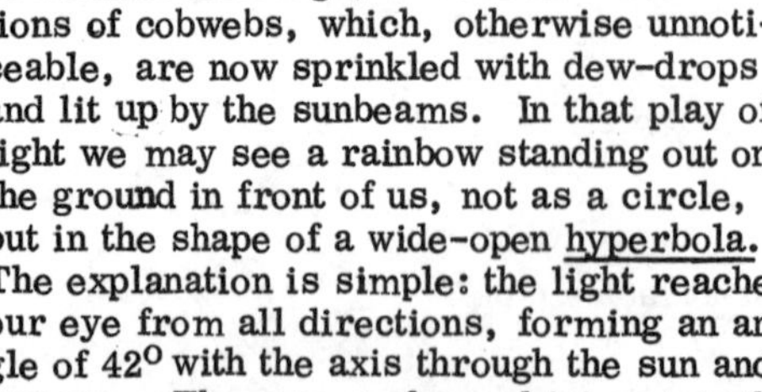

A dewbow with characteristic hyperbolic shape (X0)

the size of the drops varied from 0.1 mm. to 0.5 mm., 20 drops per cm.^{2} showing a very distinct dew-bow. (c) On a lake or on the sea, early in the morning, when the air has grown cooler but the water is still warm, so that a thin mist hangs above the surface of the water. The whole of the dew-bow is not always visible then, but only its two extremities. (d) On a frozen surface which can, apparently, become covered by dew-drops of suitable shape." (R11)

X1. February 13, 1862. Glasgow, Scotland. "There was seen today by myself and some other persons in this neighborhood, a very beautiful phenomenon, of which the cause is obvious, and of such a nature that one would expect the phenomenon to occur frequently; but I do not remember to have yet seen any instance of it recorded in any scientific publication. I refer to a prismatically-coloured hyperbolic iris, or bow of the first order, exactly resembling that sometimes seen on a field of dewy grass; but in this case it was displayed on the muddy surface of a by-road near Glasgow, and on the less-trodden parts of an adjoining turnpike road, throughout a distance of more than a mile. The time was between 12h 30m and 1h P.M. Greenwich time; the morning had been hazy, but the mist had cleared away, and the sun was shining brightly. The angular dimensions of the iris were obviously the same with those of a rainbow of the first order; its colours were complete, from red to violet, and very bright and distinct, especially where the mud was softest and moistest; where a sheet of water, how thin soever, covered the mud, the iris vanished. No trace of an iris could be seen on the grass, in the sky, or anywhere but on the mud; and on those parts of the turnpike road where the mud had been much disturbed no iris was visible. The necessary conclusion from this appearance is, that the surface of the mud must have been thickly covered with globules of pure water, perfectly spherical, and not in absolute contact with the mud, although resting on it; but those globules must have been extremely minute, for they were invisible to the closest inspection with the naked eye." (R1) Mud bows are obviously different from sandbows (GEB12)

X2. May 5, 1867. Indian Ocean. "I was in lat. 25°19' S. and long. 54°13' E., the weather was very fine and bright. Observing signs of a squall blowing up from S.S.E. while I was sitting reading on deck, I got up and went to the port gangway to watch its approach. It was about half-past one P.M., and the sun was therefore in the N.N.W.,

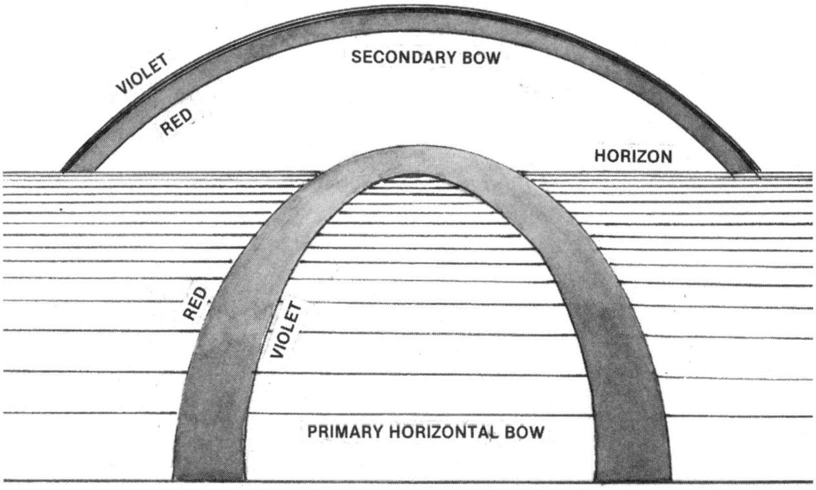

Primary rainbow apparently projected on sea surface (X2)

exactly opposite the approaching squall, upon which was already developed a rainbow of low altitude (12° to 15°). While gazing at it my attention was arrested by a yellowish brown haze upon the horizon immediately under the centre of the arc, which, although very faint, appeared from its position to have some connexion with the squall or with the rainbow; and I was thus induced to watch it attentively. At first it was a mere indefinite tinge of colour on the distant horizon, and for two or three minutes it seemed to undergo no change; but at length by slow degrees it increased in intensity, and then appeared to spread over the water, looking as though a cloud of reddish dust was hanging over the sea. For some minutes I was quite at a loss how to account for it, but carefully watched to see what would be the upshot. It now became rapidly intensified in brightness, and presently prismatic, and then slowly spread forward across the sea towards us, and at length presented the appearance of a brilliant horizontal bow lying upon the sea, its apex just capping the horizon, and its limbs seeming to fade away upon the water halfway between the eye and the horizon. As the horizontal bow increased in intensity the vertical one gradually faded away, and quite vanished immediately after the former had reached its greatest brilliancy, which was most marked about the centre or apex. In the horizonal bow the red colour was upon the outer or convex side, while in the vertical bow the red was on the inner or concave side." (R3, R4) The hyperbolic shape and horizontal position of the "primary" are anomalous. The lower bow is not a normal dewbow because it is inverted, with the apex away from the observer. The so-called "primary" and "secondary" do not have the usual geometrical relationship with each other. (WRC)

X3. 1874. Wisconsin. "I was living on an east and west ridge of high ground sloping both to the north and south. South of me was farming land, visible as far as three and a half miles. North was timber; also on the east and west was timber for one-fourth mile. There had been a shower of rain about 4 P.M. in midsummer, and it was still 'drizzling' when I went out into the clearing looking south to see about the weather. I noticed just west of me over the timber a full-sized rainbow with its north end near me. It seemed strange to see a rainbow at that time of day, to westward, so I took particular notice of it, and it seemed to move eastward. I could tell by the color of the foliage on the trees just to westward. In a few moments the colors appeared on the open ground coming eastward and approaching where I stood, at first about one hundred yards away, then closer and closer till I saw them (the colors of the rainbow) all about me, and by extending my hand I could see them between my eye and my hand. The same rate of travel being kept up soon put the colors to the east of me, and on they moved to the end of the field eastward, and there again colored the foliage on the trees to the east, and the circle of the bow appeared smaller as it receded, and after a few moments more it was the usual customary rainbow in the east." (R6) Rainbow motion along the surface from west to east is difficult to account for. See GEB5 for other moving rainbows, although of a different sort. (WRC)

X4. 1887. Australia. "A few mornings ago, when the sun was about two hours high, I noticed a curious spectrum caused by the sun shining upon the drops of dew resting on thousands of horizontal cobwebs. The spectrum was brightly colored, and extended in two straight lines from near the feet for some 150 yards, the angle between the lines being about 90°. Query: Does the mathematical theory show these lines to be actually straight, or only sensibly so?" (R5) Some curvature of the hyperbola should have been visible near the feet, although the distant parts of the bow might well seem to be straight. (WRC)

X5. September 16, 1922. Bridgend, United Kingdom. "Standing on Ogmore Down near Bridgend in this county (Glamorgan) at 2.30 P.M. and looking northwards across the broad vale towards the Maesteg hills, there appeared to me a broadcast rainbow colouring, stretching east and west for several miles along the vale. The day was exceptionally fine, with brilliant visibility and no trace of mist. The clouds were small and scattered, with a distant bank of cumulus beyond the hills, while the colours were clear and unmistakable, covering, from red in the west to blue in the east, an angle of about fifty degrees. The height of my point of view was about 300 ft. above the sea, and the whole apparition hung, like a veil of pure, immaterial colour, at about the level of my eyes, covering the distant hills but without screening their smallest particular." (R9) No bow formation is mentioned at all.

X6. Frequent occurrence. Lake Monroe, Florida. "My yachts, 'Dawn' and 'New Era,' are anchored in Lake Monroe, Florida, off the city of Sanford at the head of navigation on the St. Johns River. Lake Monroe is approximately $4\frac{1}{2}$ miles long by $3\frac{1}{2}$ miles wide; the St. Johns River enters at one end and leaves from the other. This morning when I came out on deck soon after sunrise, I noticed toward the east a remarkable phenomenon on the water. A band of prismatic colors, familiarly known as the rainbow, seemed to extend on the water, commencing about one third of the way across the lake and reaching to the shore. The atmosphere, while hazy, was in no sense foggy, and the sun was shining brightly. This phenomenon persisted and at the present time, 9.30 A.M., is truly remarkable. The haze has almost entirely disappeared. The lake is perfectly calm. The band of refracted light seems to be much broader than the usual rainbow phenomenon.

It occurred to me to measure roughly and record the angular relation of the prismatic band with the sun, and the most ready way of doing so was to take two triangles, sighting along one pointing to the prismatic path on the water and laying the other triangle across it. The result was so geometric as to be almost startling. It will be noted from the sketch that the angle formed

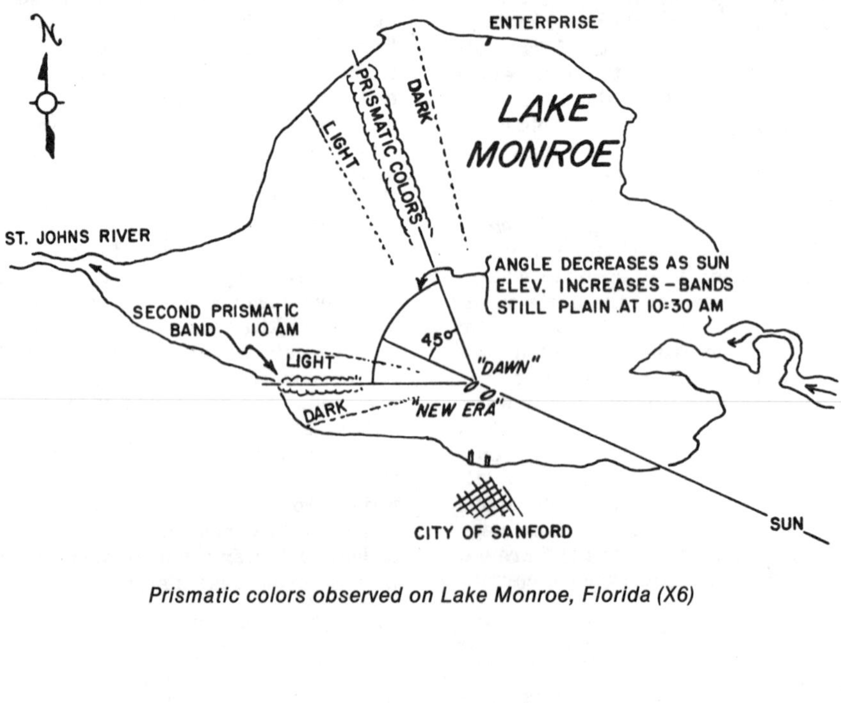

Prismatic colors observed on Lake Monroe, Florida (X6)

by the projection of the sun's rays as compared with the path of the refracted light is an angle of 45 degrees. It was perfectly simple to adjust these angles by sighting down the long side of the 45 degree angle and holding up a finger and adjusting the 60 degree engle until the shadow ran along its edge.

To me the most remarkable fact about the phenomenon is its persistency and the brilliance of the refraction all the way across the prismatic scale; an added phenomenon is a very pronouncedly dark band to the right or orange end, of the spectrum, and a bright band on the left, or purple end.

About breakfast time a young native paddled by in his canoe. When I asked him what he knew about the phenomenon, he replied that the rainbow was present every morning when the lake was calm.

Since writing the above, the other end of the rainbow has appeared, relatively as noted on the sketch. It is even more brilliant than the first---in fact, the most brilliant prismatic scale and the most extended that I have ever seen. It is now past ten o'clock; almost every particle of haze has disappeared from the lake, but still the prismatic band persists." (R8) In this example, the bands again seem straight, and the bands are flanked by peculiar dark and light bands. (WRC)

X7. No date or place given. Tiny droplets of water were floating on the surface of a lake. "The bow was seen about nine o'clock according to the daylight-saving bill, or eight by the usual local railroad time. Its appearance was about as shown in the accompanying figure. AB is the western shore-line of the lake, about 200 yards away. The bow was complete except in the following particulars: the part near S was hidden by the shadow of the observer and that of the boat in which he sat; and the part PRQ was inverted, like a reflection of what should have been the crest, the part near R being somewhat less bright than the rest. The ends of this inverted portion seemed to meet the ends of the larger arc at the shore-line, but there is no reason why such an accidental line should determine the intersection of the two branches." (R7) The inverted bow, PRQ, is probably a reflected bow. The anomaly here is that the primary is far from a hyperbola. (WRC)

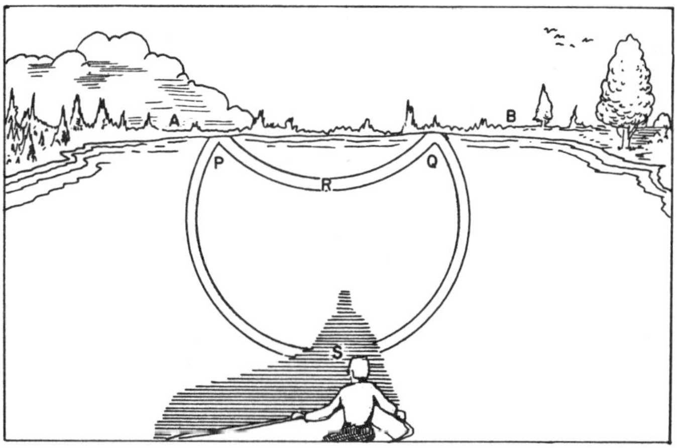

Two bows seen on a lake surface (X7)

X8. September 19, 1961. North Pacific Ocean. "Shortly after taking off from Honolulu en route to Los Angeles in a Pan American 707 jet on 19 September 1961, we attained the chosen cruising altitude of 29,000 ft. At about 10 o'clock Honolulu time, I noted from my port window that we were accompanied by a luminous arc below the aircraft. My first, wrong, impression was that I was viewing some kind of ice-crystal halo involving both reflection and refraction since it had the almost colourless luminosity characteristic of the majority of haloes. However, in a moment of further scrutiny, I noted that it seemed to originate on the cloud deck whose tops we had earlier

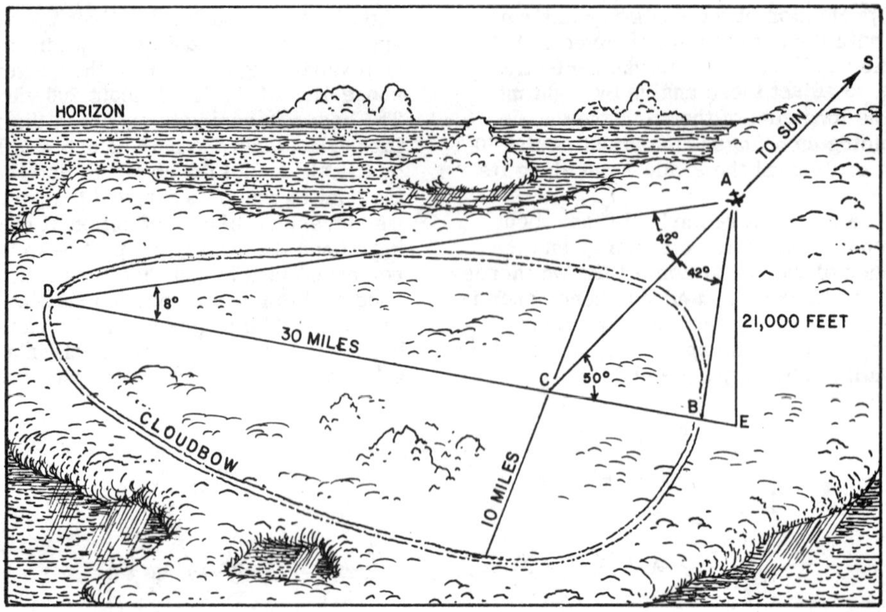

Giant horizontal cloudbow seen from an aircraft (X8)

penetrated at about 8,000 ft altitude. Its true nature was then unambiguously revealed when I noted further that here and there along its course it assumed a most vivid rainbow banding where I was looking down through breaks in the 8,000-ft deck to where the trade cumuli were raining. The near edges of the rain columns constituted for me small elements in a huge and essentially horizontal field of drops of two quite distinct size-classes: first, the tiny cloud-drops in the broken deck whose tops lay near 8,000 ft; second, the scattered patches of relatively large raindrops visible through breaks. Having recognized its real nature, I scanned along the expected arc far to the north of our flight path, and was pleased to find that the luminosity, though very pale in the distance, constituted well over half of a full bow from my vantage point.

Subsequent check on the solar altitude for that time and place permitted assigning the numerical values indicated in Fig. 1, which is drawn in a vertical plane through sun and aircraft. The sun S at altitude very near 50° sent down rays of which the one passing near an observer in the aircraft at A strikes the top of the 8,000-ft cloud deck at C. A cone of semi-vertex angle 42° revolved about that ray SC as axis intersects the deck in the rainbow-(more precisely, in the cloudbow-) locus, whose extrema D and B are indicated. Since the top of the cloud deck lay about 21,000 ft below the aircraft, the indicated angles imply that the 'summit' of the bow at D lay some 30 miles away, measured from the point E directly under the aircraft. The bow is here not the hyperbola which must always be expected with dew-bows (whose occurrence can only be expected in the morning hours of low sun, but rather an ellipse. Further calculation shows that the minor axis running perpendicular to the plane of the figure slightly to the left of C is here about 10 mi. The geometry of the small windows in the 707 precluded my seeing the full ellipse (indeed I could barely see the weak glory at C on half-standing to peer downwards), a limitation arguing the need for glass-bottomed jet transports for meteorologically inclined passengers.

Where the bow changed from cloudbow to rainbow at a break in the underdeck near a rain area, the difference in angular spread of the two parts was easily perceived. I made rough attempts to time the passage of the cloudbow over distinctive cloud turrets, estimating 3 to 4 seconds between transits of leading and trailing edge of the bow where the latter lay normal to our flight path. Assuming a ground speed of 600 mph, this gave an actual width of the cloudbow luminosity of somewhat under a half-mile, implying an angular width of about 3 to 4°. The angular width of a typical rainbow is about 1.7^{o}, so the cloudbow was about twice as wide as a rainbow, in good enough agreement with theory (Minnaert).

The phenomenon seems noteworthy on a number of grounds: here was a bow spanning a space of 10 by 30 miles, large beyond all ordinary circumstances; its locus of luminosity was elliptical rather than hyperbolic; and, finally, it was actually a mixed rainbow-cloudbow, chiefly the latter." (R12) Since this complex display is rather easy to explain, it is more curious than anomalous. (WRC)

X9. No date given. Bermuda. "I noticed that there was a mist of sorts between me and the stratified sand nearby. I looked up and as far as emotional effect went, a full charge of HE might have gone off at my elbow. I experienced a visual shell-shock, for high overhead there was etched the strongest, most materialistic rainbow I have ever seen, one end of which began in mid-air, and the other curved down, down, down, holding true from red to violet, to the rocks beside me. Once before, on a Guiana jungle river, I have been actually at the end of a rainbow, when, at my very side, one colored the bulwarks of our Akawai canoe." (R10) This must have been some sort of anomalous bow because one is never supposed to reach the end of the normal rainbow---a nearby sheet of mist may have created a rainbow in a vertical position like those seen in the spray from a water sprinkler. (WRC)

References

R1. Rankine, W.J. Macquorn; "On a Dew-Bow Seen on the Surface of Mud," Philosophical Magazine, 4:23:245, 1862. (X1)
R2. "On a Dew Bow Seen on the Surface of Mud," Scientific American, 6:352, 1862. (X1)
R3. Collingwood, C.; "Account of a Horizontal Rainbow Observed at Sea," Philosophical Magazine, 4:34: , 1867. (X2)
R4. Collingwood, C.; "Account of a Horizontal Rainbow Observed at Sea," English Mechanic, 6:298, 1867. (X2)
R5. "Cobweb Dew-Bow," English Mechanic, 46:39, 1887. (X4)
R6. Irmen, John M.; "A Curious Rainbow Phenomenon," Scientific American, 93:379, 1905. (X3)
R7. Reese, H.M.; "An Unusual Form of Rainbow," Science, 50:542, 1919. (X7)
R8. Donnelly, William T.; "An Optical Phenomenon on a Florida Lake," Natural History, 22:372, 1922. (X6)
R9. McLean, R.C.; "A Broadcast 'Rainbow,'" Nature, 110:605, 1922. (X5)
R10. Beebe, William; "Mount Bermuda," Nonsuch: Land of Water, New York, 1932, p. 11. (X9) (Cr. L. Gearhart)
R11. Minnaert, M.; "The Dew-Bow or Horizontal Rainbow," The Nature of Light and Color in the Open Air, New York, 1954, p. 184. (X0)
R12. McDonald, James E.; "A Gigantic Horizontal Cloudbow," Weather, 17:243, 1962. (X8)

GEB12 Sandbows

Description. Colored bows resembling normal rainbows---except that they are roughly twice the width---created by wind-lofted polished opaque spherules of sand. The colors of the primary and secondary have the same orders as in common rainbows.

Data Evaluation. One example only reported; but it was observed by several people and reported in detail. Rating: 2.

Anomaly Evaluation. No optical explanation has been found so far for this phenomenon. Reflection from the polished spherules certainly seems to be the basis of the sand bow, but no mechanism for dispersing the spectrum is obvious. Rating: 2.

Possible Explanations. None.

Similar and Related Phenomena. Ordinary rainbows.

Examples of Sand Bows

X1. May 16, 1901. Great Salt Lake, Utah. "On the evening of May 16, the writer was crossing the main ridge of Antelope Island ---the largest land body within the area of the Great Salt Lake. As he began the descent on the eastern slope, there appeared between the island and the mainland what seemed at first glance to be a segment of a brilliant rainbow of unusual width. It was evident, however, that no rain was falling in that direction. Clouds were gathering in the south and west, but the sun was not yet obscured. A wind setting toward the mainland had lifted from the dry flats large quantities of the 'oolitic sand,' with which the lake bottom and the recently dried patches on this side of the island are covered to a depth varying from a few inches to several feet. This so-called 'sand' consists of calcareous spherules, fairly uniform in size between the limits of No. 8 and No. 10 shot. The oolitic bodies are polished and exhibit a pearly luster.

It would seem that the outer spherical surfaces reflected the light in such a manner as to produce the bow. The colored column appeared almost to touch the lake bed, and its ends subtended with the observer an angle of about 40°. The prismatic colors were distinct, the red being outside, i.e., away from the sun. In apparent width the column was fully double that of an ordinary rainbow. A fainter secondary bow was plainly visible beyond the primary, with the colors in reverse order. The phenomenon was so brilliant as to attract the attention of all members of the party, and it remained visible for over five minutes, then, as the sun sank lower, it rapidly died away.

The production of a color bow by reflection from the outer surfaces of opaque spherules is a new phenomenon to the writer. It is inexplicable on the principle of refraction and total reflection from the interior of transparent spheroids, according to which the rainbow is generally explained." (R1)

References

R1. Talmage, James E.; "A 'Sand Bow'---An Unusual Optical Phenomenon," Science, 13:992, 1901. (X1)

GEB13 Rainbows Parallel to the Horizon

Description. Straight spectral bands parallel to the horizon and close to it, showing colors in the same order as those in a normal primary rainbow. These colored bands may stretch over a quarter of the horizon (90°).

Data Evaluation. Only a few of these rainbow-like horizontal bands have been described in the scientific literature; but these reports are consistent and seem to portray a distinct physical phenomenon. Rating: 2.

Anomaly Evaluation. Once again, the conventional theory of the rainbow fails to explain a spectral display. Indeed, there seems to be no obvious way to modify rainbow theory to encompass horizontal bands of colors close to the horizon and parallel to it. It is possible, however, that this is a type of halo, created by ice crystals, rather than a rainbow phenomenon, as discussed below. Rating: 2.

Possible Explanations. Since all of these spectral displays are close to the horizon, low-level mists may contribute to produce iridescence like that observed around the edges of some clouds. Two halo phenomena, circumhorizontal arcs and circumzenithal arcs, are also parallel to the horizon, although generally at higher altitudes.

Similar and Related Phenomena. Cloud iridescence; circumhorizontal halo arcs; circumzenithal halo arcs.

Examples of Rainbows Parallel to the Horizon

X1. July 6, 1899. Cromlyn, Ireland. "On July 6, about 4 p.m., I noticed a strange phenomenon which had never presented itself to me before, and perhaps may be new to others also. There had been showers all day, and now dark clouds and blue sky alternated, when suddenly, low down in the North-east, appeared a beautiful horizontal rainbow. There was no curvature whatever, the ends terminating abruptly, if colours so delicate can be called in any sense abrupt; and the rainbow was well defined, occupying a space about equal to sixteen (apparent) moon's diameters in length. I watched the rainbow most closely, and could not account for its peculiar shape." (R1)

X2. January 25, 1955. Indian Ocean. "0600 G.M.T. At an altitude of 27° there appeared a band of coloured light. Sextant readings were obtained as 91° in length and approx. 1° in width. This band contained all the colours of the spectrum and resembled a rainbow in every respect except that it appeared perfectly straight. The weather had been showery but at that time and for some time previous the only precipitation sighted was distant (approx. 8 miles). (R2)

X3. May 5, 1963. Bridgewater Canal, England. "At 1515 hr on 5 May 1963 I saw from the towpath of the Bridgewater Canal an unusual rainbow effect. There was a heavy cumulonimbus cloud to the eastward and the spectrum appeared to follow its lower edge. It was not a bow at all, but a more or less horizontal band. The violet portion appeared very broad and seemed to be below the edge of the cloud. The red band, which was narrow, followed approximately its lower visible edge. Part of the band was obscured by trees, but I estimated that it subtended an angle of 30°-45°. I observed it for about 10 minutes. The altitude of the sun was 38° and its azimuth 243°. The elevation of the middle of the band from a point 100 ft above sea level was 1.5°. Weather at the time: Wind W force 4, visibility good. During the afternoon there had been frequent showers of wind and hail." (R3) This may be a circumhorizontal arc. (WRC)

References

R1. Battersby, C. Maud; "A Curious Rainbow," English Mechanic, 69:510, 1899. (X1)

R2. Bury, J.E.; "Unidentified Phenomenon," Marine Observer, 26:18, 1956. (X2)

R3. Rylands, Geoffrey K.; "Coloured Band," Weather, 18:379, 1963. (X3)

GEB14 Purple Rainbows

Description. Rainbows in which purple predominates to the near exclusion of other colors. Purple rainbows have been seen only before and during sunrise.

Background. Red rainbows (GEB4) are occasionally noted during and shortly after sunset. They occur because the atmosphere has filtered out all colors except the long wavelengths.

Data Evaluation. Very few observations exist; and these contain sparse detail. Rating: 3.

Anomaly Evaluation. Just as at sunset, one would expect sunrise rainbows to be red. Instead, they are purple---the opposite end of the spectrum! Rating: 2.

Possible Explanations. The explanation may involve short wavelengths scattered from high clouds.

Similar and Related Phenomena. Red rainbows (GEB4); other low-sun phenomena (GES).

Examples of Purple Rainbows

X1. August 8, 1899. Przemysl, Austria. At an altitude of 270 meters. "During the night a heavy dew had fallen; near morning the temperature was 12°C. At the northeast appeared somber nimbus clouds, when at 3:45 A.M. the zenith assumed a strong reddish tint, this color increasing toward the northeast, and at 4 o'clock appeared a rainbow of exceptional width and of a pure dark purple; its summit was 35° above the horizon. After four minutes a second rainbow appeared, about 10° from the first, of an equal width, having a pale rose color. In the principal rainbow the purple color dominated, but the outer border was tinged with violet, and the inner edge was of a cherry red hue; its summit gradually ascended to 50°. At 4:16 the exterior secondary rainbow took an orange tint, then disappeared. The outer edge of the other became lilac, and the interior a pale cherry; by degrees, the latter color extended itself over the whole width of the rainbow, which finally assumed an orange tint." (R1)

X2. November 21, 1902. South Croyden, England. "On the morning of Thursday, 21st November, I noticed gorgeous colouring in the south-east part of the heavens, preparatory to sunrise. Near the horizon the sky was comparatively clear, but higher up was a large patch of cloud, tinted brilliantly with red and orange, the intervening spaces being blue with shades of green. There was at the same time a shower of rain advancing from the west, which caught the bright light reflected from the cloud, with the result that a purple rainbow became visible. The time of sunrise for this day is 7.29, and the rainbow was visible from 7.15 (or possibly sooner) to 7.30, when it began to fade away; before 8 o'clock rain was falling overhead. The bow was of very large dimensions, with its two extremities at right angles to the horizon." (R2)

References

R1. "Purple Rainbow," Scientific American, 82:38, 1900. (X1)

R2. Johnson, E.W.; "Rainbow before Sunrise," Knowledge, 25:11, 1902. (X2)

GEB15 Supernumerary Rainbows

Description. Additional bows or, more commonly, segments thereof, appearing just inside the primary and immediately outside the secondary bows. Supplementary bows have the same color sequences as the bows they are adjacent to, but usually red and green predominate.

Data Evaluation. A rather common though unappreciated rainbow phenomenon. Rating: 1.

Anomaly Evaluation. No anomaly exists here. The only unusual aspect worth of recording is the rare observation of large numbers of supernumeraries, as in X1. Rating: 4.

Possible Explanations. Supernumerary rainbows are the consequence of optical interference.

Similar and Related Phenomena. Multiple rainbows (GEB1).

Examples of Supernumerary Rainbows

X0. General observations. "Supernumerary bows---sometimes called spurious bows---are the bands of prismatic colors often occurring just inside the lower or primary rainbow, and less frequently outside the upper, or secondary bow. Tait describes them as having the appearance of ripples. They repeat the colors of the spectrum---sometimes several times over ---green and red being usually predominant; the other colors are often indistinguishable. As stated above, they are apt to be overlooked in a cursory observation; but as a rule they are easily seen by anyone who looks for them, especially those attending the primary bow, and they are sometimes quite conspicuous...... The Newtonian theory of the rainbow took no account of the supernumerary bows. They were first explained by Young in 1803, and more fully by Airy, in 1836-38. They are interference phenomena, due to the fact that, in addition to the rays of light producing the colors of the ordinary bow, there are other mutually parallel rays emerging from a raindrop after traversing unlike paths within the drop." (R2)

X1. June 1, 1840. Killness, Scotland. A spectacular array of ten complete bows including the primary with five supernumeraries, and the secondary with three supernumeraries. (R1)

References

R1. Fisher, J.; "On Remarkable Rainbows," Report of the British Association, 1840, p. 12. (X1)

R2. Talman, C. Fitzhugh; "Little-Known Rainbows," Scientific American, 106:305, 1912. (X0)

GEB16 Prismatic Pillars at the Foot of the Rainbow

Description. A vertical band of rainbow colors, in the same order as those in the primary, rising from the intersection of the primary with the horizon. Pillars with reversed colors may occur in conjunction with secondary rainbows. Prismatic pillars generally manifest themselves near or over water.

Background. As discussed in X1 below, a slight curvature to the pillar probably means that it is only a segment of a rainbow formed from sunlight reflected from a water surface (GEB1).

Data Evaluation. Observations of high, precisely straight pillars are very scarce. The few that have been found are also fairly ancient. Rating: 3.

Anomaly Evaluation. The only explanation profferred for this phenomenon (see below) has been termed inadequate by experts. The generally accepted theory of the rainbow does not account for tall, straight, vertical pillars. Rating: 2.

Possible Explanation. Sunlight shining on a uniform train of water waves might create a series of secondary sources of light---perhaps approaching a line source---that form the lower segments of ever-higher rainbows that meld to give the appearance of a uniform prismatic pillar. According to this theory, the arches of the several reflected-light rainbows are too weak to be seen; only the segments near the horizon are discerned. See X4.

Similar and Related Phenomena. Multiple rainbows (GEB1).

Examples of Rainbow Pillars

X1. September 3, 1821. Off the coast of Ireland, North Atlantic Ocean. This phenomenon "consisted of two beautiful segments of primary and secondary rainbows, (called by the sailors 'weather-galls,' when, as in this case, they consist only of the portions next the horizon) with some supernumerary bows within the arch of the former, and likewise, which is the extraordinary part, another spectrum rising almost vertically from the base of each of the common arcs, at its apparent termination in the horizon of the sea, so as to form two figures nearly resembling the Greek ν ." (R1) Since the arcs are slightly curved, it is probable that this particular example can be explained in terms of intersections of the lower portions of normal and reflected-light rainbows, with the primary and secondary each contributing a ν . See the GEB1 illustrations. (WRC)

X2. Circa 1840. River Clyde, Scotland. After describing brilliant primary and secondary rainbows with supernumeraries. "I first observed a faint perpendicular light on the north side of the primary bow, and apparently in contact with its outer margin, as though diverging from it in a tangent. By degrees this shot up higher, and acquired the

High, straight, vertical pillars at the ends of a rainbow (X2)

prismatic colours, which were clearly in the same order as the primary bow, though fainter. Shortly afterwards a similar tangent was seen on the opposite side, in all respects corresponding, both extending upwards to about the height of the large outer bow. My friend Dr. Black, of Manchester, who was present, first suggested that the lengthened image of the sun on the smooth water might probably produce these upright lateral columns, the image in this case acting the part of the luminary himself. We had soon an opportunity of testing the soundness of this solution, for in a little time the upper limb of the sun came in contact with the lower edge of the cloud, and as he very gradually disappeared behind it the primary bow became fainter and fainter, till at length it faded away entirely. But the upright lateral columns remained, nay, for a short time appeared brighter and higher, and only disappeared when the more complete obscuration of the sun destroyed the reflexion of his light upon the water." (R2) Although no curvature was evident, the phenomenon seems definitely associated with reflected sunlight. (WRC)

X3. July 7, 1936. Isle of Skye, United Kingdom. After describing a photograph taken of primary and secondary rainbows. "What seems to be noteworthy is that from the northern foot of both primary and secondary rainbows sprang a vertical shaft of light, each shaft exhibiting the colours in the order proper to its bow. The shafts extended to approximately three-fifths of the height of the tops of the bows. Their bases coincided with the bases of the bows. No shafts were seen at the southern feet of the bows which were not 'in' the sea as were the northern feet. The shafts were visible for about 40 minutes. When, however, we arrived at Flodigarry about 22h. the shafts and the secondary bow were no longer to be seen, but the primary bow persisted with brilliant colouring. The shafts did not appear to shew any trace of curvature." (R3)

X4. General observations. The following explanation of X2 and X3 has been given by A. Bravais: "If the sun is near the horizon, and if the surface of the sheet of water is slightly ruffled, the bow produced by the reflected image assumes a remarkable appearance; the lateral parts alone remain visible in the form of vertical columns tangential to the right and left parts of the ordinary bow. The reflected image of the sun is then greatly elongated in the vertical direction; the reflected bow is the envelope of a series of bows of the first order having their centres in the vertical through the sun; these bows are almost superimposed on each other in the parts near the horizon; but they are separated from one another towards their uppermost parts, and the dispersal of brightness resulting from this separation of the various bows tends to render them invisible in this part of the sky." (R4) This explanation is doubted by F. J. W. Whipple (R4) and M. G. J. Minnaert (R5) on the grounds that the ripples would surely reduce the light intensity so much that the rainbow colors could not be formed.

References

R1. Scoresby, William; "A Description of Some Appearances of Remarkable Rainbows," Edinburgh New Philosophical Journal, 2:235, 1827. (X1)

R2. Brewster, David; "On a Remarkable Rainbow Observed by Mr. Bowman," Report of the British Association, 1840, p. 12. (X2)

R3. Druce, Francis; "Rainbow with a Vertical Shaft," Meteorological Magazine, 71:230, 1936. (X3)

R4. Besson, L., and Whipple, F. J. W.; "Rainbow with Vertical Shaft," Meteorological Magazine, 71:259, 1939. (X4)

R5. Minnaert, M. G. J.; "Unusual or Neglected Optical Phenomena in the Landscape," Optical Society of America, Journal, 58:297, 1968. (X4)

GEB17 The Dark Space Between Primary and Secondary Rainbows

Description. A decided lack of illumination of the sky separating primary and secondary rainbows. The adjectives chocolate-colored and purplish have been applied to this phenomenon. However, in the great majority of rainbows, the dark space is not noticed.

Data Evaluation. The phenomenon is apparently rare. Nevertheless, several seemingly sound reports have been uncovered. Rating: 2.

Anomaly Evaluation. Classical rainbow theory cannot account for the dark space. No one, however, seems to have attempted to explain it employing optical interference effects and/or psychological contrast effects. Rating: 2.

Possible Explanations. Light from this region may be reduced by optical interference from light refracted from rain droplets elsewhere, similar to the manner in which supernumerary bows are created. Further, the brightness of the adjacent primary and secondary bows may make this region seem dull in comparison.

Similar and Related Phenomena. Rainbows that appear to divide sky colors (GEB19); supernumerary bows (GEB15).

Examples of the Dark Space between Primary and Secondary Rainbows

X1. August 23, 1866. Washingtonville, New York. The space between the primary and secondary was "between violet and purple; so that a portion of the mountains behind, when seen through the medium of this colored space, was considerably darker than the adjacent parts." (R1)

X2. August 19, 1875. Leeds, England. "Between 7 and 7.20 this evening, after about an hour's gloom from dense clouds, a double rainbow appeared, in diameter about 40 to 45 degrees on the horizon. The colours of the inner rainbow were unusually brilliant; those of the outer comparatively faint. The entire plane enclosed by the inner rainbow, except a portion towards the north, was of a uniform light brownish-yellow colour, unchequered by any other colour or shade. On the other hand, the plane between the two rainbows, especially towards the south, was of a uniform dark chocolate colour, forming a conspicuous and somewhat startling contrast to the lighter-coloured expanse enclosed by the inner rainbow. The lines of demarcation were sharp and defined, adding to the force of the contrast." (R2)

X3. July 27, 1897. Lausanne, Switzerland. The space between the rainbows "appeared to be a dusky shadow." (R3)

X4. July 20, 1947. Urbana, Illinois. "A remarkable set of rainbows was seen from 8:00 until about 8:15 P.M. (C.D.T.) on July 20, 1947, at Urbana, Illinois. The main primary rainbow showed a continuous band of clear color. The most unusual part of this bow was the brilliance of the violet band, which at times exceeded in intensity that of the blue and green bands. From top to bottom the colors in a primary rainbow run red, orange, yellow, green, blue, indigo, violet. Usually the indigo and violet bands are quite faint, whereas in this case they were unusually bright. A secondary rainbow was very plain, above and outside the arc of the primary bow. In this, as is usual, the order of colors was reversed, with the red at the bottom and green above. It is uncommon for the blue, indigo, and violet bands to show above the green in the

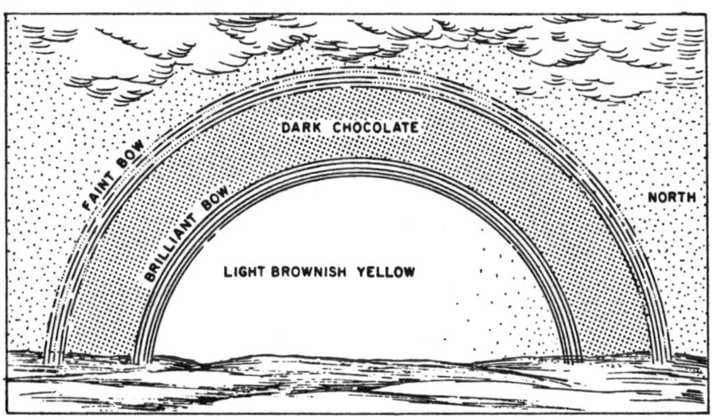

Dark space between primary and secondary rainbows (X2)

secondary rainbow, and this was no exception. However, the width of the secondary bow, even without the blues, was about equal to that of the whole primary bow. Between the primary and secondary bows the sky appeared to be lacking in light. It was dark, leaden gray in color." (R4)

References

R1. Leute, J.R.; "A Meteorological Phenomenon," Scientific American, 14:200, 1866. (X1)

R2. "Singular Rainbow," English Mechanic, 21:616, 1875. (X2)

R3. "A Remarkable Rainbow," English Mechanic, 65:575, 1897. (X3)

R4. Quirke, Terence T.; Science, 106:225, 1947. (X4)

GEB18 Grossly Distorted Rainbows

Description. Prismatic colors arranged in irregular shapes but still possessing some vestiges of the rainbow, such as arc-like shape, color sequence, and proper sun-observer-display geometry.

Data Evaluation. All observations collected so far are one-of-a-kind, although they are in general described with precision. Rating: 2.

Anomaly Evaluation. Distortion is common in optical systems, particularly one in which there are so many natural variables. The occurrence of "sports" is to be anticipated. Explanations are elusive because of the presence of so many unknowns. Rating: 3.

Possible Explanations. Some possible sources of distortion are: (1) abnormal refraction, especially when the sun is low; (2) sun glint off the sea surface; (3) intervening showers and low-lying mists; and (4) varying raindrop sizes.

Similar and Related Phenomena. Low-sun phenomena (GEL); abnormal halos (GEH); and the other rainbow irregularities in this section (GEB).

Examples of Distorted Rainbows

X1. September 13, 1955. North Atlantic Ocean. "At 1115 G.M.T. a light rain shower passed overhead and travelled in a w'ly direction. When about a mile away, bearing 280°(T), a brilliant coloration was observed on the cloud (sketch 1), approximately 300 yd long. In the sketches the letter a denotes a brownish-orange colour, b a greenish tinge and c turquoise blue. After about 3 min the left-hand side of the ellipse seemed to merge into the other side until only a 'hump' of one colour remained (sketch 2). After a further 2 min the coloration was observed to be flattening and lengthening into a thin band (sketch 3), which completely disappeared after a further 2 min. The phenomenon lasted for 7 min 1120 to 1127.....Note. The observers suggest that this was a rainbow. The sun's altitude at the time of observation was between 41° and 42° and its azimuth approximately 117°. With this altitude the primary rainbow could not have been seen, as its apex would have been on the horizon. The observation does not state the altitude of the apex of the phenomenon, but a secondary rainbow could have been formed with its apex about 13° above the horizon, bearing 297°. The coloration observed does not correspond to that of a secondary rainbow, which has blue and violet on its upper side and red and orange on its lower side. Furthermore, in normal rainbows the arcs of colour are circular and concentric. No explanation can thus be given of this very interesting phenomenon." (R1)

X2. August 22, 1958. Mozambique Channel, Indian Ocean. "While moderately heavy showers of rain were falling at 1200 G.M.T., a faint rainbow was seen bearing 110°. Vertically below the apex of the bow, in a seemingly 'compressed' state, was a remarkable 'block' of light showing the colours of the spectrum, with yellow predominating. As the bow disappeared, the 'block' of light rose and expanded until it in turn became a

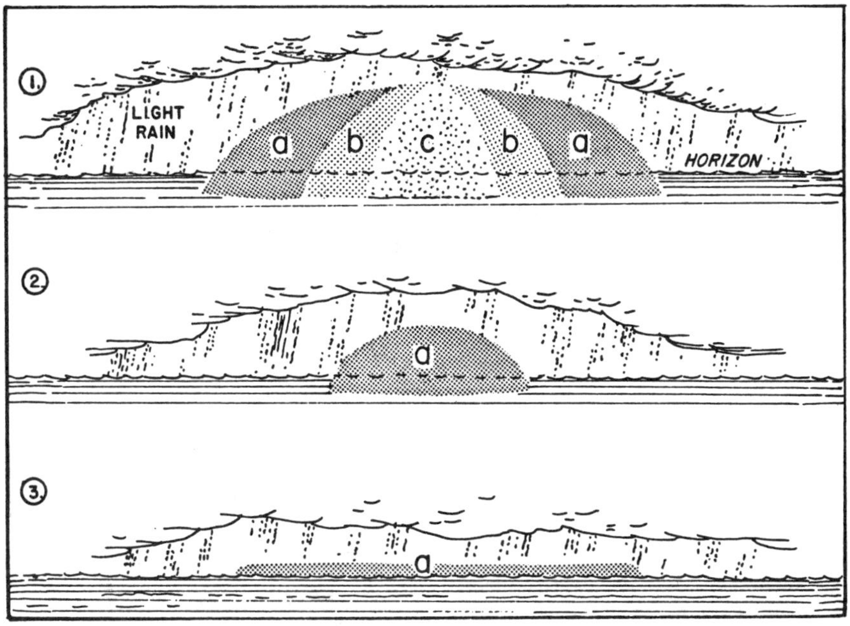

Distorted rainbows in the North Atlantic (X1)

bow, but the intensity of the colouring was by now very much reduced, and there was no predominant tint." (R2)

X3. January 15, 1963. Brazilian waters, South Atlantic Ocean. "At 1855 GMT part of a rainbow was seen forming between the ship and the eastern horizon; the arc, which was small, became complete by 1858. None of the rainbow was projected against the sky, it had the sea as a background. This condition persisted for 3 min. after which the arc appeared to straighten out, forming a low 'bank' of light, the centre of which was green; the ends---which had become slightly tilted up---were orange. During the next few minutes the ends of the 'bank' continued to tilt upwards and inwards until they met, forming an orange arc over the patch of green light. The final shape lasted for two or three minutes and then faded out completely. The sequence is shown in the sketches.....Sun's azimuth 265° (approx.); altitude 43° (approx.). Note. This is an interesting example of the manner in which the altitude of the sun (or moon if appropriate) affects the shape of a rainbow. When the sun is on the horizon a rainbow has the largest possible altitude and span. On the other hand, no primary bow can be formed when the altitude of the sun exceeds 42°. When the altitude is 42°, the uppermost part of the primary bow (i.e. the red) is on the horizon, the remaining colors being projected against the earth's surface. These conditions appear to apply in the present instance, the sun's altitude being given as 43° approx. It would seem that the violet and blue of the rainbow were invisible against the background of the sea, leaving only the red (or orange) and the green to be seen. As the sun's altitude decreased during the period of observation (the time was afternoon), conditions would become increasingly favorable for the bow to develop and increase in altitude and this would account, we believe, for the formation of the orange arc, with the green area below it. The whole phenomenon turned upon the fact that the sun was at a critical altitude, as regards a rainbow being able to form or not." (R3)

X4. July 31, 1973. Coral Sea, South Pacific Ocean. "At 0250 GMT a horizontal distribution of the spectrum of colours was observed as shown in the diagram, the base of the phenomenon being at sealevel. The sky was completely clouded over with Sc and Cu. The ship's course was 191°T while the phenomenon bore 170°T. There were showers of rain visible near the vessel but no precipitation at the time of observation.....Note. This appears to be a case of a reflection rainbow. Here sunlight is reflected from the (smooth) sea surface on to the shower cloud and the rainbow produced. The defor-

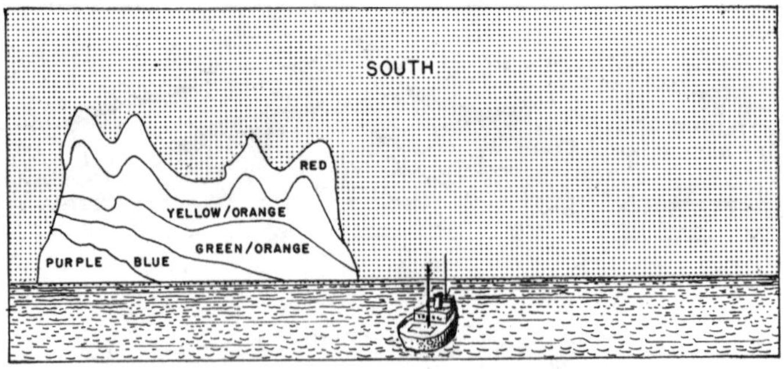

Confusion of rainbow colors over the Coral Sea (X4)

mation of the rainbow is caused by the undulation of the water surface by a smooth swell. With a perfectly calm and flat expanse of water the reflection rainbow is formed as an arc, similar to a normal rainbow, but with the centre of the rainbow being at the same elevation as that of the sun." (R4) This explanation is too easy, for the sky was completely clouded. (WRC)

References

R1. Mayne, R. J.; "Unidentified Phenomenon," Marine Observer, 26:147, 1956. (X1)

R2. Poulter, N.; "Unidentified Phenomenon," Marine Observer, 29:115, 1959. (X2)

R3. Gibbons, G. A.; "Rainbow," Marine Observer, 34:17, 1964. (X3)

R4. Thompson, A. S.; "Rainbow Effects," Marine Observer, 44:110, 1974. (X4)

GEB19 Rainbows That Divide Sky Colors

Description. The division of sky colors by rainbows, with dark hues outside the bows and lighter skies inside.

Data Evaluation. Only a few casual, rather imprecise observations exist. It is impossible to tell from the reports whether the phenomenon is merely a coincidental arrangement of clouds or a true optical effect. Rating: 3.

Anomaly Evaluation. If this is a real optical effect, similar to the occasional dark space between primary and secondary rainbows (GEB17), no inkling of an explanation has been found in the literature. Ordinary rainbow theory does not seem applicable. Rating: 2.

Possible Explanations. A psychological contrast phenomenon.

Similar and Related Phenomena. The division of sky colors by cloud arches (GWC2-X2); the dark space between primary and secondary rainbows (GEB17); halos dividing sky colors (GEH4).

Examples of Rainbows Dividing Sky Colors

X1. August 13, 1857. Belfast Lough, Northern Ireland. Sky outside the rainbow was purple-black; inside it was brighter. (R1) See complete description in GEB8-X1.

X2. August 13, 1886. Morning Sun, Iowa. "A strange phenomenon was observed here at a few minutes before sunset yesterday evening. A heavy storm had come up in the afternoon, and during a cessation, shortly before sunset, the light burst through the clouds in the western sky and a most magnificent rainbow appeared in the east, with the top of the circle about fifty-five degrees from the horizon. All that portion of the clouded sky within the circle changed to a brilliant red color, while all without the circle was of a dull bluish-gray shade. The rainbow thus formed a sharp division between the two portions of the sky which were in such striking contrast to each other. For a time the red sky inclosed by the rainbow was so deeply colored that the red portion of the rainbow, which was in itself remarkably brilliant, could not be distinguished from it, but it appeared as though the red part of the rainbow had spread out, covering the entire plane inclosed within the circle. ..... Why should the red portion of the clouded sky within the circle and the bluish-gray portion without be so sharply defined and divided by the rainbow." (R2)

References

R1. Murphy, J. J.; "Account of an Instance of Converging Rays.....," Report of the British Association, 1857, p. 35. (X1)

R2. Thomas, B. F.; "A Curious Optical Phenomenon," Popular Science Monthly, 29:841, 1886. (X2)

GEB20 The Odor of the Rainbow

Description. An undefined but apparently distinctive odor associated with rainbows and said to emanate from them.

Data Evaluation. Only vague claims propagated by the older writers. No scientific observations have been found. Rating: 4.

Anomaly Evaluation. If the so-called odor of the rainbow is simply the fresh smell encountered after a rain shower, as some believe, no anomaly exists. But, if a distinctive odor is truly created by the rainbow itself, we have a significant anomaly, because no mechanism is available to account for the supposed phenomenon. Although this phenomenon is almost certainly an instance of mistaken association, the rating assumes it to be a real physical effect. Rating: 2.

Possible Explanations. The fresh smell of rain-washed air and vegetation after a shower that coincidentally forms a rainbow.

Similar and Related Phenomena. The odor of the aurora (GLA5).

Examples of Rainbow Odor

X1. General observations. "More than half a century ago a singular controversy took place in the press regarding the existence of this supposed emanation. Its occurrence was affirmed on the authority of many writers of the past, e.g., Pliny (Lib. XII., c. 24), Aristotle (Prob. Quest. XII), a Greek writer referred to by Coleridge (in his Table Talk), Georgins de Rhodes (in his Peripatetic Philosophy), Bacon (in his Sylva), Brown (in Britannia's Pastorals), and more lately by Robert Snow (in one of his Poems); thus showing a belief held in different countries, over many centuries. Everyone is familiar at a certain season of the year with the increase of scent given by plants and shrubs on a warm evening after the air has been newly washed by rain, and this would coincide often with the appearance of a rainbow; but the suggestion of the above authors is that the perfume is actually produced by the rainbow itself. Are there any other instances of this belief on record?" (R1)

References

R1. Bentley, Richard; "Odour from the Rainbow," Symons's Monthly Meteorological Magazine, 44:92, 1909. (X1)

GEH UNUSUAL HALO DISPLAYS AND CORONAS

Key to Phenomena

GEH0 Introduction

Rings around the sun and moon are rather common; they are venerable fixtures of weather lore. The most frequently observed halos have radii of 22° and 46° and are readily explained in terms of the reflection and refraction of sunlight or moonlight by different types of ice crystals floating in the atmosphere. Rare halos with different radii as well as an immense smorgasbord of arcs are also recognized. Most of these are also fairly well understood by meteorologists. Indeed, by simulating various ice crystals with computer programs, many features of even the most complex halo displays can be reproduced. Nevertheless, even the computer combined with the power of atmospheric optics cannot account for all halo phenomena. There are, as they say in the field, "unrecognized" halos!

How can optics explain offset halos, noncircular halos, and those halo displays that are tilted away from the vertical? Even perfectly square halos have been seen, while ellipses are not particularly rare. Then there are abundant arcs and stray sun dogs and other deviations. While the variety seems endless, so are the ice crystals and orientations thereof that create halo phenomena. And of course the atmosphere itself is full of irregularities. Conceivably, if we knew all of the conditions, meteorologists might come up with reasonable explanations. Most halo displays, even the anomalous ones, pose little threat to any of our fundamental science.

Having stated these facts, two kinds of halo displays are more than ordinarily puzzling: those halos that slowly change position relative to the sun or moon and the so-called kaleidoscopic suns. It is difficult to imagine, for example, how a changing population of ice crystals can cause an offset halo to rotate about a point on its circumference. The kaleidoscopic suns are even stranger, with their weird dancing spheres and rays of colored light. Despite the secure laws of reflection and refraction, nature still has a few tricks we have not fathomed.

Halo displays are primarily products of optical reflection and refraction, but the phenomenon of diffraction also generates rings around the sun and moon---the coronas. The diffraction coronas of small size can be readily attributed to water droplets in the atmosphere and are not covered in this Catalog. The very large coronas, especially Bishop's Ring, which often (but not always) appear after large volcanic eruptions cannot be dealt with so perfunctorily. They are too variable in size and are not well-disciplined in time and geographical space. Several times Bishop's Ring has been observed in the absence of any terrestrial dust sources. We must admit that we lack complete knowledge of the origins and transportation of fine particulate matter in the upper atmosphere.

Primer on the Common Halos and Coronas

X0. General observations. "Halo phenomena are due to refraction and reflection of the light of the sun and moon by ice crystals; they are often very complicated and diverse owing to the variety of crystal forms and the large possible number of refractions and reflections. Refraction phenomena, especially those due to sunlight, are sometimes beautifully coloured.

Figure 1 gives a schematic drawing of some of the most common halo phenomena. It is well to know these so as to be able to recognize unusual forms when they occur.

A. The common halo of 22° radius, formed by refraction in hexagonal crystals at 60° angle, and which may be seen, entire or in part, on about 200 days of the year. This halo enjoys an enormous reputation as a sign of bad weather, which, as it is often seen in the cirrostratus in front of a depression, is not unfounded; it also occurs, often in fragments, in fine-weather cirrus. It is usually not well coloured, appearing as a narrow whitish ring, reddish on the inside.

B. The circumscribed halo, oval in shape, which may surround the 22° ring when the sun is higher than 29°. At lower altitudes it breaks up into upper and lower arcs of contact. Uncommon, but to be looked for, is a curved bar across the upper contact arc---the Parry arc. Anomalous contact arcs are sometimes to be seen.

C. The mock suns or parhelia of 22°, which are two bright patches, often brilliant and beautifully coloured and sometimes with long white 'tails', on a level with the sun. They often appear by themselves and arise when the axes of the ice crystals in large number are vertical. They are on the halo at low altitudes, but as the sun rises they recede from it and sometimes become connected with it by the arcs of Lowitz.

D. The halo of 46°, due to refraction at 90° in hexagonal ice crystals oriented at random. It is not common and is faint, though the colours are better than in the common halo. It has parhelia and certain contact arcs, but these are rare.

E. What has been described as the 'most beautiful of all halo phenomena', the circumzenithal arc, which appears as a miniature rainbow, red towards the sun and parallel to the horizon, a few degrees above the summit of the 46° halo when this is visible. Often this arc appears alone. It is not infrequent but somewhat transient, so it is not well known. Three or four times it has been seen as a complete circle. It arises from refraction through crystals shaped like plates or umbrellas, with vertical axes floating in a stable position.

F. The parhelic circle, a white ring passing through the sun parallel with the horizon and due, as its lack of colour shows, to reflection from the side faces of crystals floating with axes vertical. On it are the parhelia, those already mentioned and others, notably those at 90° and 120°. Other odd ones also occur and, if possible, the distances of these from the sun or moon should be carefully measured with instruments. Opposite the sun, on the ring, is the very uncommon anthelion, G, through which may pass certain arcs. The parhelic circle is often fragmentary.

Other interesting phenomena may be well seen at sunrise or sunset in the form of a pillar of light above the sun; this is gene-

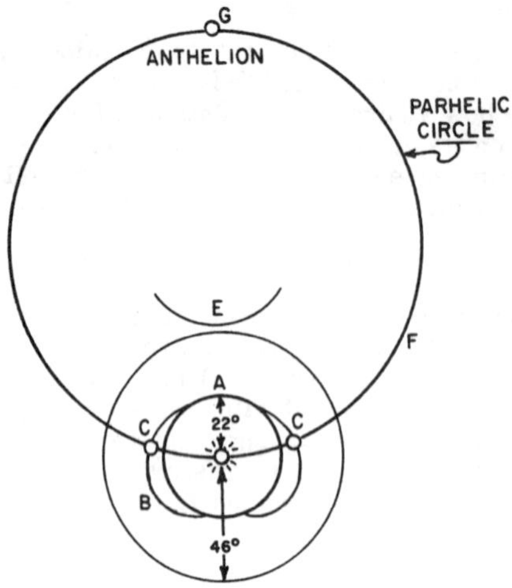

The most common halo phenomena. Letters are keyed to the text. (X0)

rally attributed to reflection of light from tabular ice crystals, but this explanation has been disputed. Pillars may be seen passing through the luminary when high in the heavens; should this coincide with the appearance of the appropriate segment of the parhelic circle, the striking appearance of a cross is produced.

....

To turn to coronae; these arise in clouds in which the drops of water or ice needles are about the same size as the wavelengths of light. In such a case diffraction occurs; the light waves run round the drops, like sea waves round a rock, and meet behind, 'interfering' with one another, which, in the case of white light, results in a separation of colour.

Most clouds show the aurole, a ring, ranging in radius from 1° to 5°, bluish on the inside, and passing through yellowish-white to red-brown. Outside this, particularly in altocumulus and fine cirrocumulus, may be one or more 'orders' of rings, blue-green-red. Coronae vary considerably in radius, the rule being the smaller the drops, the bigger the corona; the best ones are formed in young clouds where the drops are most uniform in size..... The irridescence which sometimes appears on white alto-stratus, altocumulus, and cirrocumulus, often many degrees from the sun, is probably due to fragments of very large coronae.

Following great volcanic eruptions, when the stratosphere is loaded with very fine dust, a huge aureole called Bishop's ring' may be seen. Also the vague glow round the sun in normal times may be a mixture or aureoles and coronae produced by miscellaneous atmospheric particles. This glow varies from day to day; it is worth watching since it is an index of atmospheric purity. It may be well seen by standing so that the sun is screened off by some object.

A striking relation of the corona is the 'glory', which is common in hilly country in positions where the shadow of the observer can fall on mist; the watcher sees his head (and his only!) surrounded by a series of coloured rings. The same may be seen round the shadow of an aeroplane; indeed, an aviator when passing through a cloud may see a corona in front of him and a glory behind." (R1) Note that in this Catalog glories and related phenomena are covered in GEI.

References

R1. Botley, Cicely M.; "Halos and Coronae," Weather, 18:5, 1963. (X0)

GEH1 Offset Halos and Anomalous Arcs

Description. Circular halos that are not centered on the sun or moon and arcs that do not belong to the recognized halo phenomena. Anomalous halos and arcs are usually fainter than the accompanying normal halos and rarely show colors.

Background. The most common halos and arcs are illustrated in GEH0; beyond these are many more that are "recognized"; that is, they have been observed and accepted as real, although arguments may persist as to the specific types and orientations of ice crystals that generate them. Singled out here are halo phenomena which are, as far as is known, not recognized.

Data Evaluation. Many records of anomalous halos and arcs have been found in the literature examined so far; and undoubtedly many more await collection. It is tempting when confronted with observations of anomalous configurations to claim that the observer has measured something wrong or has the diagram wrong. Often, just a slight modification in an observation can turn something anomalous into one of the myriad "recognized" phenomena. The way to combat this sort of criticism is to collect several good records of the same phenomenon, observed by different people at different times. This has been accomplished with offset halos and some of the curious arcs. Rating: 1.

Anomaly Evaluation. One's expectation is that all anomalous halos and arcs are ultimately explicable in terms of unappreciated ice crystal forms and unnoticed secondary sources of illumination. The possibilities are so profuse that the anomalies described here are very probably not threats to the present theories of atmospheric optics; it is all a matter of pinpointing the right crystal form, its orientation, and/or secondary light sources. Rating: 3.

Possible Explanations. Offset halos and arcs usually require secondary light sources for explanation. Mock suns, mock moons, intersections of halos and arcs, and possibly reflections from natural mirrors (clouds, sea surface, etc.) are among the possibilities to be considered.

Similar and Related Phenomena. Anomalous arcs accompanying rainbows (GEB3, GEB6); skewed halo displays (GEH9).

Examples of Offset Halos and Anomalous Arcs

X1. September 8, 1816. Newport, Rhode Island. "The above described phenomenon (see drawing) was observed at this place on the same day, Sunday, Sept. 8th, between the hours of two and three in the afternoon; duration of its appearance about 40 or 50 minutes. The annexed sketch was delineated at the time by actual observation. The sun at that hour bearing about south west. The halo immediately encircling the sun was about the usual diameter of those commonly seen, but uncommonly bright, being tinged in its whole circumference with prismatic colours, especially on its upper and north-eastern limb, where a bright mock sun (or sun dog, as it is usually called by seamen) was formed; the rays of light from which, formed a second halo of a smoky white colour, well defined in its whole circumference, but more faintly as it approached the primary halo on its south-western limb, where they united, the second being exactly double the diameter of the first; this second circle having a mock sun, less bright than the first, on its north-western, and another on its south-eastern limb, on a right line, nearly in that direction from the one in the primary halo; the ray of light thrown off from which, formed two other circles, double the diameter of the second, and crossing the primary halo on its south-western limb, in the brilliancy of which it was lost; and crossing the second halo on its north-eastern limb about thirty degrees from the horizon, where they were well defined, but diminished into slight pencils, until they were lost as they approached the north-eastern part of the primary halo. The rays of light thrown off from the concentration of these circles at their junction, or place of crossing the north-east, formed another and larger circle, (about 120 degrees of which was below the horizon) which crossed the two last in faint lines, and lost itself as it approached the second circle, which it would have touched, if continued on its south-western limb." (R1)
In this example, the mock suns and a point of intersection form the centers of the offset halos. (WRC)

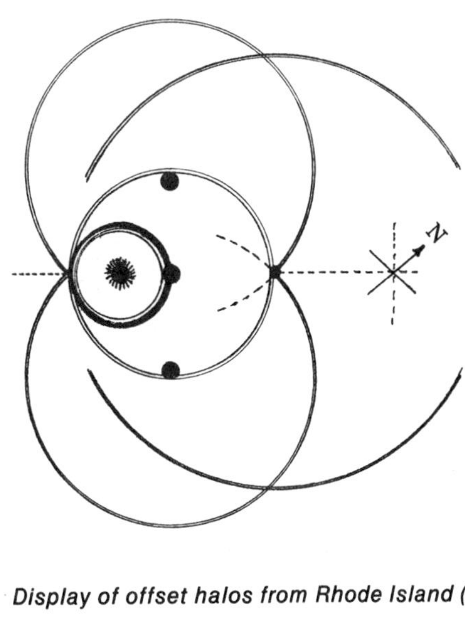

Display of offset halos from Rhode Island (X1)

X2. August 19 and 26, 1825. Tod County, Kentucky. "On Friday, the 19th of August, there appeared circles in the atmosphere, as laid down in the following diagram:

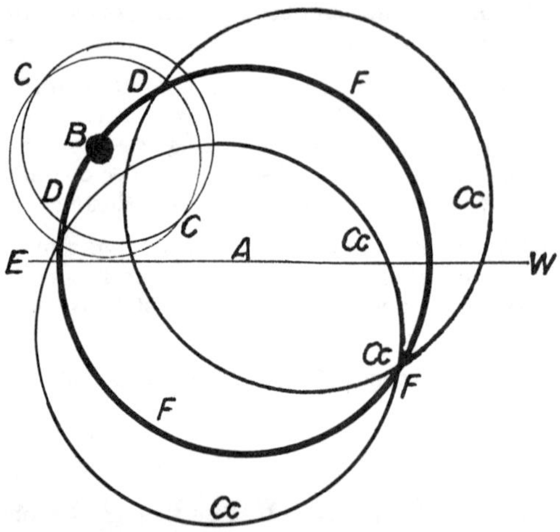

E and W represent the east and west point of the heavens; A is the zenith; B the sun; CC is a prismatic circle, of surprising brilliancy; F a very bright circle of light, passing through the sun B; Cc Cc two sections of circles, intersecting the circle F, and the point Cc F. These sections were very bright at their intersection with F, but became gradually invisible as they approached the sun. B and A and Cc F were in a right line, and the intersection Cc F was the same height above the horizon as the sun, and moved north, and approached the zenith in the same proportion as the sun moved south and approached the zenith. The circle F and the sections Cc Cc were circles of similar radii, and diminished in size, as the sun approached the zenith. These circles were first observed about 8 o'clock, and continued until 11. There was not a cloud to be seen, but so thick a haze, high in the atmosphere, that the sky appeared completely black, and the sun shone with so much splendor, and there was such a glare, that any person's eyes would be filled with tears in a few minutes, by going into the light. The same phenomena appeared again on the following Friday." (R1)

X3. August 28, 1840. Greensburg, Pennsylvania. "This phenomenon consisted of from three to five circular belts or zones of light, one of which emulated, in appearance, the splendor and magnificance of the most gorgeous rainbow. The arrangement of these rings was somewhat singular; the first or inner one, which had the sun as its center, was truly brilliant, exhibiting all the prismatic hues of the rainbow, the colors of which were so dazzling that the unprotected eye could scarcely rest upon it a moment. This, I presume, was occasioned by the sun being near the meridian, and consequently many of his rays would impinge upon the halo, without passing through the mass of vapor, to the existence of which I attributed the formation of the halo. The

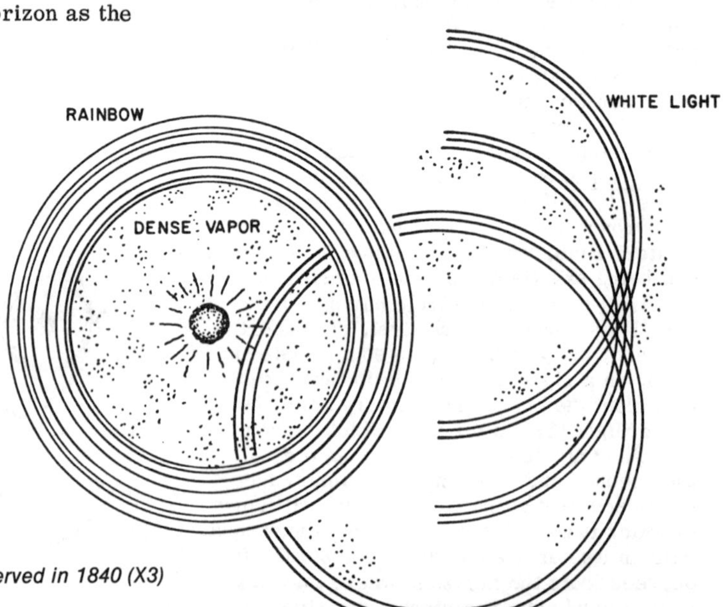

Display of offset halos observed in 1840 (X3)

outer circles, however, one only of which appeared to be perfect, were composed of pure white light, and had for their centres the circumference, or a point near it, of the inner ring. Consequently, their circumferences, if all the circles had been perfect, would necessarily have passed through the apparent situation of the sun. I mentioned, however, that only one of these rings was perfect, the others were concentric arcs of circles which crossed one another, as seen in the accompanying diagram. In the centre of the inner circle and bounded by it, a bluish mass of dense vapor was perceptible, which gave to the whole an embossed appearance, and added much to the beauty and brilliancy of the scene. Around and within the exterior circles there were also perceptible masses of vapor, though obviously much less dense than the mass which was nearer the sun. With the exception of these masses of vapor, and a large cumulus which lay to the south of us, and here and there a few scattered cirri, the sky was cloudless and the atmosphere calm and serene." (R2, R21)

X4. October 19, 1846. Nottinghamshire, England. "At 22h the phenomenon might be said to have commenced, for the sky had assumed a hazy appearance, and a halo (I) of 22°30' radius had become visible; this was of a pale straw colour and soon became very bright. The sky within the halo was considerably darker than that without. 22h 10m. An inverted arc of another halo (II), of (apparently) 33° radius, also of a pale straw colour, formed at the vertex of the other halo, appeared. At 22h 15m the halo (II) had assumed a brilliant yellow colour, and the portion (IV) which joined the halo (I) a flame-like appearance, having widened considerably, being now about 3° in width and 6° in length. This brilliant parhelion, as I shall call it, became in a few minutes intense in brilliancy, vieing with the sun, for that luminary had been deprived of his usual brilliancy by haze, and had its rays carried out considerably. No change took place until 23h 30m, when a third halo (III), of 33° radius, and also of a straw colour, was visible. This circle touched the halos (1 and 2 at IV), and its centre was 11° below the true sun. The upper half of the halo (III) was bright; but the lower portion, which is marked by the dotted lines, was not visible, owing to cumuli being very abundant on the south horizon. The lower portion of the halo (I) was also at this time nearly invisible, but the parhelion (IV) was even more brilliant than before. This appearance continued until 0h 10m, then the halo (III), together with the lower portion of the halo (I), vanished. The parhelion (IV), with the halo (II) and upper half of the halo (I), was still very bright." (R3) The arc (V) formed later. Note that halos of 33° radius are not among the standard halo phenomena. (WRC)

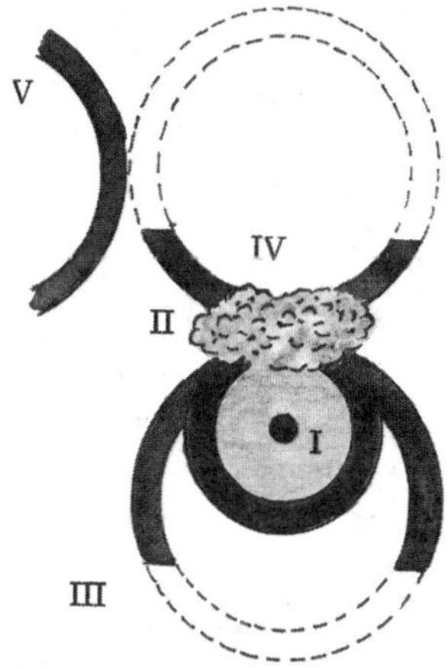

Halo complex at Nottingham, England. The sky inside the halo was darker than that without. (X4)

X5. August 13, 1861. Sydney, Nova Scotia, Canada. "This day I witnessed, as well as all our household, a remarkable phenomenon. The sun appeared as if in mourning. At twelve o'clock I first observed it. An immense dark halo surrounded the sun, as dark as a thunder-cloud; the outer edge of it was irridescent, appearing like a circular rainbow. A ring of bright white light, of greater diameter than the halo, intersected it, passing through the sun; and two other rings of the same sort, but fainter, of still larger diameter, intersected the first ring and each other, in the manner I have attempted to show in the accompanying sketch, in which I have preserved, as nearly as I could judge, the relative proportions of the halo and rings to each other and to the sun. It was a bright day, the sun blazing directly overhead, out of a clear blue sky, but there were hard, electric-looking clouds in other parts of the sky; and from these were drawn out long, attenuated, fleecy, ribbon-like appendages, which all took a circular form, having the sun apparently for a centre. The halo and rings first appeared about an hour before noon, and continued as long after

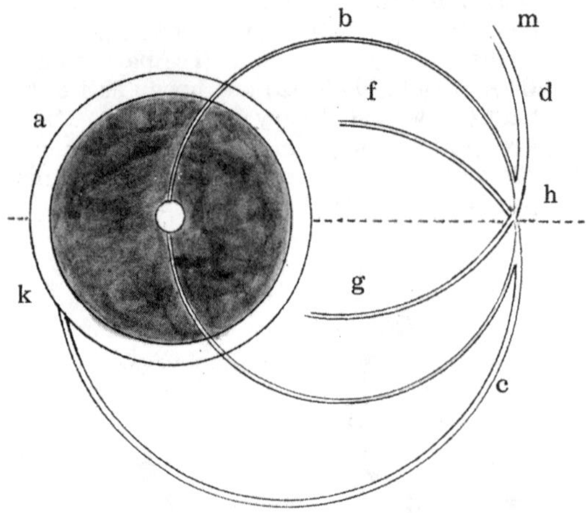

System of offset halos seen in Canada. Sky inside halo was darker than that outside. (X5)

noon. Throughout the remainder of the day the clouds were in circular tiers or ranges about the sun, which became obscured." Key for the sketch: "(a) Iridescent margin of the dark halo round the sun. (b) Ring of white light passing through the centre of the sun. (c, d) Two fainter rings of light, of which the interrupted or broken parts nearest the sun (f and g) appeared to converge toward the sun and fade away. (h) A node of light at the intersection of the three rings. (k) This end terminated here. (m) This end broken off or faded away. (R4)

X6. November 10, 1861. Muskoka River, Ontario, Canada. A halo display consisted of a circumzenithal arc, a circumhorizontal arc, plus two offset, circular halos. The ordinary 22° halo was not evident. (R8) Here, it is possible that the two offset halos might be a badly drawn ordinary halo of 22° plus the circumscribed halo. (WRC)

X7. May 26, 1865. Crawfordsville, Indiana. "About nine o'clock a series of halos or rings began to form about the sun, and continued to increase in brilliancy until near noon---then slowly disappeared during the afternoon---traces being visible at six o'clock in the afternoon. The accompanying diagram is a correct representation of the phenomenon as it appeared at twelve o'clock noon. The most brilliant ring was described about the sun as a center with an angular radius (SE) of twenty-two degrees by careful measurement. The bands were white, yellow, red and green. The white was exceedingly brilliant. The yellow and red were also very bright. The breadth of this series of bands forming the primary ring was about one and a half degrees. The arc LMH described about the sun as a center with a radius of forty-four degrees presented less brilliant, but well defined colors in the reversed order,---the white band being on the outer circumference, and green within,---with another inner band of violet. At A on the circumference of the primary circle, the center of the second circle was

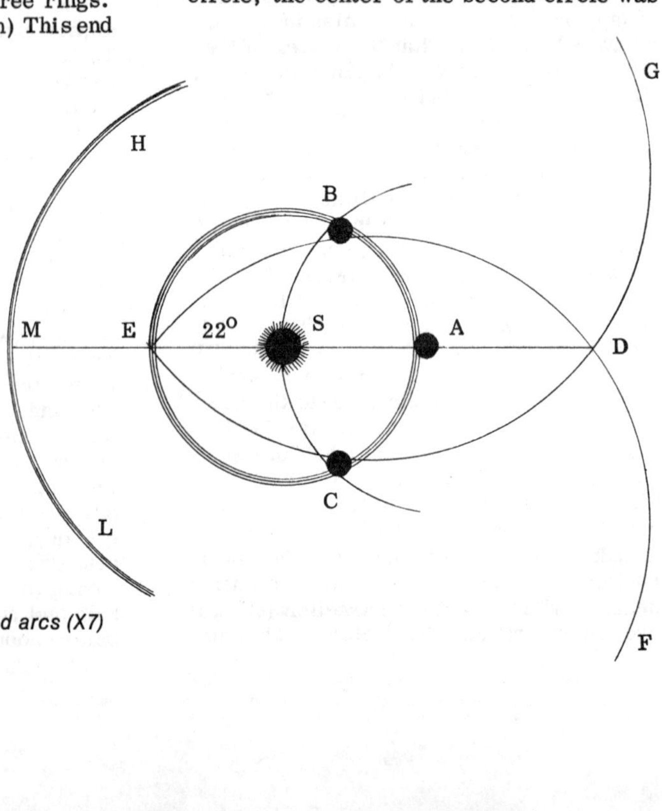

Indiana display of offset halos and arcs (X7)

located. The radius of this circle was the same as that of the primary---its circumference passing directly through the sun. The ring was chiefly white light, of less brilliancy than the primary, yet very sharply marked throughout its entire circumference. From D the arcs DF and DG branched off symmetrically with reference to the axis MSD, indicating the centers of curvature at B and C, the intersections of the secondary circle with the primary. These arcs were of white, well-defined, mellow light." (R5)

X8. January 29, 1882. Tokyo, Japan. "On January 29, between 6.20 and 6.40 p.m., I saw a very peculiar halo round the moon. I was in Tokiyo (Yedo) at the time, and the same appearance was seen by many of my friends at Yokokama (sic), 18 miles distant. It was like the figure herein. At first I was under the impression that the figure was slightly elongated on the outer circle, that is to say, the distance from the centre of the moon to the edge of the first halo was less than that from the edge of the first halo to the farthest point in the second; but on drawing it I am convinced that this was an optical illusion. The lowest part of the second halo intersected the lower horn of the moon in a peculiar way, as given in the drawing. The moon was nearly half full." (R6, R21)

with a long-armed compass, I found that the instrument confirmed what my unaided eye had detected. The inner edge of the top part of the circle was 21 1/2 degrees from the sun's upper edge, and the inner edge of the lowest part of the circle was 23 degrees from the sun's lower edge,---a difference of about three times the sun's width." (R8)

X10. March 8, 1920. Ellendale, North Dakota. "From the drawing the sun elevation appears to be about 38 degrees. This agrees with the calculated elevation of the sun at 1:30 P.M., when the display was said to be as shown in the drawing. The 22-degree halo, sun dogs, infralateral arcs, and parhelic circle are all rather obvious by now. Arc cc looks like the top of the circumscribed halo, though it is a bit surprising that the lower part is entirely missing. The supralateral arc cannot appear for a sun elevation this high, and so arc bb would appear to be the upper part of the 46-degree halo, with the rest missing. The circle centered on the zenith is not the circumzenithal arc, because it is shown tangent to the 22-degree halo. It presents an interesting problem." Many of the other arcs shown can be simulated on computers, but the arc xx represents an unsolved problem. (R20)

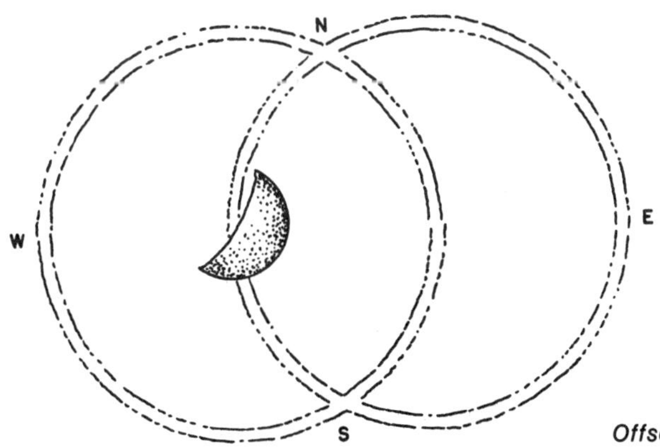

Offset lunar halo from Japan (X8)

X9. March 25, 1899. Barrie, Ontario, Canada. "On March 25th, 1899, at 10.30 a.m., I saw a solar halo at Barrie, Ont., consisting of a primary circle around the sun, but the sun was not at its centre, i.e., it was eccentric. On measuring the parts

X11. April 28, 1883. Portland, England. Two segments of slightly prismatic circular halos were seen inside the 22° halo on either side of the sun. Their centers of curvature were well outside the 22° halo. (R7) The segments were

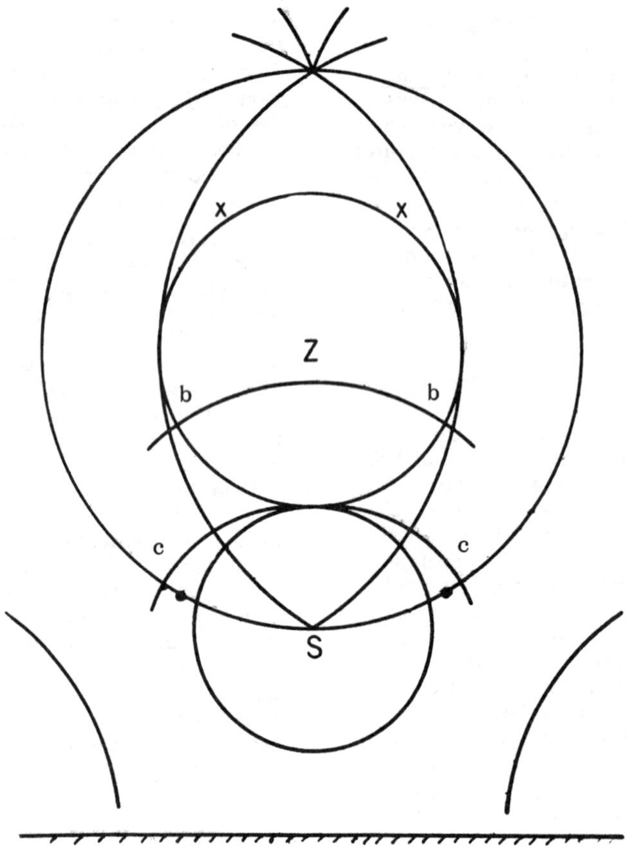

The great Ellendale, ND, halo display of 1920. S = sun; Z = zenith. Arc xx is unexplained. (X10) (Adapted from R20)

of such large radius and located so close to the sun that this does not seem to be an instance of a circumscribed halo. (WRC)

X12. March 6, 1925. Caribbean Sea, North Atlantic Ocean. "At 11.45 a.m. a halo, showing the colours of the spectrum, formed around the sun with a radius of 21 1/2°, the breadth of the spectrum subtending an angle of 3/4°. Shortly afterwards an arc of a second halo appeared to the southward, this arc being concentric with and similar to the first, while a third complete halo and an arc of a fourth were observed. Neither of these two latter showed the spectrum, nor were they concentric one with the other or the sun." (R9) No ready explanation comes to mind for the two offset halos. (WRC)

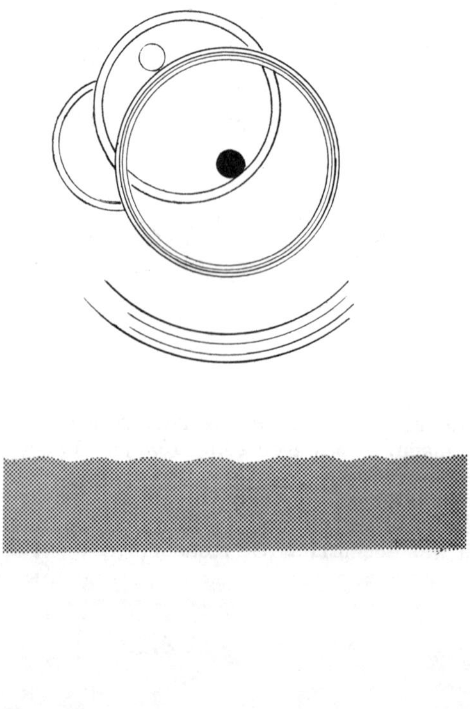

Two normal halos plus two offset halos (X12)

X13. July 15, 1928. South Atlantic Ocean. "At 4.45 a.m. A.T.S. (0740 G.M.T.) a peculiar arc of white light appeared in the sky, as shown in the illustration. On the left-hand side was a clearly defined line between the lightened portion of the sky and the dark. On the right-hand side the light gradually merged into the dark. At 5.20 a.m. (0815 G.M.T.) the moon rose in position indicated, but the arc of light remained until gradually lost in the dawn. The only clouds in the vicinity of the light were a few Ci.-St. in long strips. The moon's age was 27.1 days. Sunrise was at 6.45 a.m. A.T.S." (R10) An editiorial comment suggests that this was a display of the zodiacal light, but the zodiacal light is usually conical in shape and widest at the horizon. (WRC)

X14. February 26, 1948. North Atlantic Ocean. "0140 to 0220 G.M.T. A lunar phenomenon was observed showing three distinct arcs terminating at a central point about 5° above the horizon. Arc Z commenced in close contact with a whitish disc round the moon and all three arcs were uniform in thickness throughout their lengths, about 3/4°. Arc X equalled arc Z in length and ended about 5 1/2° above the horizon. These arcs faded out at 0220....Note.--- This is a remarkable observation, of which no explanation can be given. The arcs appear to be halo phenomena, but their positions do not agree with those of any of the known halo phenomena." (R11)

Arc of white light precedes moonrise. (X13)

Offset halo arcs in vicinity of moon (X14)

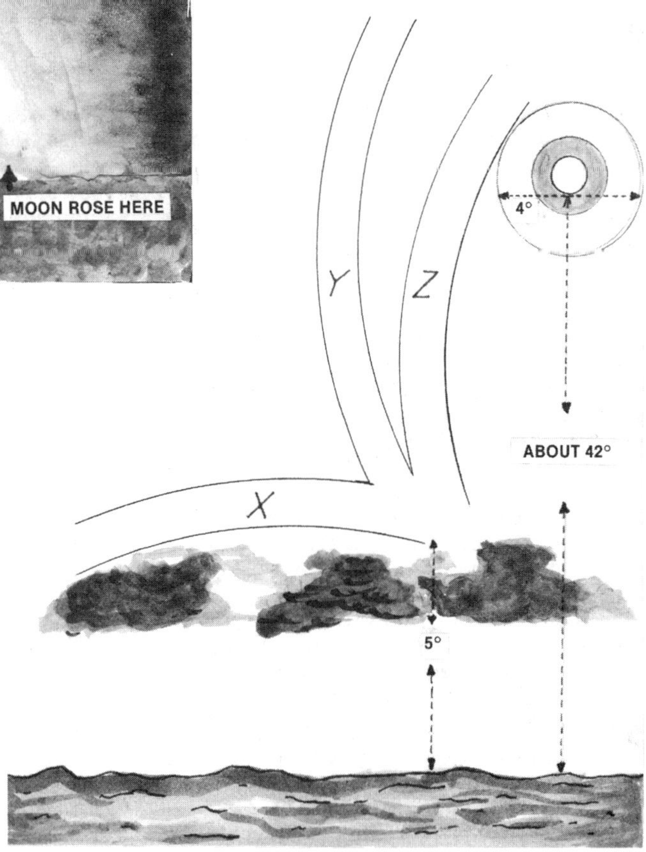

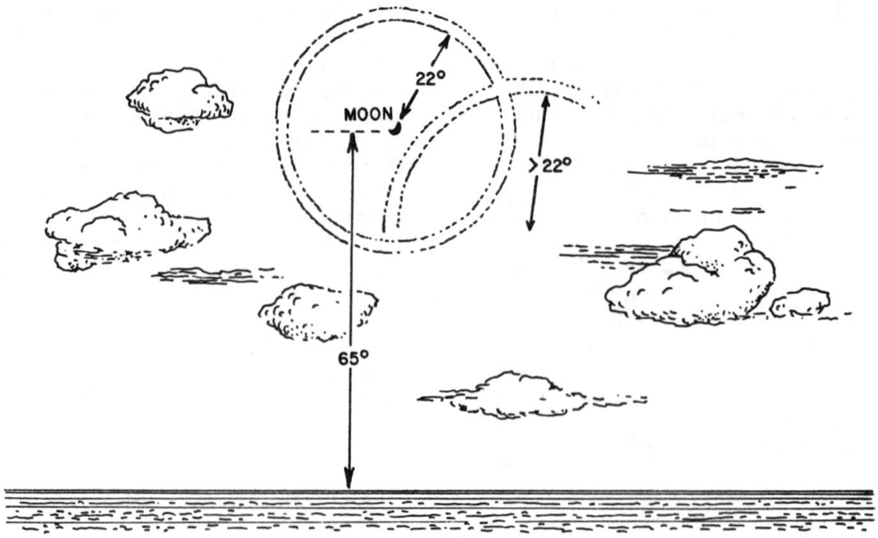

Lunar halo with larger, offset arc (X15)

X15. March 7, 1950. Caribbean Sea, North Atlantic Ocean. "0630 G.M.T. A distinct lunar halo of radius 22° was observed. The halo was obscured from time to time by Cu. At 0640 an arc of a second halo, of apparently slightly larger radius, appeared. It intersected the primary and the outer perimeter passed to within about 2° of the moon, as shown in the sketch. The secondary was visible for 5 minutes, after which the area became obscured by Cu. After a few minutes the primary was again observed, but the secondary had vanished. The moon, which was 20 days old, had an altitude of 65°.....Note. This is another interesting halo observation which cannot be fully explained. At first sight the arc of the halo intersecting the ordinary halo appears to be a secondary halo.....There are, however, two reasons why this cannot be the case: (1) that the halo is stated to be of slightly larger radius; (2) the altitude of the moon is 65°. A secondary halo cannot be formed with an altitude of the moon (or sun) greater than 60° 45'." (R12) These offset white arcs closely resemble some of those seen in association with rainbows (GEB3). R. White suggests that contrary to the Note the offset arc is actually a secondary halo about a mock moon, which could have been obscured by a cloud. (R21)

X16. October 4, 1950. South China Sea, North Pacific Ocean. The moon was ringed by the common 22° halo. A white bow of much larger radius and a center far below the moon cut through the top of the 22° halo. The angular distance between the top of the 22° halo and the top of the intersecting arc was 6°. (R13)

X17. October 20, 1950. South Pacific Ocean. "2005 A.T.S. (Zone plus 8). A lunar halo of radius 21 3/4° appeared as a plain white ring. At 2040, two arcs, meeting as in the sketch, appeared and were visible for about 12 minutes. At 2220 an area around the moon appeared green, tinged with orange red. The halo was visible until 2235.....

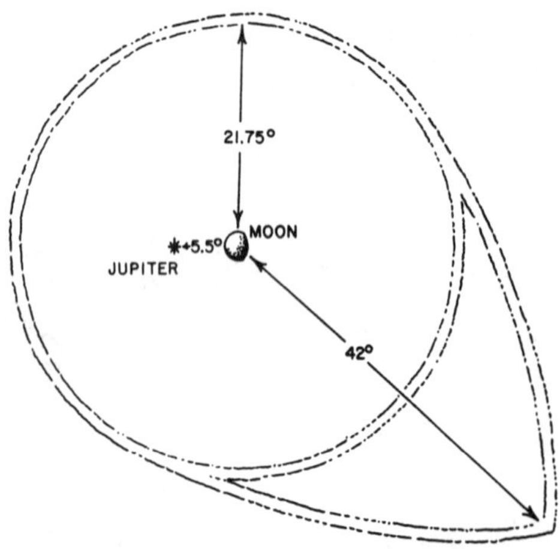

Normal 22° lunar halo with anomalous intersecting arcs (X17)

Note. This is an interesting observation; the two arcs shown inside the ordinary lunar halo do not form part of halo phenomena as at present known." (R13, R21)

X18. November 13, 1951. North Atlantic Ocean. "2345 G. M. T. Observed white lunar halo of radius 22°. At 2400 a bright white arc appeared tangent to the halo, the point of contact being very bright. Shortly afterwards another arc appeared as shown in the figure; this arc was not as bright as the former. At 0130 the halo and arcs faded. ..... Note. This is a very interesting observation, since neither of the two arcs shown are recognised halo phenomena. The upper one appears at first sight to be the ordinary upper tangent arc to the 22° halo, but there are two reasons why it cannot be: (a) with a lunar altitude of 68° the tangent arc bends downward so much as to be practically coincident with the upper part of the 22° halo, (b) the point of contact of the arc and the 22° halo, as shown in the sketch, is not vertically above the moon. The arc shown intersecting the halo below the moon is not the lower tangent arc since the latter touches the halo at a point vertically below the moon and never intersects it. Furthermore, with a lunar altitude of 68° the lower tangent arc bends upward so much as to be practically coincident with the lower part of the halo. It will be noted that the points of intersection, of which no measurements are given, are not symmetrical with respect to the vertical line through the moon." (R14)

X19. December 6, 1974. Indian Ocean. The common 22° solar halo formed with a portion of the 46° halo below it. Both showed spectral colors. In addition, a white halo of about 22° intersected the normal 22° halo. Its center was located well outside the circumference of the normal 22° halo. (R19) The intersecting halo, considering the high altitude of the sun, may have been the circumzenithal halo. (WRC)

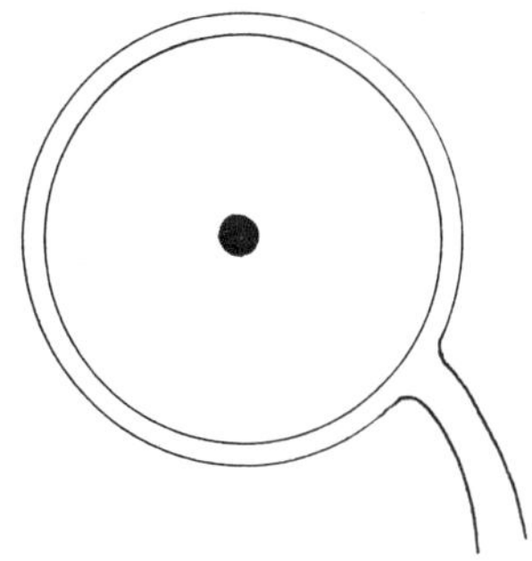

Unrecognized halo arc seen in North Atlantic (X18)

X20. September 19, 1953. Bay of Bengal, Indian Ocean. "1530 G. M. T. A lunar halo was observed with two arcs of contact. The inner radius of the halo was 20° 46'. The edges of the halo and the contact arcs were soft and merged gradually into the night sky. The arcs were slightly fainter than the halo. The phenomenon was observed for one hour, when it was obscured by Ac. Altitude of moon, 54° 35'." (R15, R21)

X21. June 13, 1954. Sistiana, Italy. "Time: 1145-1315. Description: we first noticed a circle of darker sky, concentric with the sun, at 11.45. At noon we noticed the second circle (same diameter as the first); it was paler than the first, and its circumfer-

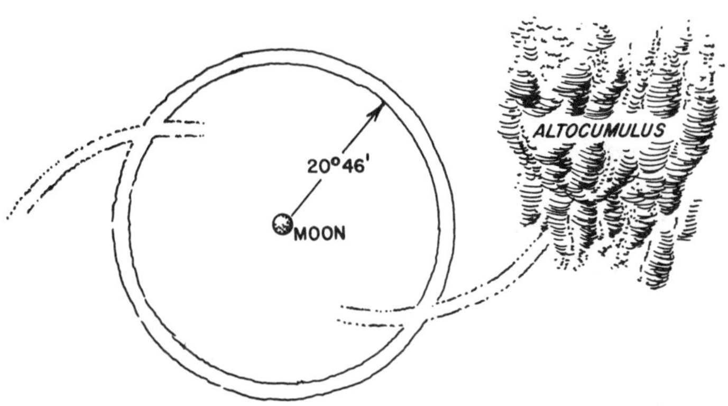

Lunar halo with two unusual arcs, Bay of Bengal (X20)

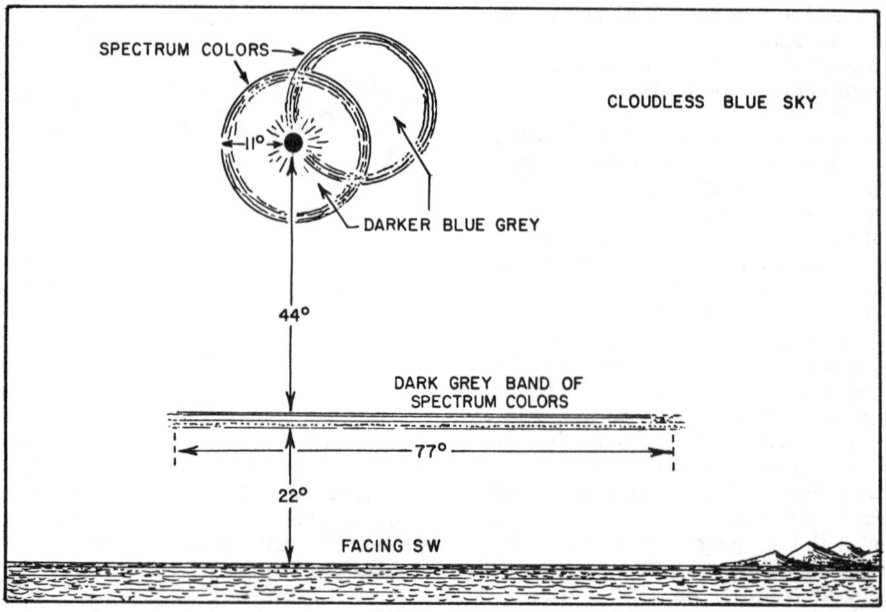

Halo with anomalous radius intersected by a second offset halo. Halos separated sky colors. (X21)

ence cut the center of the sun. The peripherae of both circles were concentric rings of spectrum colours (unfortunately I did not note the positional order of the colours). At 1210 we noticed the horizontal bar of dark grey 'cloud' in front of us, between the almost overhead sun and the horizon; it was composed of horizontal layers of the spectrum colours (again, unfortunately I did not note the positional order of the colours). Meanwhile the two circles overhead maintained their position in relation to the sun, but were gradually becoming fainter. By 1315, when we left the beach everything had practically disappeared. It was a fine day, with a cloudless blue sky and blazing sunshine." (R16, R21) The spectral colors parallel to the horizon are likely part of the circumhorizontal arc. (WRC)

X22. February 23, 1957. South Atlantic Ocean. "0710-0728 S. M. T. Halo phenomena were seen after sunrise. The radii of the two main halos were 22° and 46°. There was a brilliant vertical ray of light from the mock sun, shown as vertical lines in the sketch, and an arc of a white halo crossing this. Both the main halos were

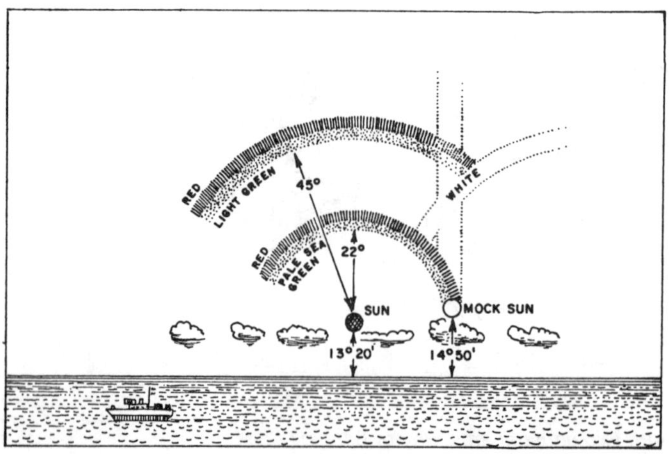

A South Atlantic halo display. The radii are normal, but the offset arc and mock sun pillar are unusual. (X22)

coloured, the 46°one red and light green and the 22° one red and pale sea-green. There was Ci cloud above the 46° halo, Ac in bands between the two halos, and Ac or high Sc above the 22° halo with a layer of separate Cu at lower altitude. There appeared to be a difference in altitude between the true sun and the mock sun and, on making careful measurement, the altitude of the lower limit of the true sun was found to be 13° 20' and that of the lower limit of the mock sun 14° 50', viz. 1° 30' greater in altitude." (R18) The offset white arc is perhaps the most anomalous aspect of this display, however, the presence of a mock sun pillar in the absence of one from the real sun is unusual, as is the difference in altitudes of sun and mock sun. (WRC)

X23. 1958. Antarctica. The arcs numbered 1, 2, 3, and 4 in the sketch are difficult to explain, although computer simulations by Greenler approximate some of them. (R20)

X24. May 29, 1963. Indian Ocean. "Between 1515 and 1520 GMT when the moon was at an altitude of 60° and bearing 302°, the halo complex shown in the sketch was observed. The halo (a) and the arc (b) formed first; (x) and (y) appeared at the time when (a) and (b) touched. The arc (z) formed when (x) and (y) moved outwards and joined; the whole system faded when (z) lay along the top of (a)." (R17) R. White has commented that this phenomenon is a dynamic one and that not all the arcs shown in the diagram are present at the same time. (R21) However, unexplained offset arcs <u>are</u> present in the description. (WRC)

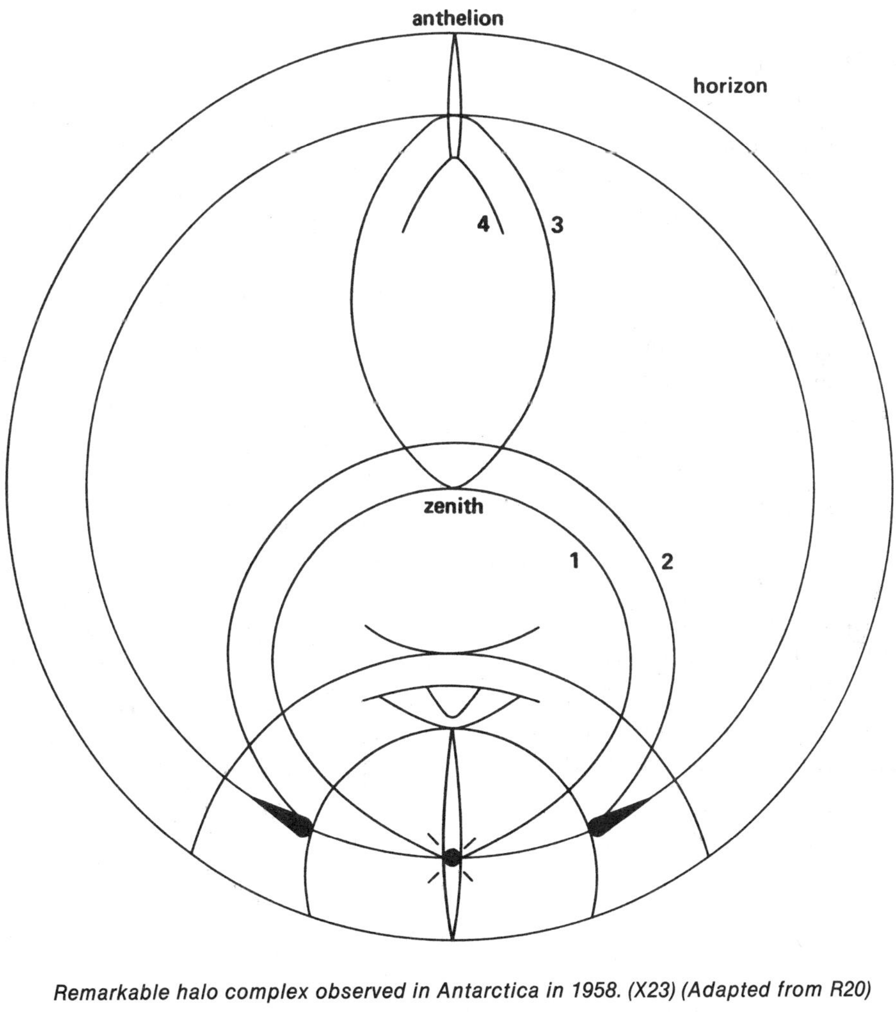

Remarkable halo complex observed in Antarctica in 1958. (X23) (Adapted from R20)

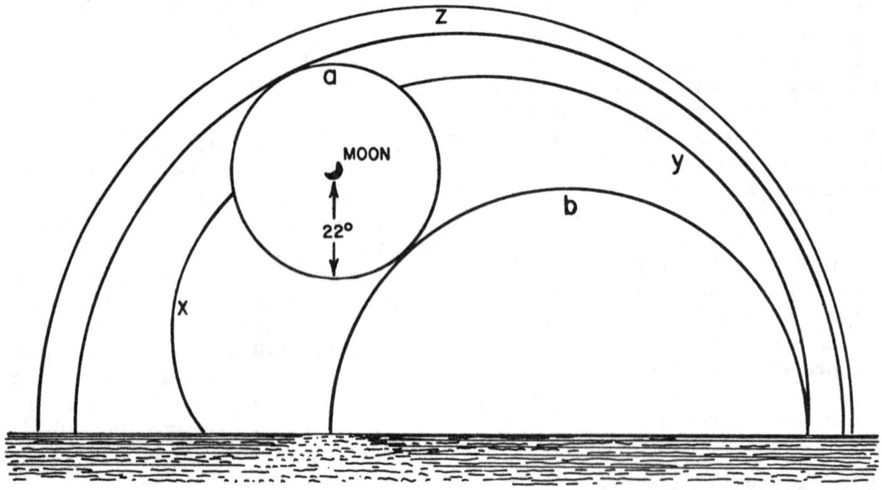

A composite drawing of the dynamic display described in X24. Letters refer to the text. (X24)

References

R1. Meriwether, Charles, et al; "History and Description of Some Remarkable Atmospheric Appearances," American Journal of Science, 1:11:325, 1826. (X1, X2)

R2. King, Alfred T.; "A Description of a Halo or Corona of Great Splendor, . . . ," American Journal of Science, 1:40:25, 1841. (X3)

R3. Lowe, Edward Joseph; "Remarkable Solar Halos Seen on the 19th of October 1846," Philosophical Magazine, 3:39: 440, 1846. (X4)

R4. Crawley, H. W.; "On a Solar Halo Observed at Sydney, Cape Breton, Nova Scotia, August 13, 1861," Report of the British Association, 1861, p. 63. (X5)

R5. Campbell, J. L.; "On a Solar Halo, Seen at Crawfordsville, Ind.," American Journal of Science, 2:40:394, 1865. (X7)

R6. Pryer, H.; "Halo Round the Moon, Knowledge, 1:521, 1882. (X8)

R7. Cardwell, E.; "Solar Halo As Observed at Portland, April 28," English Mechanic, 37:217, 1883. (X11)

R8. Hunter, A. F.; "Distorted Solar Halos," Royal Astronomical Society of Canada, Journal, 12:1, 1918. (X6, X9)

R9. Hartman, W.; "Solar Halo," Marine Observer, 3:42, 1926. (X12)

R10. Robinson, F. W.; "Luminous Arc," Marine Observer, 6:147, 1929. (X13)

R11. Lewis, J. G.; "Lunar Phenomenon," Marine Observer, 19:12, 1949. (X14)

R12. Hazelwood, H. W.; "Lunar Halo," Marine Observer, 21:16, 1951. (X15)

R13. Morgan, T. S., and Robertson, A. I.; "Lunar Halos," Marine Observer, 21: 216, 1951. (X16, X17)

R14. Davies, H., and Hobson, G. H.; "Rare Lunar Halo Phenomena," Marine Observer, 22:196, 1952. (X18, X19)

R15. Dick, W.; "Halo Phenomena," Marine Observer, 24:143, 1954. (X20)

R16. Crabbe, J. A.; "Rare Solar Haloes," Weather, 10:359, 1955. (X21)

R17. Davison, I. C.; "Lunar Halo Complex," Marine Observer, 34:73, 1964. (X24)

R18. Edward, C. M.; "Halo," Marine Observer, 28:17, 1958. (X22)

R19. Ansell, I. R.; "Solar Halo," Marine Observer, 45:166, 1975. (X19)

R20. Greenler, Robert; Rainbows, Halos, and Glories, New York, 1980. (X10, X23)

R21. White, R.; "On the Halos in the Handbook of Unusual Natural Phenomena," Journal of Meteorology, 6:133, 1981. (X3, X8, X15, X17, X19, X20, X21, X24)

GEH2 Noncircular Halos

Description. Halos that depart significantly from the expected circular shape. Ellipses are by far the most common aberration. In these instances, the major axis may be vertical, horizontal, or somewhere in between. Almost all noncircular halos are lunar.

Background. In passing one must wonder why noncircular halos are nearly exclusively the province of the moon!

Data Evaluation. Several elliptical lunar halos have been observed and measured carefully. Other shapes---squares, tear-drops---are one-of-a-kind. Rating: 1.

Anomaly Evaluation. Halo theory insists that all halos be circular under ideal conditions. Conceivably abnormal refraction and changing ice-crystal population and orientation may distort halos, but the effects of these variables would be expected to be patchy, producing irregular and poorly defined halos. Rating: 2.

Possible Explanations. Abnormal refraction; changing crystal species and/or orientation due to varying meteorological conditions.

Similar and Related Phenomena. Circumscribed halos are obviously noncircular but are readily identified and explained; many arcs in halo displays are not circular but may be easily distinguished from noncircular halos because they do not enclose the sun or moon.

Examples of Noncircular Halos

X1. April 4, 1898. Bloomfield, New Jersey. "The moon (12 days) was somewhat E. of S.E., at an altitude nearly midway between horizon and zenith. Starting from it horizontally on either side was a nebulous cloudy band, which, as the eye followed round the vault, was seen to complete a great ring in the heavens, lying in a perfectly horizontal plane, and of an average width of a little more than the moon's disc..... In the larger sketch (not reproduced) a portion of this great ring is shown, that part of it for about 20° each side of the moon being much narrower and fainter, and constituting as it were the major axis of the large elliptic halo. From above and below the moon there also extended for a few degrees 'coronal' rays." (R1) The sketch published with the letter shows an elliptical halo of strong eccentricity, with its major axis horizontal. (WRC)

X2. February 25, 1901. Media, Pennsylvania. "This phenomenon consisted of an elliptical ring around the moon with axes apparently about six degrees and nine degrees respectively. The principal axis of the ellipse was vertical while the terminator of light on the moon's surface made an angle of about 45 degrees with the horizon. The moon was about half way down in the southwest and half full. A southeast breeze was blowing and very thin fleecy clouds could be seen passing over the face of the moon. The ring persisted for fifteen minutes or more. The peculiar orientation of the terminator and major axis is the difficult part to explain. One might expect an elliptical though perhaps ill-defined ring from an elongated source of light, but why should the major axis be inclined to the terminator?" (R2) Not only was this halo elliptical, but it possessed unusual angular dimensions. (WRC)

X3. 1908. No place given. An elliptical lunar halo was observed with axes of 7° and 4°. (R4)

X4. January 21, 1913. South Atlantic Ocean. "The halo was square with quite straight sides and the angles quite sharp, and I estimated the length of each side of the square at 3 diameters of the moon. The halo did not seem to vary in size as the moon rose higher, though I think at times it became

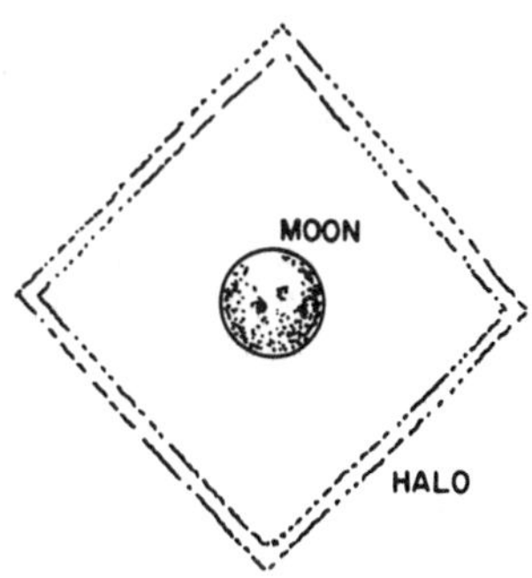

Square halo with sharp angles seen in the South Atlantic (X4)

less distinct, or possibly became rather less bright as the moon got higher, but that was hard to judge. I pointed out the halo to some of the ship's officers, none of whom seems to have seen one like it before, so it may be very uncommon. The following is an extract from the ship's log:---'Lat. 4° S. Long. 8° W. January 21, 1913. 8 p.m. The moon (aged 14 days) is framed in a dull, yellowish halo of square shape, which is about 3 lunar diameters in extent and is placed with a corner towards the horizon after this manner. (See sketch.) January 22. 5 a.m. The halo is still visible.'" (R3)

X5. June 9, 1917. No place given. An ordinary 22° halo was located well within a quadrant of a large ellipse. (R4) The elliptical arc does not seem to have been one of the common contact arcs and was well-separated from the 22° halo. (WRC)

X6. March 18, 1918. New Malden, England. "The accepted explanation of the haloes of 22° radius which are seen surrounding the sun and moon implies that they are exactly circular in form. About two years ago, I noticed a halo which appeared to be elliptical with the major axis vertical. I was unfortunately unable to take any measurements on that occasion, but on March 18 last a lunar halo, which was visible for a considerable time during the evening, also appeared to possess a decided, though slight, ellipticity. That this deviation from the circular form was not an illusion I was able to verify by noting the positions of Capella and γ Geminorum relative to the ring. At 7.30 p.m. Capella appeared to be exactly upon the inner edge of the halo, while γ Geminorum was within the ring at a distance from it, which, as nearly as I could judge, was a quarter of the moon's diameter. From these data I find that the radii of the halo as measured from the centroid of the illuminated disc of the moon through these two stars were 22.8° and 21.4° respectively. Assuming that the halo was elliptical with the major axis vertical, I deduce values of 23.3° and 21.4° for the semi-major and semi-minor axes. I am aware that a more or less complete halo the major axis of which is horizontal is occasionally seen surrounding the 22° halo, but records of haloes elongated vertically are rare." (R5)

X7. December 14, 1926. Vita, Manitoba, Canada. "The observer was located near International Boundary, south of Vita, Manitoba, on Tuesday, December 14 (when morning temperature showed coldest this winter, approximately 30 below zero, evening temperature about 10 below); lunar halos observed 5 to 5.30 p.m., Central Time. The moon was a little over half of full phase, not in meridian, but approximately south-east. One halo was about the usual size but distinctly flattened on top, and drawn down near the horizon, a clear egg-shape. An interesting halo passed through the moon's position and was as nearly as could be estimated a perfect zenith circle, being thus very large, complete and distinct; also quite permanent." (R6) The halo passing through the moon was probably the parselenic circle. (WRC)

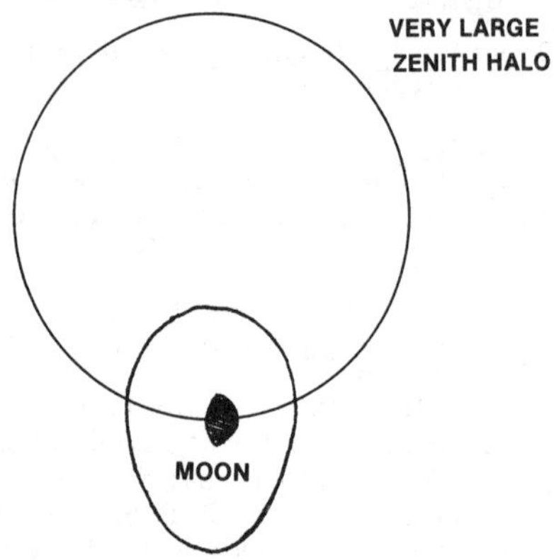

Egg-shaped lunar halo from Manitoba (X7)

X8. October 25, 1928. North Atlantic Ocean. "9.50 p.m. A.T.S. (0313 G.M.T.) Latitude D.R. 34° 08' N., Longitude D.R. 56° 33' W. A lunar halo was formed by the passage of a patch of cirrus clouds across the moon. This phenomenon was specially interesting in that the halo took the distinct form of an ellipse. Careful measurements were taken and found to be as follows, viz.; Moon 3/4 full. Altitude 48° bearing S. 1 W. Halo Major axis 25°, minor axis 20°." (R7) The drawing submitted with this report shows that the major axis of the ellipse was tilted 30° from the vertical. (WRC)

X9. 1958. Antarctica. The arcs 3 and 4 in the illustration for GEH1-X23 form two concentric figures which approach ellipses in shape. (R11) These near-ellipses are not true halos because they do not enclose the sun. (WRC)

X10. July 8, 1958. North Atlantic Ocean. "The halo shown in the sketch was seen from 1200 to 1230 G.M.T. It was silvery in appearance, against the 7/8 layer of

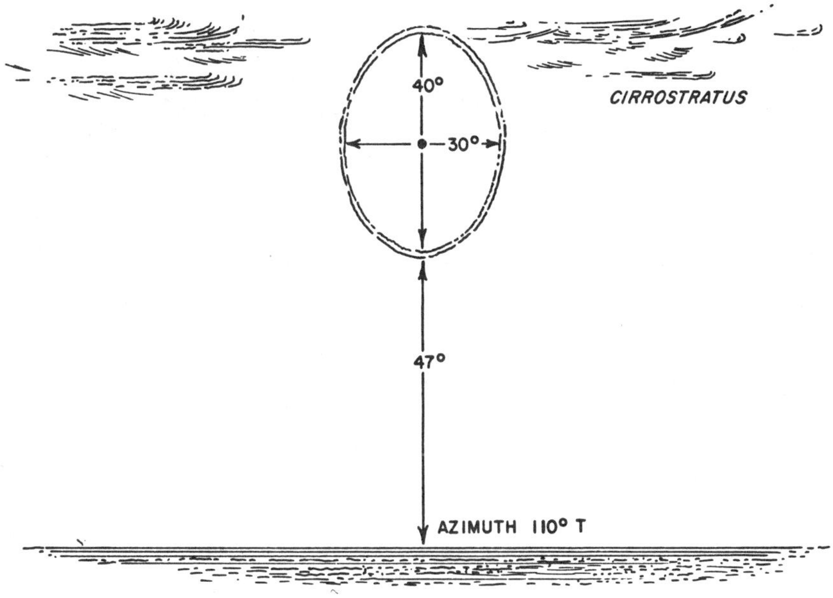

Elliptical halo, major axis vertical (X10)

white Cs cloud present." (R8) This was a solar halo. R. White states that this type of halo would result from two reflections, one in a horizontal face, another in a face inclined at 10° to the horizontal. (R12)

X11. March 8, 1963. South Atlantic Ocean. An elliptical lunar halo centered on the moon, which was 45° high. The major axis of the ellipse was vertical and measured 52° from top to bottom; the minor axis measured 40° from side to side. (R9)

X12. September 15, 1976. North Pacific Ocean. "At 0730 GMT an unusually shaped lunar halo was observed to the NNE. The moon, at its gibbous phase, had a 14° halo which projected outwards to a point about 28° from the centre. The appearance was

Tear-drop-shaped halo from the North Pacific (X12)

that of a very large tear drop, see sketch. The phenomenon lasted for about 20 minutes after which time the elongation gradually diminished until a perfect 14° halo was achieved. This was visible until 0830." (R10)

References

R1. Scrivener, H.; "Haloes Extraordinary!" English Mechanic, 67:246, 1898. (X1)
R2. Broomall, C. M.; "Lunar Halo," Science, 13:549, 1901. (X2)
R3. Evans, Lewis; "Remarkable Lunar Halo," Royal Meteorological Society, Quarterly Journal, 39:154, 1913. (X4)
R4. Hunter, A. F.; "Distorted Solar Halos," Royal Astronomical Society of Canada, Journal, 12:1, 1918. (X3)
R5. Dale, J. B.; "Elliptical Haloes," Nature, 101:126, 1918. (X6)
R6. Tuke, W. H., et al; "Two Lunar Halos," Royal Astronomical Society of Canada, Journal, 21:96, 1927. (X7)
R7. Gobel, C. J.; "Elliptical Lunar Halo," Marine Observer, 6:221, 1929. (X8)
R8. Chapman, E. G.; "Halo," Marine Observer, 29:116, 1959. (X10)
R9. Hort, L. V.; "Lunar Halo," Marine Observer, 34:16, 1964. (X11)
R10. Gilliat, S. B.; "Unusual Lunar Halo," Marine Observer, 47:110, 1977. (X12)
R11. Greenler, Robert; Rainbows, Halos, and Glories, New York, 1980. (X9)
R12. White, R.; "On the Halos in the Handbook of Unusual Natural Phenomena," Journal of Meteorology, U.K., 6:133, 1981. (X10)

GEH3 Extraordinary Mock-Sun and Mock-Moon Displays

Description. Halo displays boasting mock suns or mock moons in large number and/or in unusual positions relative to the sun. When large numbers of mock suns (say, more than four) are seen, they are almost always strung out along the parhelic (or parselenic) circle.

Background. Mock suns (parhelia) and mock moons (parselenes) are very common, particularly in the higher latitudes, where the requisite atmospheric ice crystals are more frequent. Single, double, and quadruple mock suns are almost always found on or close to the 22° and 46° halos.

Data Evaluation. Many of the famous halo displays of history were dotted with bright mock suns. (Some of these spectacular events are described elsewhere in this section.) Here are recorded those displays remarkable principally for their mock suns and mock moons rather than the halos and associated arcs. When the great halo displays are taken with the mock sun observations below, there is no shortage of extraordinary mock sun phenomena. Rating: 1.

Anomaly Evaluation. Mock suns, even in numbers as great as eight, are not necessarily anomalous for there are many possible intersections and contact points between known halos and arcs. It is at these intersections and contacts where mock suns normally form. The only anomalous features of the observations below are those mock suns occupying unrecognized positions along the parhelic circle and those appearing at points where halo-arc intersections and contacts are not known to occur. Even so, an unclassified mock sun or moon only means that we have not yet identified the correct ice crystal form or orientation. Rating: 3.

Possible Explanations. Anomalous mock suns and mock moons are almost certainly the consequence of unappreciated ice-crystal species or orientations.

Similar and Related Phenomena. None.

Examples of Extraordinary Mock Sun and Mock Moon Displays

X1. January 25, 1873. Independence, Missouri. A complex halo display consisting of a bright halo (diameter not given) and a complete parhelic circle. Six mock suns were located on the parhelic circle; two at the junctions with the halo;

two more at approximately 90°; and two more opposite the sun on each side of the antisolar point. Two additional mock suns, each with vertical pillars, were observed outside the halo centered in the upper left and upper right quadrants at about two halo radii from the sun. (R1)

X2. December 23, 1876. Colorado. Two halos plus a parhelic circle with eight mock suns spaced along it. (R3)

X3. December 30, 1880. Sauk Center, Minnesota. "It was first observed at five minutes before eleven o'clock, A. M. The mercury registered at the time 5° below zero. The sky was clear, save that the air was full of floating frost crystals that gave a leaden aspect to the heavens around. The sun, as I have tried to represent, was surrounded by a double halo, both very perfect and distinct in outline. To the right and left of the sun and on the rim of the first halo were very bright parhelia or mock suns. Passing through the sun and these mock suns and around the whole dome of the heavens, seemingly, at about 20° from the horizon, was a great circle of light. This had the appearance of the large ring of Saturn---very bright and about 1/4 of a degree in width. Again, about 15° from the rim of the outer or second halo, and in the path of this circle of light, were other parhelia on either side of the sun; and on the opposite side of it, from the sun, was another or third parhelion. This circle was very brilliant, describing a diameter of about 100°. Intersecting this bright circle, at the points of the two parhelia, passed a somewhat less brilliant ring of light in form of an ellipse, with its longest diameter some 130° and the short one 80. At the northern end of the ellipse were three parhelia as represented in diagram. These mock suns were all very distinct and beautiful. To complete this phenomenon there was as perfect a rainbow as we ever see in midsummer, describing an arc of about 35°, with its crown resting on the rim of the second or outer halo, directly over the sun. In fact this rainbow was of such brilliancy that it was too dazzling for the uncovered eye to look at, unlike the soft mellow tints in our summer rainbow phenomena. The duration of this most rare spectacle was an hour and fifteen minutes." (R2) The lower elliptical ring in the diagram is probably the parhelic circle with an antisun. The upper elliptical ring may be the circumzenithal halo, since the sun was high in the sky. (WRC)

X4. December 21, 1934. Quebec, Canada.

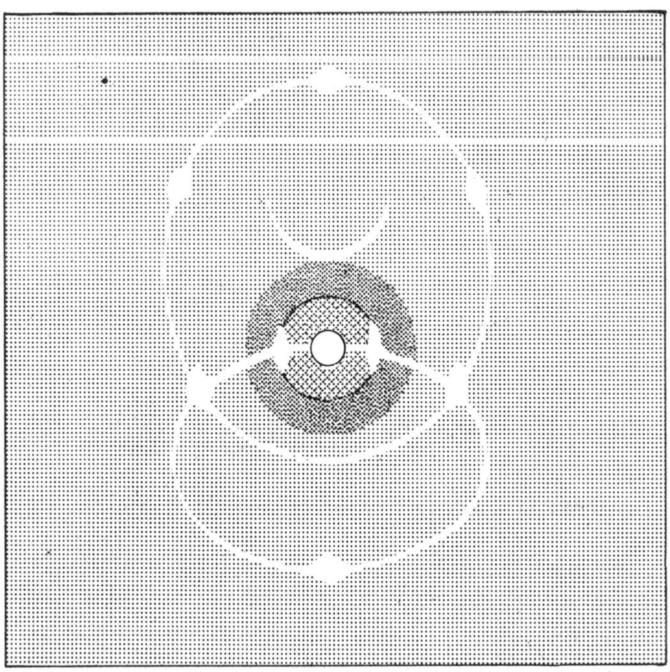

Sauk Center, MN, halo complex with eight mock suns (X3)

Lunar halos of 22° and 46°, a parselenic circle, with three pairs of mock moons at 22°, 46°, and 134°. A lunar pillar intersected both halos, producing an impressive "gunsight" effect. (R5)

X5. April 24, 1935. Wick, United Kingdom. "On April 24th, 1935, Mr. T.J. McGuiness of Wick (Caithness) observed the optical phenomenon shown in the attached diagram. A_1 and A_2 were the first parts to appear then the small halo marked $B_1B_2B_4B_5$ and finally the arcs of the halo C_1C_3 (probably the 22° halo) with the upper arc of contact C_2. Mr. McGuiness reports that the whole was brightly coloured 'like a rainbow with the shaded part white' and that the radius of the inner halo and the difference between the radii of the two halos were both equal to approximately 4 1/4 inches measured with a pencil at arm's length. The brilliant colours at so many point are exceptional and confirm that this was not an ordinary display. The halos faded from left to right and in the reverse order to their appearance. A_1 faded about 5 minutes before A_2 which finally faded out at 7h. 25m." (R4) If C_1 and C_3 were at 22°, the inner halo and mock suns possessed much smaller, and very rare, dimensions. (WRC)

References

R1. Woodson, D. M.; "Extraordinary Parhelia Observed at Independence, Mo.," Scientific American, 28:132, 1873. (X1)

R2. Morgan, F. M.; "The Meteor," Scientific American, 44:68, 1881. (X3)

R3. Wangelin, Otto H.; "The Sun Dogs of Colorado," Scientific American, 44:116, 1881. (X2)

R4. "Unusual Halo Phenomena Seen at Wick," Meteorological Magazine, 70:114, 1935. (X5)

R5. Middleton, W. E. Knowles; "A Rare Halo at Quebec," Royal Astronomical Society of Canada, Journal, 29:177, 1935. (X4)

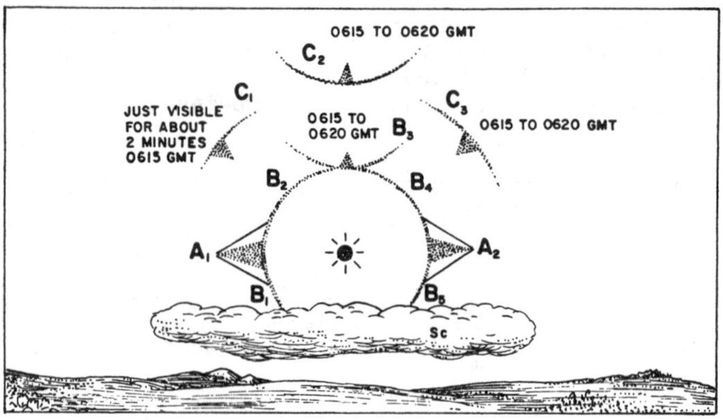

Mock sun arrangement observed at Wick. See text for letter key. (X5)

GEH4 Halos Dividing Sky Colors

Description. Halo displays in which the sky inside the halo(s) is considerably darker than that without.

Data Evaluation. Several casual observers have commented on this phenomenon, but no scientific investigations have been discovered. Rating: 2.

Anomaly Evaluation. From the scant data it is impossible to decide if this is a real atmospheric phenomenon. It may be merely a matter of human psychology, with the region inside the halo only seeming to be darker due to the very bright surrounding halo. Rating: 3.

Possible Explanations. In the absence of instrument readings objectively recording brightness levels inside and outside the halo, one is tempted to write this phenomenon off as purely one of psychological contrast.

Similar and Related Phenomena. Rainbows dividing sky colors (GEB19).

Examples of Halos Dividing Sky Colors

X1. October 19, 1846. Nottingham, England. The sky within the halo was very much darker than that without. See the description and illustration for GEH1-X4. (R1)

X2. August 13, 1861. Sydney, Nova Scotia, Canada. The sky within the normal halo was dark. See description and illustration for GEH1-X5. (R2)

X3. July 15, 1928. South Atlantic Ocean. An arc of white light clearly defined the border between light and dark parts of the sky. See description and illustration for GEH1-X13. (R3)

X4. June 13, 1954. Sistiana, Italy. Two halos of 11° radius, one encircling the sun, the other offset and intersecting the sun. The areas within both halos were darker than the surrounding sky. See description and illustration for GEH1-X21. (R4)

References

R1. Lowe, Edward Joseph; "Remarkable Solar Halos Seen on the 19th of October 1846," Philosophical Magazine, 3:29: 440, 1846. (X1)

R2. Crawley, H. W.; "On a Solar Halo Observed at Sydney, Cape Breton, Nova, Scotia, August 13, 1861," Report of the British Association, 1861, p. 63. (X2)

R3. Robinson, F. W.; "Luminous Arc," Marine Observer, 6:147, 1929. (X3)

R4. Crabbe, J. A.; "Rare Solar Halos," Weather, 10:359, 1955. (X4)

GEH5 Bishop's Ring and Other Rare Coronas

Description. Rings of color around the sun or moon either of very large radius (15° and more) or grossly distorted in terms of shape, eccentricity, or color pattern.

Background. The stipulation of very large radii eliminates from consideration the very common and well-understood small coronas with radii of only a few degrees. Such coronas are created when sunlight or moonlight is diffracted by tiny water droplets in the atmosphere. Very large coronas were first recognized by science following their worldwide appearance after the eruption of Krakatoa in August of 1883. These coronas are called Bishop's Rings after their discoverer, Sereno Bishop. In passing, one wonders why these large coronas were not recorded following major eruptions occurring before 1883!

Data Evaluation. Large coronas or Bishop's Rings have been widely observed after several volcanic eruptions since Krakatoa and are well-described in the scientific literature. The handful of other corona idiosyncracies are one-of-a-kind. The following rating applies only to Bishop's Ring. Rating: 1.

Anomaly Evaluation. The size and structure of Bishop's Rings can be explained in terms of the diffraction of sunlight by very fine dust or possibly ice-coated dust in the upper atmosphere. Bishop's Rings, however, are rather undependable. They appear after some major volcanic eruptions but not after others. Their appearances are spotty in time and geography. Some investigators have claimed correlations with lower atmosphere weather. Several times Bishop's Rings surrounded the sun when there were no known sources of terrestrial dust. Such variability, though, like the weather, challenges no scientific laws; it simply indicates how little we know about the origins and distribution of high-altitude dust. Rating: 3.

Possible Explanations. Bishop's Rings originate in the diffraction of sunlight by fine dust in the upper atmosphere lofted by volcanos or perhaps injected from extraterrestrial sources.

Similar and Related Phenomena. Other diffraction phenomena include the glories seen around

Brocken Specters (GEI1). It is also possible that the white sky phenomena of 1912 may be related (GWC1).

Examples of Bishop's Ring and Other Rare Coronas

X1. November 2, 1827. New Haven, Connecticut. "On the evening of the 2d of November, 1827, between the hours of seven and eight, there appeared around the moon, (a little more than its width in diameter), a very luminous saffron colored light. On the outer edge was a circle of bright red, which was graduated into a dark purple; around the purple was a circle of bright blue which faded into a yellowish green, increasing towards the outer edge, to a very vivid green. There appeared to be faint white rays passing from the moon across these columns, whose circles formed, around this lunar glory, a larger circle of a dark leaden color, which gave the whole a very beautiful appearance." (R1) In most respects the above observation is that of a typical corona with the red outermost. It is also small and highly colored. To familiarize the reader with the terminology, the colored region immediately surround the sun is called an "aureole." A representative aureole has been reproduced from a sketch made in the Indian Ocean (R23). A minor anomaly in the X1 description is the presence of radial white lines similar to those observed in some halo displays. (GEH11) (WRC)

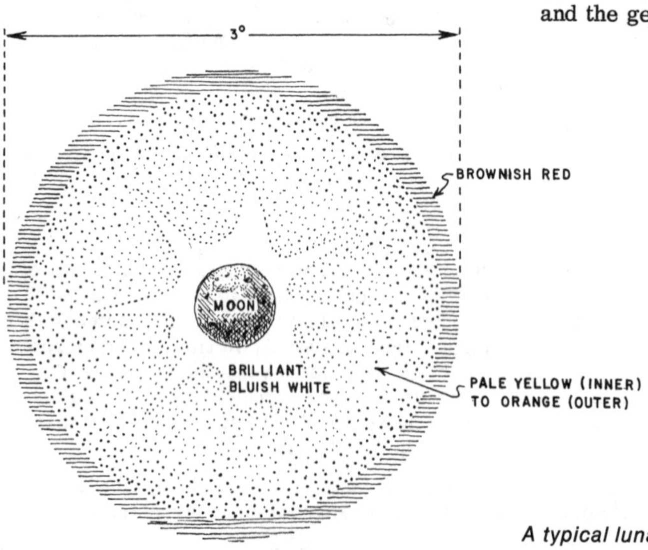

A typical lunar corona (X1)

X2. Late August 1883 through 1885. A worldwide phenomenon. A remarkable corona around the sun first reported by Reverend Sereno Bishop, from Honolulu, on September 5, 1883, just nine days after the explosion of Krakatoa. The ring, dubbed Bishop's Ring, was seen from time to time during the remainder of 1883, all of 1884, and less frequently in 1885. (R13)

November 1883. United States. "In November, 1883, at the time of the remarkable after-glows, I noticed that there was a broad, reddish ring around the sun even at mid-day. Soon after, I briefly described the appearances in Nature. Since then, I have constantly observed this phenomenon. The sky is very bright for about ten degrees from the sun; then comes the ruddy zone about twenty degrees wide, the deepest color being at about the natural distance of halos. My observations show that at this place there are but few days of the year when the chromatic glow is not visible; but it varies in intensity not only from day to day, but even from hour to hour. About a year ago I discovered that an increase in the depth of color preceded a fall in the temperature, and the formation, first of a structureless haze in the upper atmosphere, and, soon after, of cirrhus-clouds (sic). At other times storms came on with no increase in the depth of color. Soon it became evident that the latter cases were when rain fell, and the general temperature was not low.

Hail and sometimes snowstorms were accompanied by great depth of color. During the summer of 1884, I passed several weeks in Maine. On two occasions the colored zone appeared around the sun as distinctly as it ordinarily does here. Both times the appearance of the glow was followed by violent thunder-storms, with high winds and hail. While temperature would not affect the diffractive power of particles of volcanic dust directly, yet it is possible that at low temperature the dust particles, on account of the condensation of the air, may be enough nearer to each other to give a perceptibly greater diffractive power to the mass of air in which they are suspended. But so often has an increase in the depth of the circumsolar glow preceded the formation of clouds, that it seems far more probable that the glow is caused by the precipitation of atmospheric moisture at low temperatures. If dust is involved in the process, it is probably only by its increasing the depth of color, or by its facilitating the precipitation of moisture." (R5) Of interest here are the great variability of Bishop's Ring and its possible correlation with weather. (WRC)

Late 1883 through 1885. Europe. "The ring of reddish brown color surrounding the sun, and enclosing a disk of glowing whitish light, to which Prof. G. H. Stone called attention in Science for May 22, p. 415 (R5), has been most carefully studied by Kessling of Hamburg, who has shown clearly that it is due to diffraction on minute particles suspended in the air. Careful observers agree that it was not seen before November, 1883; but since then it has attracted much attention. It is manifestly produced in the upper regions of the atmosphere; for it is best seen from elevated mountains, where it is continually visible in clear weather. It is often hidden at low stations by the plentiful reflected skylight that comes from the coarser dust of the lower atmosphere, even though the sky seems tolerably clear. It has been astonishingly distinct here in Cambridge through the past winter, on the clear anti-cyclonic days, with north-west winds, following the withdrawal of the cyclonic cloud disk; and it attains its greatest visibility between clouds, because much of the lower dusty air is then in shadow, and does not out-shine the delicate colors of the ring. On some recent cloudless but slightly hazy days it has been entirely invisible. I have not observed the connection between the visibility of the ring and the changes of temperature and formation of clouds noted by Professor Stone, and should be glad to learn more details as to date of observations, and as to closely of the connection in point of time.....The most remarkable point in connection with the ring is its persistence long after the cessation of the brilliant twilights with which it began. How is the volcanic dust or the ice dust that it causes it supported so long? It seems incredible that dust could simply float for a year and a half in so thin a medium as the atmosphere at a height of ten or more miles." (R6)

January 12, 1884. Lafayette, Indiana. "On the evening of Jan. 12, at 9.20 (90th meridian time), my attention was called to a peculiar appearance about the moon. The sky was quite clear at the time, and there appeared around the moon several colored circular bands. The first was of a bright silver-gray shade, and about two diameters of the moon in width. The next was yellow, the next faint orange, and the next violet. The three bands were each about one-half a diameter in width. The outermost band was of a green shade, and about two diameters in width. At ten o'clock the innermost light band remained, but all the others had been replaced by a blue band lighter than the surrounding sky." (R2) This is the only example found of a lunar corona observed during the period of the Krakatoa Bishop's Ring. One would expect more since the solar corona was so prominent. (WRC)

Late 1883. Berkeley, California. No Bishop's Ring seen here; only a whitish glare extending 20° to 25° from the sun. (R3)

Summer 1885. Europe. Although spectacular sunsets were observed, Bishop's Ring did not appear until summer, when it took the form of a yellowish brown halo around the sun. (R4) Why the long time delay---almost two years after the eruption ---especially when the dust produced remarkable sunsets? (WRC)

Late 1883 through 1887. Pike's Peak, Colorado. "The reddish ring about the sun first appeared here (at the base of Pike's Peak) on November 22, 1883. For several days before that date, a faint discoloration of the region about the sun had attracted my attention. This gradually grew more intense, and, on the day mentioned, became unmistakable. The subsequent history of Bishop's ring as seen at this place is, in brief, as follows:---The colour was most intense during the winter of 1883-84, and diminished in brightness from that time until its disappearance. At first it was visible almost all the time. Later, it appeared only at the time of cold storms, which were accompanied by great vertical movement of air, or when, for any reason, the clouds reached to great height. It was, on the average, brighter during the winters than in the summers; also, it was brighter when the sun was near the horizon. Many times in cold weather there has been not a trace of the ring, although the air was so clear that peaks a hundred miles distant were distinctly visible from the heights behind this city. At other times the ring has been very bright when the air was so hazy that the mountains only ten miles away were hardly visible. During the later months of 1885 it was invisible most of the time, but suddenly flamed out in wonderful intensity at the time of the great norther of January 9-11, 1886. Then for about two months it frequently appeared in the morning, or towards evening. During the warm months of 1886 it was not seen. On October 15 it appeared distinctly. About a week later it appeared very faintly a few times, and since then I have not been able to see a trace of it. My observations have been made at elevations of from 6000 to about 13,000 feet, and there was but little apparent difference in intensity at the different elevations. It is well known that the atmosphere here is, in general, very dry and transparent. The diffraction-ring was often more coppery, almost rosy, in tint at the time of the northers, and in the thickening haze in the upper air preparatory to hailstorms. The great intensity of the colour at such times, and its peculiar tint, and that, too, irrespective of the amount of haze in the lower atmosphere, makes it probable that the ring was in part due to diffraction on ice-particles. If so, the ice particles themselves may have been due to precipitation on dust-particles. The fact that no diffraction-ring has been seen around the sun during the past winter is not conclusive, for we have had no great northers, the season being unusually mild. But the disappearance of Bishop's ring for so long a time makes it certain that, even if there can be a circumsolar glow due to diffraction on ice-particles, yet the proper conditions for such a ring are realised only rarely, except when there is a great amount of volcanic dust in the air." (R7) Once again, many variables seem to affect the prominence of Bishop's Ring, including weather patterns in the lower atmosphere. (WRC)

1886-1887. England. Bishop's Ring has not wholly disappeared. It is still seen at sunset. Whitish wisps can be seen in and near the ring. (R8) No one seems to have speculated on the nature of these wisps. (WRC)

X3. January 17, 1894. Finchley, England. "On Wednesday last (17 inst.) a finely coloured lunar corona was seen from here. The colours appeared to me brightest when the misty clouds which floated across the sky were sufficiently dense to obliterate all but first-magnitude stars. The inner diameter was about 3 1/2 degrees, and the colours commenced with a red circle inside, and it seemed to me that the rainbow hues could be traced <u>outward</u> as far at the blue; but the violet was not perceptible to my eyes. Following this there was a dark space, with another red circle of about 10 degrees diameter. I thought I had observed the whole carefully, but I see by one of the books I have since consulted, that the red is the <u>outermost</u> colour. Is my observation at fault, or will theory allow of the colours being reversed?" (R9)

X4. Summer-Fall 1902. England. This item begins the discussion of the so-called "second appearance" of Bishop's Ring, which was generally associated with the eruption of Mt. Pelee in May 1902. "The phenomena connected with the volcanic dust are undergoing distinct changes. In common with observers in the south of England, I noted the fresh appearance of the dust phenomena in the end of June, especially on June 26, but they were not very striking until August 1. At first the most decidedly volcanic feature was the great corona round the sun, known in the case of the Krakatoa effects as 'Bishop's Ring.' Whether this name should be applied to the corona this year is doubtful, as its radius has been fully double that of the Krakatoa corona, having until recently averaged about 70°, measured from the sun to the middle of the reddish part. Yesterday and this morning, however, it averaged only about 40°, and its reddish part was a yellowish-brown rather than a red. The colour of the corona this year has always been much less decidedly pink than was the case with Bishop's Ring; indeed, it has sometimes been an absence of blueness in that part of the sky rather than any positive redness. The pink glows after sunset were very strong in the end of June, but stronger still in November, and on November 1, 17 and 18 there was also a faint second glow, a phenomenon I had not previously seen since the Krakatoa sunsets." (R10)

August 1903. Switzerland. M.F.A. Forel reports a reappearance of Bishop's Ring. (R11) Angular measurements of the ring in Switzerland put the radius of the visible outer limit of the reddish ring at between 20° and 25°. (R12)

Winter 1902-1903 through the summer of 1903. Blue Hill Observatory, Massachusetts. Bishop's Ring observed in conjunction with brilliant sunsets. (R12, R14)

X5. January 1, 1909. Honolulu, Hawaii. Sereno Bishop notes the reappearance of the ring named after him. He suggests that it is due to volcanic activity in Italy. (R15)

X6. May 19, 1910. Europe. As the earth passed through the tail of Halley's Comet. "According to reports from various parts of the world no magnetic nor electric phenomena were noted that appear necessarily, or even very reasonably, attributable to the comet; and the same statement would be more or less true of all other observed phenomena, except possibly certain halos and coronas seen on the 19th.

While the unusual appearances of some of these rings can not with certainty be referred to the comet, nevertheless they and one or two special conditions of the sky are recorded here for whatever they may be worth. Similar phenomena have been reported from other places. Professor E. B. Frost, Director of the Yerkes Observatory, states that from noon till about 1 p. m. of the 19th there were seen from that Observatory iridescent clouds of unusual brilliancy, and also the rare phenomena of a halo of 15° radius.

Professor Max Wolf reports from the Koenigstuhl Observatory that late in the afternoon of the 19th a Bishop's ring was seen around the sun, and that this was followed by a twilight of unexpected intensity, extension and duration; that three twilight purple glows succeeded each other, and that all the particular sky-phenomena following the eruptions of Krakatoa and Mount Pele appeared in an intensified manner. Also that a Bishop's ring more intense than any he had ever seen before, and of 28° radius, indicating particles of 1.5μ diameter, appeared about the moon.

In the same number, 4414, of the Astronomische Nachrichten, W. Krebs, reports from Grossflottkek that from 4 to 4:30 p. m. a parhelium was seen, and that at the same time a peculiar shimmering white corona of 5° radius appeared around the sun.

It is reported by Emile Marchand that observations made on Pic du Midi and at Bagneres-de-Bigorre showed, on the morning of the 20th, a lunar corona and a bright tinted sky such as was seen after the eruption of Mount Pele. These phenomena were even more pronounced on the evening of the 20th. There was also an anlogous solar corona of 3° to 4° diameter on the 20th. that was still seen on the 2nd of June.

The halos, coronas and other phenomena listed above were both widely scattered and, in some respects, distinctly unusual; and their occurrence coincident, as near as can be determined, with the passage of the earth through the tail of the comet suggests for them a cosmical origin. Still there certainly far from universal, and besides they have all been seen before when there was no comet to which they could be attributed; and, therefore, while admitting the possibility, in this case, of a cometary influence, it would seem rash, without additional evidence, to conclude that the comet was the principal or even partial cause of any of the appearances mentioned above." (R22) One would expect cometary and volcanic dusts to be at widely different altitudes, producing significantly different effects. (WRC)

X7. January 16 and 19, 1914. Switzerland. "While at Kandersteg, Canton Berne, recently this interesting appearance was twice clearly seen, namely, on Friday and Monday, January 16th and 19th, in each case from an Alp at 5000 to 5500 ft. elevation, and to less advantage on the 16th from Kandersteg itself, at 4000 ft. It could not be distinguished when the sun was in sight, but directly that was behind a peak the orange-tawny ring round a markedly greenish circle of 15° to 20° radius became sufficiently marked for others of the party to be interested when their attention was called to it. The ring itself stretched over another 15° or so, as roughly measured by the eye." Correlating these sightings of Bishop's Ring with the Krakatoa-like sunsets in 1913, the writer concludes both phenomena were probably the result of a volcanic eruption in the Aleutians in May 1912. (R16)

X8. July 1916. Switzerland. "Dr. Julius Maurer, director of the meteorological service of Switzerland, reports that since the middle of last July the atmosphere of the high Alps has shown marked optical deterioration, as indicated by the extension of the glow around the sun to which Mauer applies the term 'aureole,' reaching a diameter of 140 deg. by the end of August. A Bishop's ring, or large corona due to dust, was visible August 3rd-4th. Phenomena of this character seen in 1883-1884, 1902-1903, and 1912 were attributed to volcanic dust from the explosive eruptions of Krakatoa, Pele and Katmai, respectively, but no violent volcanic eruptions have been reported this year. The question has been raised whether the cumulative effects of gunfire in the course of two years of war may be equivalent to those of a volcanic eruption in charging the upper air with dust." (R17, R19)

July 1916. South Africa. A dark gap or ring observed around the sun. (R18)

X9. October 11, 1924. Aberdeen, Scotland. "A lunar corona of elliptical form was observed here at 19h. 30m. G.M.T. on October 11th. and remained visible unchanged for a period of about twenty minutes. The phenomenon consisted of an elliptical ring eccentric about the moon. The colours were distinctly marked, and graduated from a warm creamy glow near the moon through lemon-yellow to the orange-red-brown of the ring. The upper half of the ellipse was bounded by a band of violet light of a very pure colour. This band, thin at the top, widened at the sides of the ring. Above this came a broader arc of light of a tint resembling 'electric-blue,' but of a less pronounced brilliancy than the violet band. Next this, a band of dull olive-green formed the outer top edge of the display. All three bands terminated at the sides about level with the horizontal diameter of the ellipse. The sky round the moon was covered with a coarse streaky type of cirro-stratus; wisps of fracto-stratus were seen at times.

The corona became irregular by 20h., but later assumed the usual circular form, showing a faint tint of cream colour with very little, if any, indication of the brownish red ring. Afterwards the corona disappeared, and a lunar halo formed." The elevation of the moon was 30°. The horizontal diameter of the ellipse between the outer edges was 12 lunar diameters. From the bottom of the moon to the lower edge of the ellipse, the distance was 3 diameters; it was 5 diameters from the top of the moon to the upper edge. (R20)

X10. July 23, 1932. Broken Hill, Australia. Bishop's Ring visible at 5 P.M. This date was 10.5 weeks after a volcanic eruption in the Andes. (R21)

X11. May 1982. Houston, Texas. Bishop's Ring was seen almost every evening about 10 minutes after sunset. Attributed to the eruption of El Chichon in the spring of 1982. (R24)

References

R1. "Singular Appearance of Circles Around the Moon," American Journal of Science, 1:14:397, 1828. (X1)

R2. Huston, H.A.; "Halos Round the Moon," Science, 3:115, 1884. (X2)

R3. LeConte, John; "Upperglow of the Skies in Relation to Halos and Coronas," Science, 3:143, 1884. (X2)

R4. Walter, J.G.; "Luminous Phenomenon Round the Sun," Journal of Science, 22:45, 1885. (X2)

R5. Stone, G.H.; "The Ruddy Glow Around the Sun," Science, 5:415, 1885. (X2)

R6. Davis, W.M.; "The Reddish-Brown Ring Around the Sun," Science, 5:455, 1885. (X2)

R7. Stone, G.H.; "Disappearance of Bishop's Ring in Colorado," Nature, 35:581, 1887. (X2)

R8. Backhouse, T.W.; "Bishop's Ring," Nature, 36:102, 1887. (X2)

R9. "Lunar Corona," English Mechanic, 58:510, 1894. (X3)

R10. Backhouse, T.W.; "Volcanic Dust Phenomena," Nature, 67:174, 1902. (X4)

R11. Nature, 68:396, 1903. (X4)

R12. Rotch, A. Lawrence; "The New Bishop's Ring," Nature, 68:623, 1903. (X4)

R13. "Bishop's Ring," Observatory, 26:431, 1903. (X2)

R14. Clayton, Henry Helm; "A Second Bishop's Ring Around the Sun and the Recent Unusual Twilight Glows," Science, 17:150, 1903. (X4)

R15. "The 'Bishop Ring' Is Seen Again for the First Time Since Its Discovery," Scientific American, 100:103, 1909. (X5)

R16. Clark, J. Edmund; "Bishop's Ring," Symons's Meteorological Magazine, 49:33, 1914. (X7)

R17. "Alpine Haze Due to the War?" Scientific American, 115:405, 1916. (X8)

R18. Holcroft, A.V.; "Dark Gap or Ring Round the Sun," English Mechanic, 104:48, 1916. (X7)

R19. Shaw, Napier; "Optical Deterioration of the Atmosphere in July and August, 1916," Nature, 98:90, 1916. (X7)

R20. Jarrold, P.F.; "An Elliptical Lunar Corona at Aberdeen," Meteorological Magazine, 59:261, 1924. (X9)

R21. Paterson, J. Fraser; "Bishop's Ring and the Andean Eruption," Nature, 130:843, 1932. (X10)

R22. Hogg, Helen Sawyer; "Halley's Comet and Unusual Atmospheric Phenomena," Royal Astronomical Society of Canada, Journal, 59:89, 1965. (X6)

R23. Finch, E.R.; "Lunar Halo and Aureole," Marine Observer, 47:111, 1977. (X1)

R24. Livingston, W., and Lockwood, G.W.; "Volcanic Ash over Arizona in the Spring of 1982: Astronomical Observations," Science, 220:300, 1983. X11)

GEH6 Halos of Unusual Radii

Description. Solar and lunar halos with radii other than 22° and 46°.

Data Evaluation. Although our literature survey has located very, very few reports of unusual halo radii, both Minnaert and Greenler (two modern authorities on halo phenomena) are persuaded that several other halos exist with radii between 8° and 90°. They must be exceedingly rare. Rating: 2.

Anomaly Evaluation. Actually it is not particularly surprising to find odd-sized halos because there exists such a wealth of ice crystal forms and possible orientations. Rating: 3.

Possible Explanations. The few halos with unusual radii probably owe their origin to ice crystals of pyramidal form.

Similar and Related Phenomena. None.

Examples of Halos of Unusual Radii

X1. General observations. "Circles round the sun of 8°-9°, 16°-19°, 23°-32°; always be careful to screen off the sun when looking for these faintly luminous rings! They are due to refraction in pyramidal ice-crystals, fortuitously oriented. This is why they often occur several at a time." (R1) Note that the literature surveyed so far has yielded no examples of faint, multiple halos with odd radii. (WRC)

"A scattering of reports in the scientific literature indicates a halo with angular radius near 90 degrees. It is called Hevel's halo, and suggestions for its origin involve combinations of reflections and refractions in ice crystals having the form of right hexagonal prisms.....It seems unlikely, however, that such a crystal can account for the variety of other rare halos that have been reported, including a halo of radius 8 or 9 degrees, another of radius 17 or 18 degrees, some with radii slightly larger than 22 degrees, and one with a radius between 30 and 35 degrees." The author has measured photos of halos with radii of 8 and 18 degrees. (R3)

X2. May 19, 1910. Europe. A rare 15° halo was reported when the earth passed through the tail of Halley's comet. (R2) See more complete description in GEH5-X6.

X3. April 30, 1977. Manitoba, Canada. A 15° halo suddenly changed into a 22° halo. See details in GEH7-X3. (R6)

X4. 1629. No place given. "A halo of 26°-29° was seen by Scheiner in 1629, and afterwards by Greshow and by Whiston." (R4)

X5. 1661. No place given. "A halo of about 90° was seen by Hevelius in 1661, afterwards by Erman, Sabine, and others." This halo has always appeared as a white circle in contrast to the other halos of abnormal radius, which are usually colored like the 22° halo. (R4)

X6. 1796. No place given. "A halo of 4°-6° radius was seen by Hall in 1796." (R4)

X7. 1831. No place given. "A halo of measured radius 19°25' was seen by Burney in 1831, and afterwards on three occasions by Hissink." (R4)

X8. 1839. No place given. "A halo of measured radius 14° was seen by Heiden in 1839 and a halo of measured radius 16° by the same at the same time." (R4)

X9. 1892. No place given. "A halo of 9°-10° radius was seen by Van Buijsen in 1892 and afterwards by Hissink and Russell." (R4)

X10. 1898 and 1899. No places given. "A halo of 7°-8° radius was seen by Arctowski in 1898 and by Hissink in 1899." (R4)

X11. 1899 and 1905. No places given. "A halo of measured radius 17°55' was seen by Hissink in 1899 and 1905, and afterwards by Krcmar." (R4)

X12. 1900. No places given. "A halo of about 17° was seen by Besson and Dutheil in 1900. (R4)

X13. 1904. No places given. "A halo of 34°-38° was seen by Fueillee, by Parry, and by members of the Charcot expedition in 1904. (R4)

X14. April 14, 1974. England and Holland. A remarkable display of halos of unusual radii were seen and photographed. Measurements of the photos yielded radii of about 22°, 19°, 18°, and 9°. (R5)

References

R1. Minnaert, M.; "Very Rare and Very Doubtful Halo Phenomena," The Nature of Light and Color in the Open Air, New York, 1954, p. 203. (X1)

R2. Hogg, Helen Sawyer; "Halley's Comet and Unusual Atmospheric Phenomena," Royal Astronomical Society of Canada, Journal, 59:89, 1965. (X2)

R3. Greenler, Robert; "Rare Ice-Crystal Halos," Rainbows, Halos and Glories, New York, 1980, p. 61. (X1)

R4. Besson, Louis; "The Different Forms of Halos and Their Observation," Monthly Weather Review, 42:436, 1914. (X4-X13)

R5. Tricker, R.A.R.; "Arcs Associated with Halos of Unusual Radii," Optical Society of America, Journal, 69:1093. (X13)

R6. Wadsworth, D.J.; "Solar Halo of 15 Degrees---30 April 1977," Weather, 33:113, 1978. (X3)

GEH7 Jumping and Moving Halos

Description. Halos that change position or radius by rotation, translation, or sudden, discontinuous "jumping."

Data Evaluation. Although the observations of jumping and moving halos are very rare, several good quality sightings are on record. Rating: 2.

Anomaly Evaluation. The discontinuous jump of a halo from one radius to another is startling but not particularly anomalous, for it may be explained in terms of a sudden change in ice-crystal population or orientation. However, the slow, continuous translation or rotation of a halo of fixed radius is much more puzzling. Rating: 2.

Possible Explanations. All examples of translating and rotating halos involve offset lunar halos. Offset halos are not readily explained even when stationary (GEH1), but they may, like offset rainbows, be a consequence of secondary light sources. Bright clouds, especially at night, may serve as secondary light sources. Since clouds do move, the halos they create may also move.

Similar and Related Phenomena. Offset halos (GEH1); offset rainbows (GEB3, GEB6).

Examples of Jumping and Moving Halos

X1. February 26, 1934. Halifax, Nova Scotia, Canada. 0200 GMT. "Whilst leaving Halifax an unusual lunar halo was observed. The sky seemed to be cloudless when the phenomenon appeared. It consisted of a halo round the moon 23 $1/2^o$ in radius with a second ring of greater brilliance passing through the moon and lying to the west of it. This second ring measured 47° in diameter, and slowly revolved till it assumed a vertical position over the moon (this moved east) when a third arc appeared showing fairly distinct spectrum colours, red on the upper edge of the arc. This latter arc was 23 $1/2^o$ above the horizon and parallel with it. The primary ring was whitish and well defined, the second ring was also white but more pronounced than the first. The third arc was very brilliant directly under the moon and decreased outwards. The phenomenon lasted 20 minutes, disappearing suddenly. The moon at the commencement was 45 min. west of the meridian bearing 209° True, at an altitude of 68° 20'." The remarkable feature is, of course, the rotation of the second halo by 90°, keeping its circumference on the moon, and using the moon as the axis of rotation. The spectral arc parallel to the horizon was probably the circumhorizontal arc. (WRC) (R1)

X2. November 17, 1961. Indian Ocean. "At 1500 GMT a white halo was observed, cen-

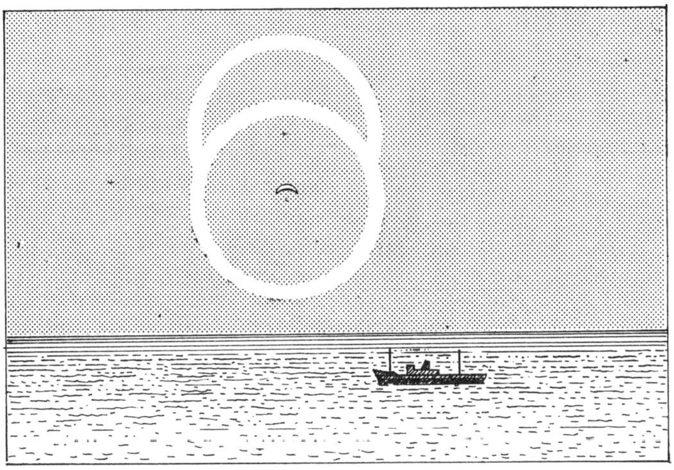

Second lunar halo separates from normal halo and moves into an offset position. (X2)

tred round the moon. During the following hour, a second halo, separating from the original one, moved slowly into the position shown in the sketch. At 1600, when the upper part of the second halo was coming very close to the moon, the whole of this halo disappeared; a few min. later the original halo also vanished. The cloud in the vicinity of the moon consisted of parallel bands of cirrus, producing a great deal of light..... Note. We cannot explain this very interesting phonomonon. It is possible that there was a gradual change in the distribution, nature and behavior of the ice particles within the cirrus cloud. The halo with the moon at its centre was probably normal, while the second displaced halo was a very interesting abnormality." (R2)

X3. April 30, 1977. Manitoba, Canada. "...at 1358 Central Standard Time plus an hour of daylight saving time I was attempting to observe the sun. Conditions were very violent with evidence of cross currents in cumulonimbus between the cumulus itself and the glaciated upper portions. As conditions deteriorated I looked up to ascertain what my chances of further solar observations might be. The cumulonimbus were moving from west to east and the glaciated tops from north to south. In the glaciated sheet a weakly coloured halo was visible. Using a home-made theodolite I undertook a routine measurement of the angular radius of the halo. To my surprise the value was 15 degrees from the sun to the inner edge of the halo with about a half degree spread of the spectrum, blue to the outside and red to the inside. I checked the instrument and made a second measurement, having doubted my first. This confirmed my first observation but, still not sure, I attempted a third measurement. Whilst making this I saw the halo jump to a radius of 22 degrees. The jump was over a period of seconds only. The 15-degree halo faded quickly and a 22-degree halo replaced it." (R3) This jumping phenomenon is perhaps not as startling as the slow, progressive rotation or translation of halos, as in X1 and X2. (WRC)

References

R1. Rothwell, A.; "Lunar Halos," Marine Observer, 12:8, 1935. (X1)

R2. Davies, T.; "Lunar Halo Complex," Marine Observer, 32:186, 1962. (X2)

R3. Wadsworth, D.J.; "Solar Halo of 15 Degrees---30 April 1977," Weather, 33: 113, 1978. (X3)

GEH8 Kaleidoscopic Suns

Description. A puzzling spectral display of multiple rays and spheres of light in all imaginable colors. Although concentrated in the direction of the sun, the colors seem to flash on and off both in the adjacent sky and amidst terrestrial surroundings.

Background. Kaleidoscopic suns resemble in some respects the luminous aerial bubbles described in GLD7. The two phenomena should be compared closely.

Data Evaluation. Remarkably, several fairly consistent accounts of this phenomena, from various times and places, have been found in the meteorological literature. Although the observers are genuinely startled at the bizarre event, their accounts are generally matter-of-fact. Still, the data are rather imprecise and also testimonial. Rating: 2.

Anomaly Evaluation. Spectral phenomena like the kaleidoscopic suns are probably the consequence of unobserved objects or inhomogeneities in the atmosphere; but we can only guess at the nature of these color-makers. Rating: 2.

Possible Explanations. One possible source of such spectral phenomena could be large ice flakes in the atmosphere or even seeds buoyed by the wind. Light diffracted, reflected, and refracted by these impurities could be highly colored. When the sun is low, inhomogeneities in the atmosphere may also contribute, especially close to the horizon.

Similar and Related Phenomena. Aerial bubbles (GLD7); sunset shadow bands (GEL10); low-sun phenomena in general (GEL); shadow bands seen during solar eclipses (GES1).

Examples of Kaleidoscopic Suns

X1. June 1885. Birmingham, Michigan. "The occurrence herein related took place from twenty to thirty minutes before sunset in the latter portion of June of the year 1885. The weather was more than usually fine. The sky was clear with the exception of a few clouds of the cumulo-nimbus order a few degrees above and to the northward of the sun. Suddenly there appeared a peculiarly weird and hazy condition of the atmosphere. There was an indescribable commingling and general diffusion of all the hues of the rainbow. During this state of things there appeared in the sky, on the earth, and on the trees, innumerable balls of decomposed light, presenting all imaginable colors and apparently of about the size of a bushel basket. They were uniform in size and appearance. This phenomenon was confined to that region of the sky about the sun, extending but a few degrees each side of it. It lasted about twenty minutes, when it disappeared as suddenly as it came." (R1, R2)

X2. May 10, 1902. South Devon, England. "A friend informs me that, on looking at the sunset sky on May 10 last from about 6.50 p.m. to 7.5 p.m., or perhaps a little later, he saw numberless 'suns' in various parts of the sky, coloured 'magenta to violet' principally, although they also changed to yellow and green. He called his sister out of the house; after that, his farm manager; and then again referred to the local postman; all of whom verified the curious sight, the lady remarked they (the 'suns') looked 'like toy air balloons.' The locality is near Ivybridge, South Devon. Weather was clear at the time. I understand the 'suns' were visible at a moderate altitude, and in the direction of the sunset. I have questioned my friend personally on the subject, with a view to ascertaining whether or not he was the victim of some complementary effect, after looking at, or towards, the bright image of the sun. He is a man of the world, has travelled largely, is a good sportsman, in excellent health, and is the last person, one would consider, to make such a report, unless something unusual had really been seen. Moreover, he was thoroughly corroborated by the individuals to whose attention he directed the phenomenon, and who saw the 'suns' quite plainly." (R3)

X3. October 13, 1917. Fatima, Portugal. The optical phenomena reported during the famous Miracle at Fatima included a 'dancing sun' and rotating beams of colored light. (R5) Although no report has been uncovered in the scientific literature, the popular reports resemble the other entries in this category. (WRC)

X4. January 6, 1923. County Cork, Ireland. "A very remarkable optical (?) display was seen here on January the 9th, about 1.30 p.m. local mean time. During the entire forenoon an almost clear sky and sunshine prevailed and up to 1 p.m., when a slight snow shower fell, clearing again a little later. A friend shouting to me to hurry out, I saw the sun behaving in a most unusual fashion, now surrounded by bright red, flashing rays in all directions, then changing to yellow in which the body of the sun, though more clearly visible, appeared to dance and shift about here and there in a radius of about 5°; again, changing to green, the rays flashing as in the red---all these changes taking place in less time than it takes to write. Could this have been caused by a cloud of snow particles crossing the sun in otherwise clear air? Of all the beauties seen I should think the quickly changing mock suns the most beautiful as they flashed here and there faster than it was possible to count them. The colours were so brilliant and dazzling, that even after I had come indoors anything I looked at appeared a mixture of all the colours seen. The only thing which I can compare it is Venus 'boiling' in the telescope on a hazy windy evening, on a very large scale; but no description is able to convey what was seen in these ten or twelve minutes." (R4)

References

R1. Alexander, S.; "Remarkable Optical Phenomenon," American Meteorological Journal, 3:486, 1887. (X1)

R2. Alexander, S.; "Remarkable Optical Phenomenon," Royal Meteorological Society, Quarterly Journal, 13:308, 1887. (X1)

R3. Markwick, E. E.; "Curious Phenomenon," English Mechanic, 75:417, 1902. (X2)

R4. Mintern, Joseph; "A Kaleidoscopic Sun," Meteorological Magazine, 58:10, 1923. (X4)

R5. Osborne, Charles; "The Dancing Sun," Catholic World, 169:208, 1949. (X3)

GEH9 Skewed Halo Displays

Description. Halo displays that are not symmetric around a vertical line drawn through the sun. The most common manifestations are tilted arcs of contact and mock-sun lines that do not intersect the sun.

Data Evaluation. A few qualitative reports plus one careful observation with approximate measurements. Rating: 2.

Anomaly Evaluation. For a halo display to be tilted from the vertical, all of the ice crystals creating it must be similarly inclined, which means that some vertically skewed force must be acting. However, it is not unreasonable to blame the motion of air for such tilting. Rating: 3.

Possible Explanations. Air currents tilt ice-crystal axes away from the vertical.

Similar and Related Phenomena. Offset halos (GEH1); rotating halos (GEH7).

Examples of Skewed Halo Displays

X1. May 19, 1910. No place given. Light pillars seen below the moon and, after sunrise, below the sun, were inclined to the vertical. Observations were from a balloon. (R1)

X2. July 18, 1913. No place given. Upper arc of contact was 5° to the right of the solar vertical. (R1)

X3. September 4, 1915. No place given. A photo of a 22° halo with an upper arc of contact showed the contact point to be 5° or 6° to the left of the solar vertical. In addition the mock sun ring was below the sun. (R1)

X4. March 6, 1941. Great Weldon, England. Refer to the sketch. "The portions of the arcs which were brilliant are shown as continuous lines. The visible portion of the mock sun ring was dead straight and it was considerably out of the horizontal. The line of the ring when produced, passed approximately 5° below the true sun. The mock sun on the left was a brilliant white, showing no trace of colour. The glow on the right was of a yellowish tint; but this may have been due to the fact that it was on the edge of a large cumulo-nimbus, which later drifted across the glow and obscured it completely. The portions of the halo of 46° and its arc of contact were undoubtedly of similar radius, a rough check having been obtained by holding lengths of wood at arm's length and pivoting them about the sun and the apparent centre of the arc of contact. The arc appeared to be refracted out of the vertical by the same amount as the mock sun and its associated ring." (R1)

References

R1. Satow, P. G., and Whipple, F. J. W.; "A Skewed Halo Complex," Royal Meteorological Society, Quarterly Journal, 67:294, 1941. (X1-X4)

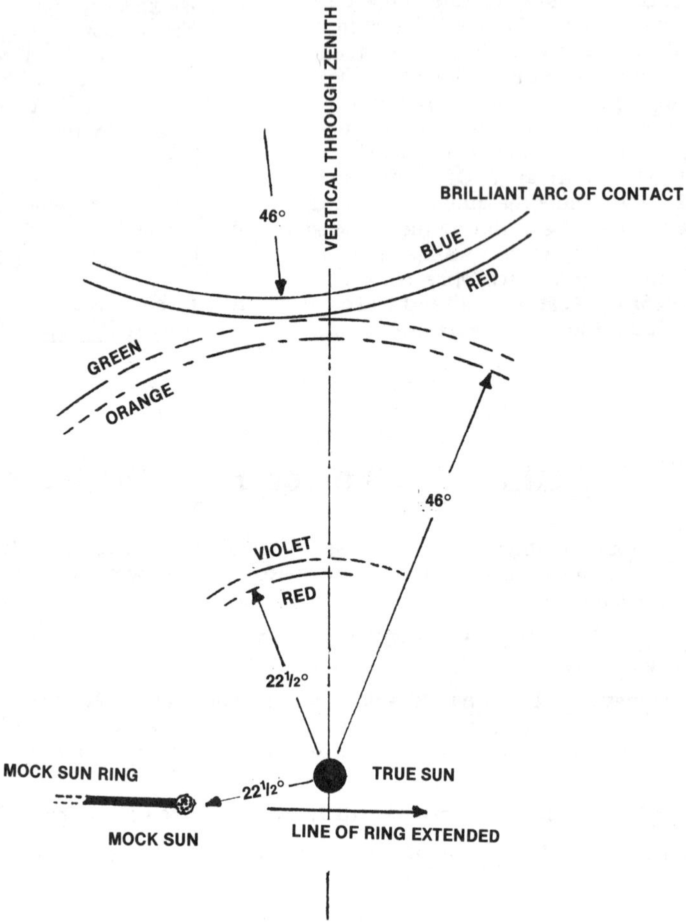

A skewed halo complex observed in England (X4)

GEH10 Bottlinger's Rings

Description. Faint rings of white light concentric with the subsun. If the subsun is elongated, Bottlinger's Rings will be likewise. Often several rings are present.

Background. The subsun is a bright spot of light reflected mirror-like from plate-type ice crystals. Obviously it is seen in the direction opposite the sun. Since it is as far below the horizon as the sun is above it, the subsun is more often observed from aircraft flying over clouds containing abundant ice crystals.

Data Evaluation. No specific examples are available, but the phenomenon is mentioned (with a photograph) in a well-known book on atmospheric optics (R1). Rating: 2.

Anomaly Evaluation. No generally agreed upon explanation exists for Bottlinger's Rings, but the basic mechanism probably involves nothing more than repeated reflections and refractions from ice crystals. Rating: 3.

Possible Explanations. Multiple reflections and refractions from ice crystals.

Similar and Related Phenomena. Solar halos (GEH).

Examples of Bottlinger's Rings

X1. General observations. "The rings surrounding the subsun spot are called Bottlingers rings, and I do not understand their formation. If they were diffraction rings resulting from the small reflecting ice plates, I would expect to see them colored like the coronas.....One of the more plausible explanations is that they are produced by multiple reflections between flat-plate ice crystals, each with a small tilt from the horizontal. Thus the subsun results from one reflection, the first ring from three reflections, the second from five, and so on. It seems that the form of the rings might be explained this way; yet I suspect that such a mechanism would produce too low an intensity in the rings to be observable. The question is yet to be answered." (R1)

References

R1. Greenler, Robert; "Summary of Ice Crystal Effects," Rainbows, Halos, and Glories, New York, 1980, p. 104. (X1)

GEH11 Transient Lines Superimposed on Halo Displays

Description. Bright or dark lines or rays that intersect halo displays. The lines usually move relative to the display and, especially in the case of dark lines, seem to produce a ripple effect. Stationary but transient line structures have also been reported.

Data Evaluation. Systems of moving dark ripples are the most common variety here; and several good reports exist. The other transient linear features are one-of-a-kind and their existences are questionable. There is a strong possibility that two or more separate phenomena have been lumped together in this entry. Rating: 2.

Anomaly Evaluation. Promising explanations (see below) have been profferred for the moving dark and bright lines, so that a low anomaly rating is in order here. Rating: 3.

Possible Explanations. The moving dark lines might well be indirect evidence of travelling disturbances in the atmosphere that deflect ice-crystal axes, temporarily depriving part of the halo display of light (from the standpoint of the observer). Gravity waves propagating through the atmosphere are likely candidates. The slowly moving bright rays crossing halo displays could be a manifestation of the so-called banded sky (GLA23).

Similar and Related Phenomena. Visible sound waves (GES2); rainbows disturbed by thunder (GEB9); the banded sky (GLA23).

Examples of Transient Lines Superimposed on Halo Displays

X1. October 11, 1899. England. "On Oct. 11 a curious phenomenon occurred at about 1.30 p.m., and continued with variations for over an hour. The curious part of this 'solar halo' was that it did not appear as a simple circle or semicircle, but as a series of ever-changing wavy lines, above which was an upright and inverted rainbow of brilliant colour." (R1) The inverted rainbow is probably the upper contact arc. (WRC)

X2. November 23, 1910. Hampstead, England. "A curious lunar phenomenon was seen at the observatory of the Hampstead Scientific Society, near the flagstaff on the Heath, on the 23rd ult. About 10.30 p.m. the writer, who with three others, was engaged in observing the comet, noticed a bright patch in the sky immediately above the moon, which was a few hours short of full. Some twenty minutes later his attention was again directed to this appearance, and it was then seen that the moon was encircled with a nearly complete 22° halo, with mock-moons east and west, and an arc of the rarely seen inverted halo touching the primary halo at its topmost point, the junction of the two halos forming the bright patch first seen. At this point there were distinct indications of colour. The sky was piled with thin cirro-stratus cloud. It was then noticed that, in addition to the halos and mock-moons, there was a system of straight, bright rays, radiating from a point on or near the horizon immediately below the moon, well defined and easily fixed in position by bright stars over or by which they passed. The longest of these rays were those lowest down in the west, where they were easily traceable for quite 90°, two of them passing beyond Regulus. A sketch was made on the spot, a tracing of which accompanies this letter. The sketch was begun at 10.55. Five or six minutes later it was observed that the whole system of rays was rotating toward the right, the ray next to the left of the moon being now to the right, and the ray which passed over Spica, now passing just below it. At 12 o'clock the inverted halo was still conspicuous, the mock-moons and the rays visible with difficulty, but still radiating from the same point. The last observation was made by the writer about a mile away, at the town end of the Heath, so that any terrestrial explanation of the phenomena is rendered difficult. The comet had set for over half an hour, and the position of the rays in the south seems to negative the possibility of any connection. The appearance was noted by three observers...." (R2) The rotating system of rays resemble the banded sky phenomenon (GLA23). (WRC)

X3. August 6, 1916. Fort Worth, Texas. For a period of 4 minutes, the upper contact arc of a 46° halo was divided longitudinally by a single dark line. The band widths were in the ratio 2:3, with the nar-

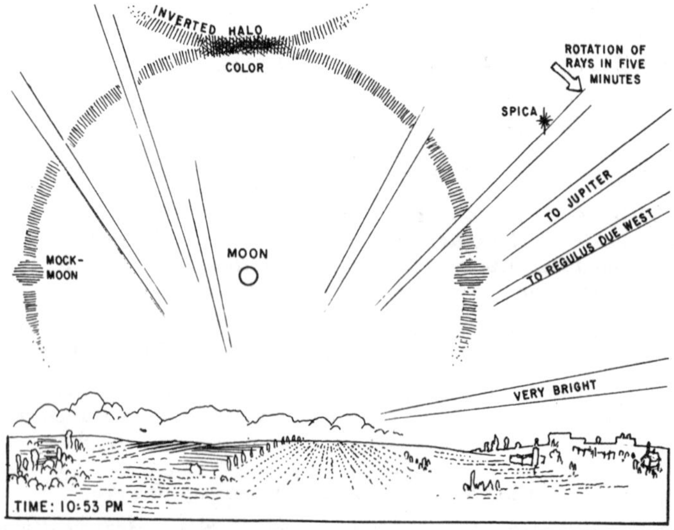

Lunar halo crossed with diverging rays (X2)

row part being toward the sun. The arc may have been the circumzenithal arc. (R3, R4)

X4. August 9, 1944. Cambridge, England. "A phenomenon which I have not seen described before occurred in a part of the horizontal ring of a solar halo on August 9, 1944. It consisted of dark bands moving like waves through the very bright ring. The sky was very clear at Cambridge on the morning of that day when a cloud trail was formed by an aeroplane. The trail moved from northwest in an easterly direction. At 11.20 a.m. a section of the horizontal mock sun ring appeared in that part of the cloud which in the northeast had the same altitude as the sun. As the cloud moved on, being steadily deformed and taking the shape of the letter ~ the halo shifted slowly from left to right. About a minute later, I noticed dark bands crossing quickly through the brightness of the halo. They were of various grades of darkness and could not be seen continuously. Each band was about $1/2^{\circ}$ in width and seemed to be perfectly straight. The distance between two adjacent bands was approximately $1/2$-1°. The middle one of a batch of bands was usually the darkest, the brightness of the brilliant mock sun ring was being reduced by more than 75 per cent, while the accompanying bands were less dark. The best description of the phenomenon I can give is by comparing it with the rippling of the smooth surface of a lake by an occasional breeze. The bands moved from right to left and their speed seemed very high, approximately 5° per second, that is, very much faster than any cloud seen in the sky on that day. They gave the impression of being lower than the halo. I called a second observer who independently noted the phenomenon after I had directed his attention to the spot. A description was given in the words: 'It is as if the ether waves have become visible.' At 11.25 a brilliant mock sun appeared on the right of the sun. The 22° halo developed and at 11.30 a mock sun could be seen on the left. However, no dark bands were visible in them. Later, a part of the horizontal ring appeared in another cloud trail at 11.45 a.m. The bright patch was this time in the north-west. First one dark band moved quickly through the patch in a northerly direction, taking about 1/2 sec. to cross it. A few seconds later a whole batch of waves followed, going in a slightly different direction. Then lower clouds covered this part of the sky." (R5)

X5. June 17, 1979. Bolton, England. "On Sunday 17 June 1979 several areas of cirrus clouds had developed from aeroplane contrails in the sky over Horwich near Bolton in Lancashire. In these contrails various forms of haloes were observed. I first noticed the ordinary mocksuns of 22° at 1825 BST and later in the evening the upper tangent arc of the 22° halo and parts of the solar pillar were visible. At 2100 BST my attention was fixed on the upper tangent arc because of its straight line appearance which was probably caused by the striations of cirrus cloud. Suddenly at 2103 narrow dark vertical lines passed quickly through this arc. The distances between the lines and their intensity were quite irregular. The whole appearance lasted for a few seconds only, but after about half a minute the same happened again. The lines moved from west to east at a speed which I estimated at roughly 2° per second. There were about four lines per degree." (R6)

X6. July 30, 1944. Arromanches, France. "The entry in my diary for 30 July 1944 states in part: 'Plentiful tufts, bands, and patches of cirrus and cirro-cumulus moving from the west, seemingly very high. In mock sun at 1930 see concussion waves from firing on front passing through cirrus.'.... The phenomenon took the form of dark parallel bands moving rapidly from bottom left to top right of the sun dog, i.e., from south to north. I think that the ripples showed themselves most clearly in the red part of the parhelion, but this association with red in my mind may be a caprice of memory. At any rate I can recall a clear picture of moving bands within which the luminous pink colour was momentarily extinguished allowing the blue of the sky to show through the tenuous cloud. Through binoculars, in fact, the striations of cloud appeared to flicker on and off as a ripple passed. Ripples crossed the mock sun at irregular intervals every few seconds. I cannot recall whether they came in closely associated groups, like ripples from a stone dropped into a pond, or singly. I do remember that most were quite faint, but that every now and then there would be a stronger impulse apparently originating from a point further east than the others." (R7)

X7. Spring 1976. No place given. "It occurred one evening at about 6 pm during April or May 1976. The cirrus was in patches and a brilliant mock sun appeared at an elevation of about 10° to 12°. It was brightly coloured and had a moderate length of tail. Suddenly dark bands passed across the parhelion. These were vertical from the direction of the sun and as mentioned by Blake the bands did not go beyond the tail. At first, I thought it was smoke, but soon it became apparent that ripples were present in the atmosphere. The phenomenon lasted for about 30 to 45 seconds and finally disappeared completely leaving only the parhelion showing." (R8)

X8. December 15, 1976. Chelmsford, England. "At 1505 G.M.T., while observing an upper arc of contact to the 22 degree halo, a number of closely spaced grey bands were seen rapidly crossing it from left to right, roughly south to north.....The bands were first seen for about ten seconds. Then there was a period of about 15 seconds when no bands were seen. More bands then appeared for about 5 seconds. During the interval the attention of three colleagues was summoned, two of whom are trained meteorologists. All three witnessed the second sighting and confirm my own observations. During the first sighting an estimated 30 or so bands crossed the arc of contact, there being about 5 to 10 bands visible at any time. The bands appeared greyish in colour and did not have a sharp outline. They were narrow, appeared to be straight, parallel and regularly spaced. They moved steadily, taking an estimated two seconds to cross the arc of contact. During the second sighting the appearance and behavior of the bands was (sic) identical to the first sighting. As the bands crossed the arc of contact, their length varied so that they were only visible within the confines of the arc." (R9) Supersonic aircraft were suspected to be the cause, but they would not usually produce so many closely spaced ripples. (WRC)

References

R1. Horner, D.W.; "Remarkable Mock-Sun Phenomena," English Mechanic, 70:253, 1899. (X1)

R2. Hepburn, Patrick H.; "Curious Lunar Phenomena," English Mechanic, 91:425, 1910. (X2)

R3. "A Novel Optical Phenomenon," Scientific American, 116:301, 1917. (X3)

R4. English Mechanic, 105:209, 1917. (X3)

R5. Archenhold, G.H.; "A Solar Halo Phenomenon," Nature, 154:433, 1944. (X4)

R6. Archenhold, G.H.; "Solar Halo---Moving Dark Lines," Weather, 34:476, 1979. (X5)

R7. Blake, M.; "Rapidly Moving Ripples in Cirrus," Weather, 32:276, 1977. (X6)

R8. Scutt, R.; "Rapidly Moving Ripples in Cirrus Cloud," Weather, 33:38, 1978. (X7)

R9. Burton, B.J.; "Fast-Moving Dark Bands Crossing the Arc of Contact," Journal of Meteorology, U.K., 2:233, 1977. (X8)

GEH12 Optical Effects Where Halo Displays Touch the Horizon

Description. Bright patches of light, often with spectral colors, occurring where lunar halos and mock moons touch the horizon. This phenomenon may be confined to snow- and ice-covered terrain.

Data Evaluation. In addition to some older observations by Nansen, recent Antarctic expeditions have yielded a host of good observations. Rating: 1.

Anomaly Evaluation. Some enhancement of halo brightness would be expected where displays meet a snow/ice horizon due to simple reflection. The spectral colors, though, are not handled as easily. Unfortunately, Nansen did not provide enough detail. If the colors are seen well above the horizon, there is no ready explanation; if they occur below the horizon, they probably come from halo-like reflections from ice-crystals on the ground, just as in terrestrial halos (GEH16). Rating: 3.

Possible Explanations. Reflection from ice and snow and, possibly, from terrestrial ice crystals, as in terrestrial halos. Mirage effects, too, may be involved, because low-lying mist is frequently a factor and many of the effects are seen below the true horizon.

Similar and Related Phenomena. Hyperbolic halos formed from hoarfrost (GEH16).

Examples of Optical Effects When Halo Displays Touch the Horizon

X1. November 1893. Arctic Ocean. The Nansen Expedition. "These last days the moon has sometimes had rings round it, with mock-moons and axes---accompanied by rather strange phenomena. When the moon stands so low that the ring touches the horizon, a bright field of light is formed where the horizon cuts the ring. Similar expanses of light are also formed where the perpendicular axis from the moon intersects the horizon. Faint rainbows are often to be seen in these shining light-fields; yellow was generally the strongest tint nearest the horizon, passing over into red, and then into blue. Similar colours could also be distinguished in the mock moons." (R1)

X2. Frequent occurrence. Antarctica. "On numerous occasions a type of mock-sun was observed, to which, for want of a more euphonious name, I refer as an 'hypohelion.' It was also seen five times in the Antarctic by Arctowski (1899), whose diagrams are reproduced in Pernter-Exner's Meteorologische Optik (1922); it is mentioned briefly in Bower's diary (Cherry-Garrard, 1922). Its characteristic is that it always occurs on or near the visible horizon, either with or without other optical phenomena, although as far as I am aware it has not been seen alone previously. It is usually large and bright (cf., Bowers, loc. cit. 'Below was a dome-shaped glare of white light which contained an exaggerated mock-sun, which was as dazzling as the sun himself.') and always diffuse and does not show prismatic colors, being as a rule white, except that when the sun is very low it may become yellow or even red. It was observed in North East Land as follows (only the four most interesting observations have been reproduced here):

March 5, 12.00 h. to 14.00 h. Temperature -13°. ---A pair of mock-suns near the halo of 22°, very definitely below the horizontal circle through the sun, but well above the horizon, remaining about equally distant from both the whole time.

March 10, 12.00 h. Temperature -16°. ---The halo of 22° was seen, ice-crystals being clearly visible in the air. On it were three pairs of mock-suns viz.:---

(a) On the horizontal circle, which was visible for a few degrees outside the halo. These were the usual prismatic parhelia, showing the colors red, yellow, green, blue, and were fairly well defined.

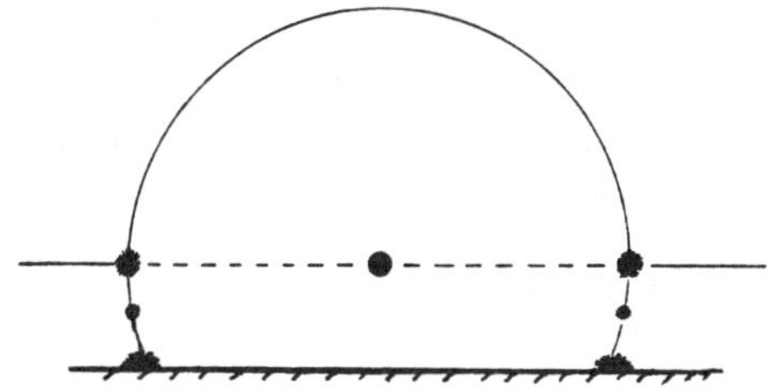

(b) On the visible horizon. This pair was large and diffuse, of an orange-yellow color, but showing a little red on the 'inner' side, i.e., that towards the sun.
(c) Half-way between (a) and (b). These showed the same color sequence as (a) but were slightly less bright, and were smaller.

The sun was 5° or 6° above the horizon. The halo was white and of small intensity towards its highest point.

March 17, 13.00 h. Temperature -23°. ---Three rather faint and misty hypohelia of a reddish color were seen below the topographical horizon. They lay in a straight line, one of them being vertically below the sun and the others about 22° from it.

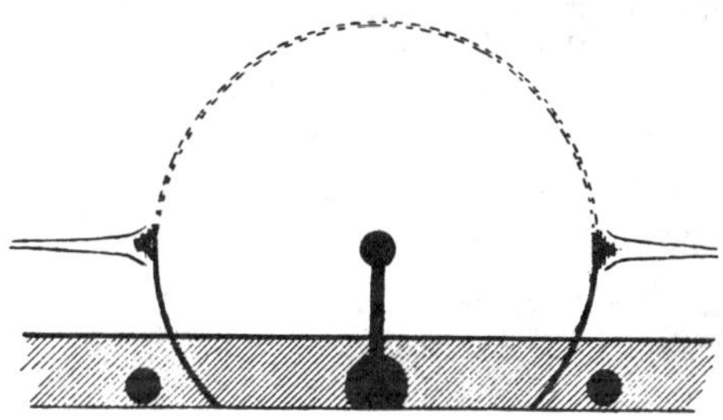

April 20, 16.00 h. Temperature -5°.--- A remarkably fine and brilliant halo system with mock-suns was seen. The 22° halo was brightly colored below the level of the horizontal circle, part of which developed outside the halo, and on it were two parhelia. Above this level the halo was of less intensity, but still brightly colored. On the horizon, directly below the sun, was a large, very bright, glittering, white mock-sun, and between it and the sun a rather faint pillar. On the same level and vertically below the prismatic parhelia were nebulous, white hypohelia somewhat smaller than the central one. These were quite distinct from the halo which was colored down to ground level, a definite clear space separating them. Slightly above the level of the white parhelia, the upper edge of a horizontal bank of cloud could be distinguished through the general mistiness.....

Thus it seems that the hypohelia and the more usual parhelia are distinct phenomena, although they may often occur together; in fact they generally do. The former seems to be connected with the rather thick low-lying mist, while the latter are produced by the larger ice-crystals which may become visible to the naked eye, as when the circumzenithal arc was seen. The air temperature when hypohelia were seen was always below zero Fahrenheit and more usually in the neighborhood of -20°F." (R2)

References

R1. Nansen, Fridtjof; "The Winter Night," Farthest North, London, 1897, vol. 1, p. 185. (X1)

R2. Moss, R.; "Atmospheric Optical Phenomena in North East Land," Royal Meteorological Society, Quarterly Journal, 64:451, 1938. (X2)

GEH13 Close, One-Sided Mock Suns

Description. A single mock sun appearing 10° or less on one side of the true sun at approximately the same altitude. No traces of the halos usually associated with mock suns are detectable.

Data Evaluation. Two excellent observations, well-measured, one with a good photograph of the phenomenon. Rating: 2.

Anomaly Evaluation. Mock suns are common at 22° and not rare at even larger distances from the true sun. The mock suns of this entry, however, are not located at the radii of any recognized halos. They therefore constitute one of the more profound anomalies in atmospheric optics, assuming of course that they are actually created by the reflection and refraction of light by ice crystals. Rating: 2.

Possible Explanations. It is possible that these close mock suns may be spurious halo phenomena, and that we are dealing merely with lateral mirages.

Similar and Related Phenomena. Lateral or sideways mirages (GEM3); double suns (due obviously to refraction near the horizon) (GEL4).

Examples of Close, One-Sided Mock Suns

X1. November 5, 1896. Wales, United Kingdom. A mock sun a few degrees from the true sun. True sun was dark red; the mock sun looked like the moon when seen during the day. (R3)

X2. February 16, 1926. North Atlantic Ocean. "The phenomenon was first seen at 15h. 46m. G.M.T. when the true sun, with an altitude of 14° and bearing 227° was

Real sun and mock sun separated by 10° (X2)

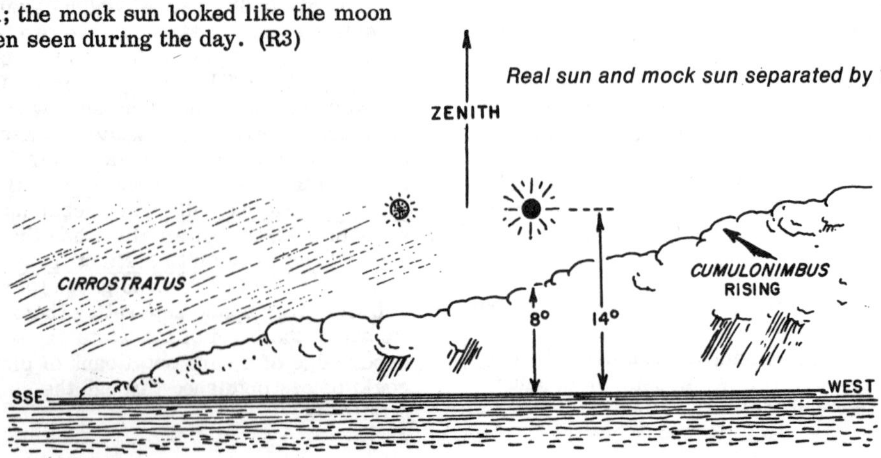

shining brightly with good limb in medium sextant shade, and a heavy squall was moving from west to south-southeast with the upper edge of the cumulo-nimbus cloud at a point vertically below the sun and having an altitude of 8°. The phenomenon was an uncoloured image of the true sun at the same altitude and at an angular distance of 10° to the left, certainly not 22°. It was fairly well defined, sufficiently so for an azimuth bearing, but not enough for anything but a doubtful sextant altitude. The image had the appearance of the true sun shining through denser layers of cloud, although the cirro-stratus was fairly uniform in density; but with a slightly mottled appearance. No image was seen to the right of the true sun, and probably, though not certainly, the corresponding position to the right of the sun was already occupied by the moving cumulo-nimbus. No halo or mock-sun-ring was observed. The image dimmed as the squall approached the true sun and disappeared when the true sun became covered." (R1)

X3. June 15, 1965. Sea of Timor, South Pacific Ocean. "On 15 June 1965 the motorship Fairstar navigated in the Sea of Timor. The sea was very quiet, the sky almost perfectly clear. A few passengers were on deck, in the open air, looking at the sea, no glass pane or any other object being interposed between them and the prospect. About 10 minutes before sunset, Mrs. N.S. noticed to her surprise a 'second sun', much less bright than the real one, somewhat to the left, but at the same height above the horizon. There was nothing at the symmetrical point to the right. She pointed this out to two other passengers, who then also noticed the phenomenon. She ran to her cabin and made a colour photograph, which by a piece of good luck had almost the right exposure. After development, it was found to confirm fully the visible observation..... The original was measured with a plate measuring machine and gave results, accurate to about 0.01 mm and well reproducible. The focal distance of the camera objective amounted to 45 mm, so that 1° corresponds to 0.786 mm, and 1 minute of arc to 0.0131 mm. The results are summarized in the diagram, where all dimensions are given in units of 0.01 mm. The following conclusions may be drawn.

(a) The sun and the mock sun are indeed precisely at the same height above the horizon: 1° 8'.

(b) The horizontal diameter of the sun of 31' 32" (after the ephemeris) should correspond to 41.2 of our units on the photograph; the mock sun is found to have precisely the geometrical diameter, the real sun is slightly larger (irradiation?).

(c) The vertical diameter of the mock sun is 38, that of the real sun is 42 (irradiation); for a solar altitude of 1° 8' the observed ratio between the vertical and the horizontal diameter corresponds to the normal flattening by atmospheric refraction.

(d) The mock sun was at a distance of 3° 30' to the left of the sun.

(e) Below the mock sun there is a slight brightening of the sky just as it is seen below the real sun, only much weaker. It looks as if the mock sun was simply a faint duplicate of the real sun.

The phenomenon remained visible till the sun was setting. It seems impossible to explain the observation by any reflexion or by instrumental effects inside the camera, since it was first noticed visually by three persons. Mrs. N.S., who reported the observation, makes a very reliable impression, however she has no special scientific education." (R2)

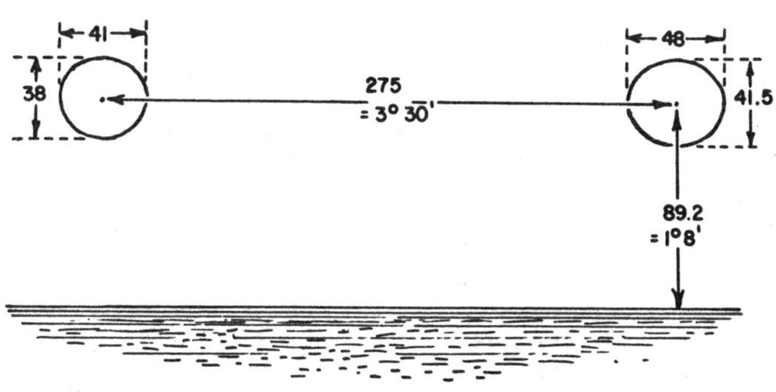

Close, one-sided mock sun (left); real, slightly flattened real sun (right) (X3)

References

R1. Vigurs, J. T. C.; "An Unusual Mock Sun," Meteorological Magazine, 61:91, 1926. (X2)

R2. Minnaert, M.; "An Exceptional Phenomenon of Atmospheric Optics: A One-Sided Mocksun at 3°30'," Weather, 21: 250, 1966. (X3)

R3. "Mock-Suns Close to the Sun and Lateral Mirage," Weather, 22:260, 1967. (X1)

GEH14 Halo Displays Formed by Terrestrial Ice Crystals

Description. Hyperbolic strips of colored light stretching along the ground away from the observer towards a low sun. The apex of the hyperbola appears just in front of the observer, while the arms maintain a constant 22° and/or 46° angle with the sun. The ground must be covered with ice crystals; for example, large flakes of hoarfrost.

Data Evaluation. Few observers have reported this phenomenon in the literature, but Minnaert (R2) treats it as a well-established phenomenon. Rating: 2.

Anomaly Evaluation. These terrestrial halos yield readily to the same rules governing atmospheric halos. Rating: 4.

Possible Explanations. Terrestrial spectral hyperbolas are formed by the same reflections and refractions in ice crystals that give rise to the ordinary atmospheric halos.

Similar and Related Phenomena. Normal halo displays (GEH0); horizontal rainbows (GEB11); which are very similar in form but stretch away from the observer in the direction away from the sun and depend upon liquid water droplets for their formation.

Examples of Halo Displays Formed on Terrestrial Ice Crystals

X1. March 23, 1845. Amherst, Massachusetts. "On the 23rd of last March I witnessed an optical phenomenon in nature, which was to me entirely new. The sun had ascended eight or ten degrees from the horizon, and I was stepping from my door, which opens to the east, when I was surprised to notice a luminous curvilinear band, three or four feet wide, thickly studded with shining points of every prismatic hue, strecthing over the dead grass of the wide street before me. Its form seemed to be that of a parabola or hyperbola; its nearest point, which was the vertex of the curve, was not more than twelve or fifteen feet distant, and its two branches extended several rods, one to the right and the other to the left of the sun,---the axis being the intersection of the ground plane with a vertical through the sun and the eye. Though the band could thus be distinctly traced, yet the illumination was not continuous, as in the rainbow, but pencils of intense light were seen to come from innumerable but separate points; and these exhibited, without much order of arrangement, the countless shades of the prismatic spectrum. As I moved, the luminous arch moved also in the same direction, so as to retain its relations to my eye and the sun, while the individual points changed from hue to hue and were extinguished, and others started into sight, to twinkle for a moment in their turn, and then disappear. My first thought was, that it was the lower limb of a rainbow, such as I have often traced in dew-drops, as they rest on vegetable leaves and are strung upon spider lines that are stretched along the grass. Its form appeared the same as the rainbow must assume when seen in such circumstances. But it was in the wrong direction; I stood facing the sun, and the colors plainly came from points not more than thirty degrees from it. I perceived in a moment, however, that there were no dew-drops, but that the spires of dead grass were feathered over with frost crystals of unusual size. The true nature of the phenomenon instantly flashed upon my mind; and I was delighted to witness a most interesting confirmation of the truth of the generally received theory respecting the large solar and lunar halos. In the first place, though the colors were scattered promiscuously, yet there was evidently a prevalence of the red on the most distant, that is, the concave edge of the band. And in the next place, I ascertained, by a rude measurement, that the distance of the curve from the sun was about twenty two degrees. These two particulars characterize that species of halo, most frequently encircling the sun and moon, respectively called the parhelion and the paraselene."(R1)

X2. General Observations. When an observer faces the sun and the ground is covered with hoarfrost, just after sunrise or before sunset, hyperbolic halos of 22° and/or 46° may be seen scintillating with "the most wonderful colors." (R2)

References

R1. Snell, E.S.; "Singular Case of Parhelion.....," American Journal of Science, 1:49:73, 1845. (X1)

R2. Minnaert, M.; "Halo Phenomena on the Ground," The Nature of Light and Color in the Open Air," New York, 1954, p. 207. (X2)

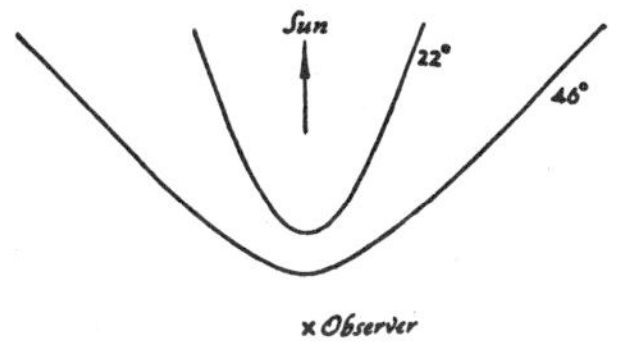

Hyperbolic halo geometry applicable to terrestrial halos seen facing the sun (X2) (After R2)

GEI OBSERVER-CENTERED PHENOMENA

Key to Phenomena

GEI0 Introduction

A wild tale has long circulated about a giant specter seen by mountain climbers on the Brocken, a German mountain. As the story goes, there was once a climber working his way along a precipice who suddenly saw an immense human figure rise out of the mists towards him. In his fright he lost his footing and fell to his death. Doubtless just a story, but the Brocken Specter does exist, not only on the Brocken but wherever shadows are cast upon mist, fog, and veils of fine water droplets.

That one's shadow can be seen, even in giant proportions, on a wall of white mist is not surprising. Some features of the phenomenon, however, remain to be fully explained. A common feature of the Brocken Specter is the "glory" or concentric rings of color centered on the observer's head. As many as five separate, multicolored rings have been observed. There is considerable controversy as to whether glories have the same origin as solar coronas, which everyone agrees are formed by the diffraction of sunlight by fine water droplets. The problem is that glories are seen with the sun at one's back and no adequate mechanism has been uncovered that reflects enough sunlight from the mist back toward the observer to form the glories. Thus, this class of phenomena is but a small one---merely three entries where the observer himself is actually a part of the display. Of course, many other optical phenomena seen in the open air require specific angular relationships between the observer, the light source, and the phenomena. The rainbow is an obvious case in point; but the observer never sees his shadow in the center of the rainbow, a halo, or mirage; he is only a passive observer of nature. Whereas the viewer of a Brocken Specter can play an active role in the phenomenon.

Curious things happen inside the glories. Sometimes an observer sees only his own shadow; at other times, nearby persons and terrain are also visible. Some glories are elliptical; others are more grossly distorted, especially when another person approaches the observer. Another unusual feature is the occasional appearance of dark radial streaks that seem to emanate from the observer's arms or even features of the terrain. Such facets of the phenomenon are probably not intractable to atmospheric optics, but no convincing explanations have yet been found.

Two related phenomena are recorded here, not because they are anomalous, but because they are also observer-centered and most interesting in their own rights. Heligenschein is a luminous halo seen about one's shadow cast upon wet grass. Then, there are the bright spokes of light seen rotating wheel-like around the shadow of the observer's head cast upon a ruffled pond or swimming pool. Both phenomena have easy explanations; but their close cousin, the Brocken Specter, retains some puzzling features.

GEI1 Puzzling Features of the Brocken Specter

Description. Shadows of observers cast upon a veil of mist by the sun, usually surrounded by one or more rainbow-like rings called "glories." Red is almost always on the outsides of the rings, with yellow, blue, or another color inside. Several features of the Brocken Specter and its accompanying glories are intriguing and not always easy to interpret: (1) The shadows of the observers may be gigantic or very small; (2) The order of the colors in the glory is variable; (3) The observer generally sees only his own shadow, but the exceptions are legion; (4) The glory may be round, elliptical, or more complex shape; (5) Dark, radial rays often project from parts of the shadows; (6) Some displays apparently incorporate true rainbows and fogbows; (7) Sometimes shadows of distant objects appear outside the glory; (8) The center of the glory may suddenly shift to a different part of the shadow.

Background. Brocken Specters received their name from their frequent appearance on the Brocken, a European mountain, where huge shadows suddenly and eerily appear before climbers on a curtain of mist. Not all Brocken Specters boast glories; and not all glories have well-defined shadows at their centers. These days, glories frequently appear to airplane passengers around the shadows of their aircraft projected on the clouds below.

Data Evaluation. Reports of "standard" Brocken Specters are abundant; and they are not cataloged here. Data supporting the existence of the anomalies listed above are surprisingly frequent, except for (7) and (8), which are one-of-a-kind. Overall rating: 1.

Anomaly Evaluation. The reader should not think that excessive space is being devoted to a phenomenon that is merely curious save for a few minor features. Optical theory does not satisfactorily account for the basic glory. At one time, glories were attributed to the diffraction of light scattered back toward the observer by mist droplets, but actually most sunlight is scattered away from the observer. Beyond this fundamental problem, atmospheric optics must contend with all the idiosyncracies of this phenomenon, many of which will probably succumb to careful analysis using ray diagrams and incorporating the varying distances of the mist screen and water droplet diameter. Rating: 2.

Possible Explanations. Optical diffraction is the most likely candidate for the explanation of the glory's multiple rings of color, which may number as many as five! The varying sizes of the shadows seen within the glory depend strongly upon distance to the veil of mist, as one would expect, although there are some disconcerting exceptions in the reports. The dark radial rays and the shape of the glory seem to be associated with objects (and persons) close to the line of sight of the observer. Like an obstruction in a telescope tube, they intercept some of the light that would form the scene before the observer, thus creating shadows on the mist and deformed glories. These suggestions are manifestly vague, and any sound explanations will depend upon careful study.

Similar and Related Phenomena. Rainbows also occur with the sun at one's back (GEB); Bishop's Ring is a diffraction phenomenon seen in the direction of the sun (GEH5); halos are spectral displays that usually occur close to the sun (GEH). Rainbows possess such large angular diameters that they can almost always be easily distinguished from glories.

Examples of Brocken Specter oddities

X0. Ulloa, the famous Spanish scientist gave this classic description of the Brocken Specter. "He was on Pambamarca with six companions at daybreak. The top of the mountain was entirely covered with dense clouds. As the sun rose, it dispelled these clouds, and nothing remained in their stead but some very light mists, which it was almost impossible to distinguish. Suddenly, on the side opposite that in which the sun rose, each of the travellers saw, at a dozen fathoms from where he stood, an image of himself reflected in the air, as though upon a mirror. This image appeared in the centre of three rainbows shaded with different colors, and surrounded at a certain distance by a fourth arch of a single color. The tinting farthest on the outside of each arch was flesh-colored, or red, the next shade was orange, the third was yellow, the fourth straw-color, and the last one green. All these arcs were perpendicular to the horizon; they moved about and followed the person reflected in every direction, surrounding his image like a gloria. What was most remarkable was, that, although the seven travellers stood together in a single group, each of them saw the phenomenon only in relation to himself, and was disposed to deny its existence in reference to the others. The extent of these arches increased progressively in proportion to the height of the sun. At the same time their colors faded away, the spectra became paler and paler and more vague, and at last the phenomenon entirely disappeared. When this display began, the shape of the arcs was oval; and, toward the last, it was perfectly circular." (R4)

Brocken Specter with glories (X0)

September 29, 1880. Putney, England. A Brocken Specter formed with an artificial light source. "Having occasion ten days ago to go into my garden about half past ten o'clock at night I found there was a thick white fog, through which, however, a star could be seen here and there. I had an ordinary bedroom candlestick in my hand with the candle lighted, in order to find the object I wanted. To my great surprise I found that the lighted candle projected a fantastic image of myself on the fog, the shadow being about twelve feet high, and of an oddly distorted character, just as the spectre of the Brocken is said to be. It is of course usual on going into the open air to use a lantern with a solid back for any light that may be wanted, and with this, of course, such a shadow would not be seen; but in this charmingly foggy valley of the Thames, and in these days of 'Physics without Apparatus,' the effect I saw can probably be seen only too often. May not the gigantic spirits of the Ossianic heroes, whose form is composed of mist, through which the stars can be seen, be derived from the fantastic images thrown upon the mountain fogs from the campfires of the ancient Gaels?" (R5)

X1. July 31, 1858. Catskills, New York. "About half past six o'clock, P.M., as I joined again the assemblage on the rocks a little south of the hotel overlooking the valley, the valley appeared like one vast sea of mist or cloud nearly to our feet; grayish clouds were hovering over it in the south, and before me (east) I saw myself in a circular rainbow, brightening until a second pale inside circle surrounded the first, like a halo, but with a little amber-colored space between them. And what I saw of myself, each one saw of himself, and only of another when in close contact. As I moved my arms or head, the image moved; sometimes the circle deepened in tints, at others paled, as the sun shone brightly or was obscured by passing clouds. Shortly after, the sun went into a cloud and the vapors rolled away from the valley. But in a few minutes the valley and mountain side were again shrouded in heavy white clouds, and soon the sun shone out again; but it was so much nearer its setting, that when the vision reappeared it was thrown farther off, and was different in many respects. The reflection was not in a circle, but a long horizontal oval, and the rainbow tints were much paler; besides, the picture was not of an individual only, but of the group of two or three persons observing it, and of the irregular outline of the edge of the mountain precipice with the scattering wild plants growing on it." (R1) The distance to the mist has a strong effect upon just how much is included in the corona or glory. Despite what some books say about the Brocken Specter phenomenon, situations do exist in which more than one person can be seen in the glory. (WRC)

X2. 1871. Snowdon, England. "This was a short account of a curious phenomenon observed by the author two years ago on the top of Snowdon. He saw his own shadow surrounded by five concentric coloured bows, which seemed to approach as the fog came nearer, until at last he saw the shadow of his head surrounded by a brilliantly coloured ring. Similar phenomena have often been observed; but so great a number of bows has never been seen." (R2) This item is included because of the "record" number of bows. (WRC)

X3. July 12, 1875. Snowdon, England. "On Monday, July 12, I, with a party, ascended Snowdon. The atmosphere was clear until we had reached within half a mile of the summit, when a light cloud rising stealthily from amongst the southern peaks enveloped it. Drifting towards us, when very near, the cloud dropped over the eastern shoulder of the mountain just where it dips towards Capel Curig. As we stood watching, great was our surprise and delight when we beheld painted upon it, not the arc-en-ciel with which we are familiar, but a complete and brilliant prismatic circle, apparently about thirty feet in diameter, in the very centre of which we ourselves were depicted, the image being somewhat enlarged but clearly defined; as we arranged the party in groups or bowed to each other, every form and movement was faithfully reproduced in the picture. It was now about 8 o'clock, with the sun nearly in a line with us." (R3) Another instance where more than one observer, many in fact, were encircled by the glory. (WRC)

X4. August 13, 1884. The Grisighorn, Switzerland. Dark rays emanated from the observer's arms. These long extensions moved as the arms were moved. (R6)

X5. February 14, 1884. Skye, United Kingdom. "Our shadows were apparently thrown against the precipitous side of a deep corry, distant 200 feet, or perhaps more. They vanished and reappeared as thin mists passed through the corry, the sun shining continuously. We could not see each other's shadows unless close. The distance apart at which they became visible I do not clearly remember, but know it was approximately as one of my friends, Mr. W. A. Brown, writes:---'So long as we kept a few yards apart each could only see his own shadow, but when two were within arm's length a double shadow was visible to each, and on getting still nearer the shadows merged into each other.' My estimate of the angle subtended by the diameter of the rainbow was 15°, that of my friend 10°." (R6)

X6. February 21, 1886. Adam's Peak, Ceylon. On the summit awaiting the famous shadow phenomenon described in GES8. "..... suddenly at 6.30 A. M. the sun peeped through a chink in the clouds, and we saw the pointed shadow of the Peak lying on the misty land. Driving condensed vapour was floating about, and a fragment of rainbow-tinted mist appeared near the top of the shadow. Soon this fragment grew into a complete prismatic circle of about 8° diameter by estimation, with the red outside, formed round the summit of the Peak as a centre. The author instantly saw that with this bow there ought to be spectral figures, so he waved his arms about and immediately found shadowy arms moving in the centre of the rainbow. Two dark rays shot upwards and outwards on either side of the centre, as shown in the diagram fig. 2, and appeared to be nearly in a prolongation of the lines of the slope of the Peak below. The centre of the bow appeared to be just below the point of the shadow, not on it; because we were standing on a platform below a pointed shrine, and the subjective bow centred from our own eyes. If we did not stand fairly out in the sun, only a portion of the bow could be seen." (R7, R8) These ray-like prolon-

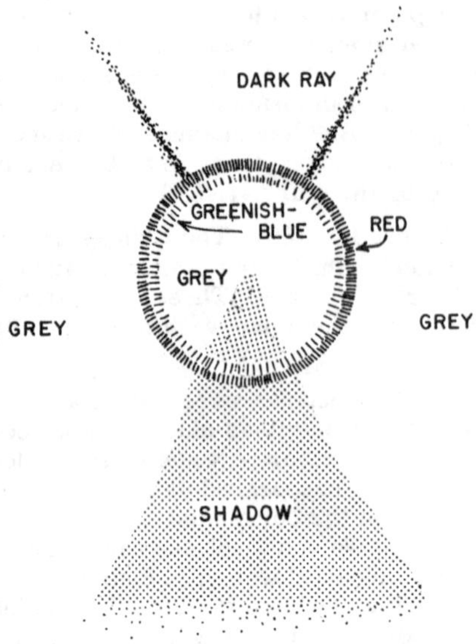

Shadow of Adam's Peak with glory and radial rays (X6)

gations of the mountain's outline are analogous to the rays extending the observer's arms in X4. (WRC)

X7. March 1, 1887. Skye, United Kingdom. As two observers of the Brocken Specter approached one another the inner edges of the hitherto separate coronas or glories disappeared, leaving a single glory shaped like a double peanut. (R6)

X8. March 1887. Teneriffe, Canary Islands, Spain. "As we stood near the edge of the precipice looking over the valley, with the setting sun behind us, we saw our shadows projected on the thin mist; each shadow of course followed every movement of its owner, keeping for example excellent time to a reel, and on one occasion (March 8th) it was so sharp, that when I went through some cuts and guards of the sword exercise the shadow of my walking stick was perfectly clearly defined. I should say that the spectre varied in distinctness on different occasions and at different spots on the same day, the variations being evidently due to the varying density and outline of the surface of the mist. A lady friend saw it again on April 12th, under similar circumstances in the same place, but it was then far less clear. When seen at its best the spectre was surrounded with three complete sets of faintly coloured rings, the head of the figure being in the centre of the system. A blue haze surrounded the head and reached about to the shoulders; the colours followed in the reversed order of the spectrum, green, yellow, orange and red; the first red ring crossed the figure at its middle, then came blue again, and so on the second red ring reaching about to the feet; outside this again a third reversed spectrum, the outside red ring extending about once and a half the height of the figure. Beyond this triple system of coloured rings, at a distance of perhaps 15° from the line connecting the spectator's eye with the head of the shadow, was an arch of white light, broad and faint, like a lunar rainbow, but sufficiently distinct. Each person could see the rings only around his spectre, or that of a person standing very close to him." (R9) Of particular interest here is the outside white ring. This appears in many accounts and is usually called a fogbow (GEB10). If such is true, and it does seem reasonable, we have here a combination of two optical effects: the glories arising (supposedly) from diffraction and the outer fogbow depending, like the rainbow, on reflection and refraction. (WRC)

X9. October 30, 1896. Juvisy, France. "On October 30 last, while engaged with M. Flammarion in measuring γ Arietis, a fog of an almost incredible density, which rendered invisible objects at 100 feet distance, sprang up all of a sudden from the valley of the Seine, with the result that we were obliged to give up our work at once. On stepping out to the terrace adjoining the dome to see what was going on, I was very agreeably surprised to see my shadow some 200 feet high, and thrown on the mist by a strong lamp, rise up to the zenith! It was a very curious spectacle indeed. Every movement of the hand or head was faithfully reproduced by the phantasm. But only the head and shoulders of the figure were neatly delineated. The remainder of the body was exceedingly indistinct. Gigantic rays of a grey colour radiated from the head (the eyes as a centre) in all directions, and especially from the projecting parts of the figure, the ears, shoulders, corners of the hat, &c., to a very great distance from the centre of emanation. MM. Flammarion and l'abbé Moreux, who happened to be present at the time, saw exactly the same appearances. A very striking feature of the phenomenon was that everybody saw his own spectre only, not that of his neighbor." (R10) Once again, the radiating rays make their appearance. From this description, there can be little doubt that they are simply shadows thrown upon the mist by projections and closely related to the crepuscular rays at sunset and sunrise (GEL2). (WRC)

X10. December 1903. North Wales, United Kingdom. "During the Christmas holiday my brother and I were in North Wales, and happened to be on the ridge that lies north of Llyn Llydaw; the sun was about 1h. from the time of setting, and was low enough to clear the lower edge of the thin clouds which came from a northerly direction. The hollow (Cwm Glas) to the north of the ridge was, every now and then, filled up with thin mist on which our shadows were projected; surrounding the shadow was a faint oval-shaped rainbow, which, as the sunlight strengthened, became brighter, and a second bow outside the one nearest to the figures appeared, though very faintly. Although my companion was within a few feet of me, we each saw our own shadows only. We also saw, when the mist was further from us, a shadow of the ridge itself with our two figures on it, in this case the figures appearing much smaller than in the other effect, and without any bow. (R11) Apparently, at greater distances, the glory disappears completely. More surprising is the reduction in size of the shadows, when one should expect giant Brocken Specters! (WRC)

X11. August 20, 1918. North Wales, United Kingdom. "In North Wales, on August 20, about two hours before sunset, I saw a rainbow effect which was quite new to me. The summit of Tryfaen (some four miles north-east from that of Snowdon) has three sharp, rocky peaks running roughly north and south. We had climbed up the eastern cliff in a south-westerly gale, which brought up much cloud with some light showers, and were sitting just below the top of the southern peak. The Holyhead road lay north-east and 2000 ft. below us. From it rose the upright portion of a brilliant rainbow. At the centre of its circle was the shadow of our peak with those of the other two peaks to the left of it, all sharply defined. Around the shadow of our peak was a most vivid and persistent bow, the smallest I have ever seen, the radius of the inner edge being about half that of the outer. The central space changed a good deal, being frequently almost filled by a diffused yellow glow, which sometimes appeared to condense towards the center until it resembled a nebulous sun on a whitish ground, while at intervals little yellow streamers seemed to radiate from it to the inner edge of the bow. Outside this bow (which had the colours in regular rainbow order, red (outside) was part of a third bow of perhaps double the diameter, but dim and intermittent. We stood up and made gestures, expecting some sort of Brocken effect, but could detect none. However, as we were not on the extreme summit, and the cloud was very distant, our shadows would at best have been extremely minute." (R12) The strange characteristics here are the appearance of

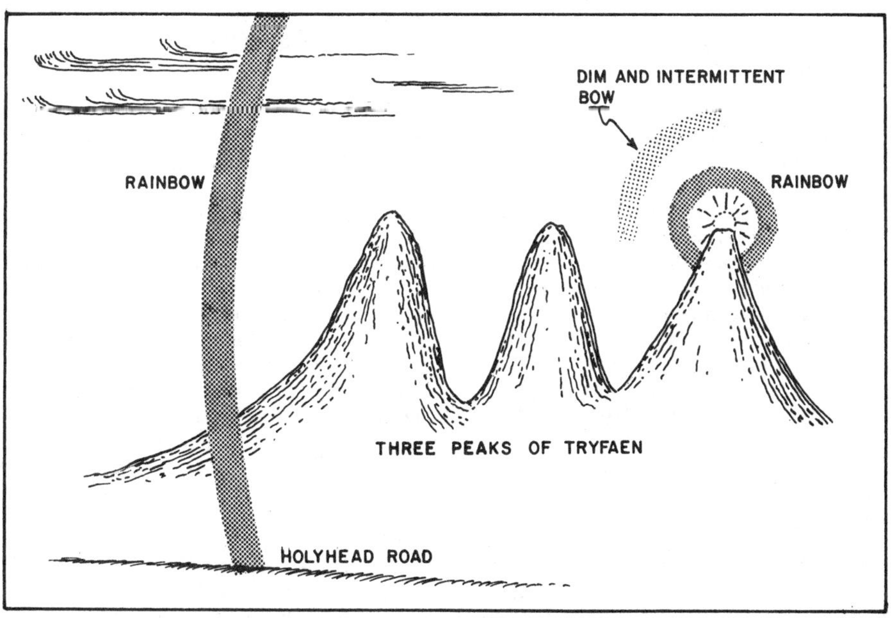

Shadows of three peaks in Wales. The observers were on the peak on the right. (The two small "rainbows" are really glories.) (X11)

images outside the corona and the simultaneous distant rainbow (or fogbow?). (WRC)

X12. December 1, 1952. Grand Canyon, Arizona. The Watch Tower spectre at Desert View. "The specter consists of a series of colors which arc or bow around the shadow of the tower. Yellow was innermost, with reds and purples forming the intermediate and outer bands. Cloud particles diffused the colors along their adjacent edges and formed many interesting blends. Later the same day, the spectre shifted and was centered over the shorter shadow of the kiva section of the tower. When the tower and the bow of colors were first noticed during the spring occurrence, it was about 9:00 A.M. The erect shadow of the tower, and the spectre were then projected on the clouds 200 feet below the north rim, which is about nine miles northwest of the tower. Elevation at the tower is 7452 feet and at the north rim it is about 8300 feet at the Canyon edge. According to Mrs. Farmer, who was on duty at the tower that day, the spectre stayed with the shadow as it was shortened by the rising sun and moved from the west toward the east. As late as 4:00 P.M., the spectre was still visible in a small side canyon lying about ENE of the tower; the shadow was then about 400 feet in length. The bows of color subtended angles (measured from the tower) of less than one degree near the north rim, to approximately ten degrees late in the afternoon of the same day. When the phenomenon was observed on December 1, it was about 3:00 P.M. The shadow at this time extended for about 500 feet NE of the 70-foot tower. Colored photographs have been taken of the phenomenon by Virgil Gipson, local Fred Harvey photographer, and Charles Farmer of Desert View. However, diffusion by the cloud particles usually causes the pictures to have indistinct lines and fuzziness. Unlike true rainbows, the outer band of color is reddish violet instead of red. To further complicate matters and defy explanation, the innermost color is yellow, hence, it is neither a true rainbow nor an inverted one." (R13) The shift in the center of the display is unprecedented, unless the observer happened to move, too. The long duration of the display is likewise most unusual. (WRC)

X13. February 9, 1967. Somerset, England. "The glories seen about the shadows of aircraft flying over layer cloud usually appear circular, but when the sun is low in the sky they appear elliptical or even hyperbolic. The light comprising the top and the bottom of the glory may then come from areas of the cloud sheet many miles apart, This effect seems to have been important in determining the shape of the glory seen by one of us (G. C. B.) when flying above extensive stratocumulus over Somerset at about 16 GMT on 9 February 1967. The sun was low and the glory basically elliptical. Its top, however, broadened towards the horizon as shown in the attached sketch of the first and second red rings. The distortion was presumably caused by the effective drop diameter in the cloud sheet far away being smaller than in the nearer parts." (R14)

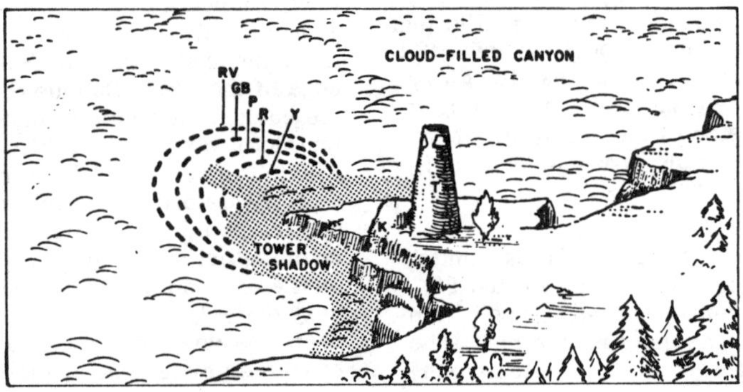

The Watch Tower specter at Grand Canyon (X12)

References

R1. "Reflections by the Clouds in the Catskills, near the Mountain House," American Journal of Science, 2:26:298, 1858. (X1)

R2. Schuster, Arthur; "On a Curious Phenomenon Observed on the Top of Snow-

Distorted glory seen from an aircraft (X13)

don, " Report of the British Association, 1873, p. 40. (X2)
R3. Wetenhall, H. J.; "Mirage on Snowdon," Nature, 12:292, 1875. (X3)
R4. Zurcher, Frederic; "The Circle of Ulloa," Meteors, Aerolites, Storms, and Atmospheric Phenomena, New York, 1876. (X0)
R5. Rogers, J. Innes; "Spectre of the Brocken at Home," Nature, 22:559, 1880. (X0)
R6. Sharpe, Henry; "The So-Called 'Brocken Spectres' and the Bows That Accompany Them," Royal Meteorological Society, Quarterly Journal, 13:245, 1887. (X4, X5, X7)
R7. Abercromby, Ralph; "The Peculiar Sunrise-Shadows of Adam's Peak in Ceylon," Philosophical Magazine, 5:23:29, 1887. (X6)
R8. Abercromby, Ralph; "The Sunrise Shadow of Adam's Peak, Ceylon," Nature, 33:532, 1886. (X6)
R9. Longstaff, G. B.; "Note on the 'Spectre of the Brocken,' as Seen on the Peak of Teneriffe," Royal Meteorological Society, Quarterly Journal, 13:310, 1887. (X8)
R10. Antoniadi, E. M.; "Aerial Spectres," British Astronomical Association, Journal, 7:79, 1896. (X9)
R11. Warner, H. M.; "Curious Shadow Effect," Nature, 69:296, 1904. (X10)
R12. "A Curious Rainbow," Nature, 102: 85, 1918. (X11)
R13. Black, Donald M.; "The Brocken Spectre of the Desert View Watch Tower, Grand Canyon, Arizona," Science, 119: 164, 1954. (X12)
R14. Bridge, G. C., and Corsford, G. C.; "Distorted Glory," Weather, 22:300, 1967. (X13)

GEI2 Heligenschein

Description. The appearance of a luminous halo about the shadow of the observer's head but not about the shadows of any companions. Each observer sees his own halo, but not that of others. Heligenschein (holy light) is best seen on bedewed vegetation, although dry surfaces often produce very faint versions of the phenomenon.

Data Evaluation. A very common phenomenon for those who are about when the sun is low and the grass wet. Rating: 1.

Anomaly Evaluation. Heligenschein is adequately explained by prevailing theories. It is included here for purposes of comparison with other observer-centered effects. Rating: 4.

Possible Explanations. The most common form of heligenschein occurs when droplets of dew reflect sunlight (or light from any source) directly back to the observer. This effect makes the regions immediately around the shadow of his head appear much brighter than adjacent areas. The so-called "dry" heligenschein is fainter and exists because the region around the antisolar point (where the shadow of one's head falls) is much brighter than surrounding areas due to the absence of shadows. Just a short angular distance away from the antisolar point one begins seeing shadows of the grass blades, and this dims the region.

Similar and Related Phenomena. Glories and Brocken Specters (GEI1); spokes around the shadow of one's head (GEI3).

Examples of Heligenschein

X0. General observations. "Haloes or glories around the shadow of the head thrown on dewy grass in early morning or evening have been called 'Heligenschein' to distinguish them from haloes of the ordinary type. The 'Heligenschein' is a real phenomenon which can be photographed, and it extends some distance from the edge of the shadow. It is not to be confused with the apparent brightness around the shadow thrown on a flat surface, which is a purely subjective phenomenon.....The Heligenschein is most clearly seen in meadows where the grass is more or less uniform in length and orientation, and is covered with small drops of dew. The height of the sun should be such that the length of the shadow is 40 ft. or more. The phenomenon is attributed to the light reflected from the spherical dewdrops both directly and after two refractions and one or more internal reflections. These effects give a maximum intensity in the direction of the incident light, the intensity falling off without alteration of colour. It is, therefore, quite distinct from haloes, coronae, and glories." (R1)

In mountainous regions, the heligenschein may appear as a circular or elliptical patch of light on grassy slopes a mile or more distant, but the shadow of the observer's head cannot be distinguished under these circumstances. (R2)

Heligenschein is seen only around the shadow of the observer's head. The shadows of people nearby do appear but without the surrounding glow. Since many dry surfaces reflect considerable light back to an observer gazing at the antisolar point, faint heligenschein can appear without the benefit of dew. (R3, R4)

References

R1. "Coronae, Glories, and Heligenschein," Nature, 91:114, 1913. (X0)
R2. Evershed, J.; "Luminous Halos Surrounding Shadows of Heads," Nature, 90:592, 1913. (X0)
R3. Minnaert, M.; "Heligenschein," The Nature of Light and Color in the Open Air, New York, 1954, p. 230. (X0)
R4. Greenler, Robert; "The Heligenschein," Rainbows, Halos, and Glories, New York, 1980. (X0)

GEI3 Rotating Spokes about the Shadow of One's Head

Description. Bright rotating spokes seen radiating from the shadow of one's head when cast upon a water surface. The effect is most pronounced when the water contains fine particles. As with heligenschein, each observer sees the spokes around his head only.

Data Evaluation. Only a few mentions of this phenomenon exist in the literature, but it must (or could) be observed just about anyone. Rating: 1.

Anomaly Evaluation. A simple and convincing explanation exists. This phenomenon is cataloged because of its curious features and close relationship to the Brocken Specter and heligenschein. Rating: 4.

Possible Explanations. See the detailed discussion in X0.

Similar and Related Phenomena. Brocken Specters (GEI1); heligenschein (GEI3).

Examples of Spokes Surrounding the Shadow of One's Head

X1. General observations. "Stand in the sunlight, when the sun is high, so that the shadow of your head falls upon the surface of a slightly turbid pond or lake, whose surface is covered with gentle waves, or ripples: you will see the shadow of your head surrounded by a halo of quivering radial streaks of light. From the deck of a steamer on the Thames I have often observed this. The waves or ripples of the surface cause the sun's rays to enter the water with unequal intensities at different points. At certain points the beams will enter almost without loss by reflexion, and will be traceable by their illuminating the particles of the turbid water. These beams will be nearly parallel to one another; but as they all retreat from the observer, they will appear to converge to a point exactly opposite the sun; to a point within the shadow of his head, in fact, giving the nimbus effect." (R1)

"One of the greatest joys of physics lies in finding explanations for common everyday phenomena which confront all of us, yet which are not quite obvious.

The following example of such a phenomenon excited my curiosity even when I was very young; standing on the end of a diving board over deep water, the sun behind me, I often wondered as I studied my shadow on the water's surface about the rays which seemed to emanate from my head. They radiated from a spot between my eyes, and the rays seemed to glisten and to rotate like the spokes of a wheel, generally rotating continuously in one direction. When a friend would join me, we would both see the rays coming from our own heads. This phenomenon is observed sufficiently seldom to elude investigation, for it is not a very striking effect and requires: (1) Deep water, over three feet, depending on the nature of the bottom; (2) Sunshine; (3) Suspended particles, such as algae or minerals; (4) Relatively quiet water.

The explanation is straightforward, though very satisfying: Sunlight produces the same effect in water containing suspended particles as it does in air, where it produces a sunbeam. In the phenomenon described above, the sun, observer's eyes, and shadow comprise the optic axis. Rays of sunlight strike the water parallel to this axis and, if the water be unperturbed, continue through the water parallel to each other. If, however, the surface be disturbed by superposed wave trains, traveling this way and that, the regions where the waves are refracted and continue as through still water (i.e., level patches of water) become spaced according to the surface wave form.

Thus shafts of sunlight within the water are the result of a ruffled surface. An observer looking down into the water sees these shafts but, because of his viewpoint on the optic axis, they appear to radiate from a common center in much the same manner as the lesser buildings appear to radiate from an observer looking down from atop the Empire State Building.

As for the apparent rotation of the 'spokes,' this is due to the effect of the moving wave trains. Although the directions of the wave trains may differ, there is generally uniformity in the sense of their travel. As the wave trains progress, they produce the observed shifting in the shafts, causing an appearance of rotation." (R2)

References

R1. Thompson, Silvanus P.; "On Certain Phenomena Accompanying Rainbows," Philosophical Magazine, 5:6:272, 1878. (X1)

R2. Jacobs, Stephen F.; "Self-Centered Shadow," American Journal of Physics, 21:234, 1953. (X1)

GEL LOW-SUN PHENOMENA

Key to Phenomena

GEL0 Introduction

As the sun sinks toward the horizon, the amount of atmosphere between it and an observer increases rapidly. The longer path that sunlight must now travel through the capricious atmosphere introduces a variety of novel optical phenomena. Images of terrestrial and astronomical objects may become grotesquely distorted, multipled, and painted with unlikely hues. The bending of light waves (optical refraction) by the different layers of air creates this type of magic. Still closer to the horizon, clouds and mountain peaks intercept and further alter the sun's rays, giving rise to the well-known, fan-shaped crepuscular rays, as well as more exotic phenomena. The crests of mountain ranges may act as a sharp edge that casts shadow bands on the earth when the sun is low through optical diffraction. Another optical phenomenon at work is that of dispersion, which spreads out the spectral colors of sunlight. The famous "green ray" or "green flash" is due to atmospheric dispersion. Even when the sun is well below the horizon, its rays may be reflected earthward from high altitude dust layers. Such reflections produce the so-called "purple light" after sunset and the Alpine Glow. And as night draws near, the shadow of the earth itself may often be seen cast upon the upper atmosphere as it advances relentlessly from the east. Most of these low-sun phenomena also occur at sunrise, but in reverse order, of course.

Many low-sun phenomena are well-understood; such as the solar pillars, the typical green flash, and distortions of sinking sun's disk. Many other phenomena, though, are curious and enigmatic; and some of the most common optical effects have puzzling aspects. The green flash and the crepuscular rays, for example, have long been considered as fully explained. The stock explanations, however, seem simplistic when the full ranges of these phenomena are considered. Some green rays seem purely subjective; others are real and can be photographed. And when crepuscular rays stretch all the way across the vault of the sky to the east in regular geometric arrays, one questions that a few random clouds make a sufficient explanation, especially when no clouds are known to exist over vast adjacent regions. Even the famous Krakatoa sunsets of 1883-1885 have a mysterious dimension. How could they have been seen months before the actual eruption in Africa---and then recorded episodically for decades afterwards? Were these spectacular sunsets and sunrises truly caused by the volcano's dust? And then we have fluorescent landscapes and unexpected spectral bows at sunset that are not centered on the sun. Low-sun phenomena are not only very beautiful but a few of them severely try the mettle of meteorological optics.

GEL1 Puzzling Features of the Green Flash

Description. Abnormal flashes of light exhibited by astronomical bodies as they sink below or rise above the horizon that cannot be accounted for by conventional refraction theory. The "normal" apparition consists of a bright bead or flash of green, more rarely blue or purple, lasting a second or so, as the upper limb of the sun (moon, Venus, or Jupiter) is covered by the horizon. The reverse order of effects occurs at sunrise. Also considered "normal" are red flashes as the sun's lower limb appears below a sharply defined cloud bank near the western horizon. Abnormalities include: (1) Multiple flashes; (2) Long rays of colored light; (3) Structures resembling Bailey's Beads; (4) Complex fine structure of the flash; (5) Green edges around terrestrial objects viewed against the sun prior to the green flash; (6) Green flashes before the sun has set; (7) The apparent psychological origin of some green flashes; (8) The failure of the green flash to appear under ideal conditions; and (9) The apparent inadequacy of normal refraction at the horizon to account quantitatively for observations.

Background. Most scientists and the books they write portray the green flash as an easily explained atmospheric phenomenon. As the sun sinks below the western horizon, the tip hanging on the horizon turns green for an instant because the refractive power of the atmosphere spreads (disperses) the spectral colors vertically so that the shorter wavelengths are the last to be seen. Green, blue, and finally purple (violet) should appear in sequence. In practice, the complete succession of colors is rare because the atmosphere preferentially scatters the blue and purple. Theory thus predicts a green flash.

The problem posed by the vagaries of the green flash is the adequacy of this theory.

Data Evaluation. Reports of normal green flashes, as well as blue and purple flashes, are abundant. Good photographs also exist. The vagaries, on the other hand are hard-to-find in the literature. Nevertheless, each abnormality listed above, except (5) and (6) is represented by more than one observation or discussion. For the vagaries alone, rating: 2.

Anomaly Evaluation. Even if all the puzzling features described below are legitimate, they can almost certainly be explained in terms of atmospheric refraction---with some modifications, as suggested below. The changes required are minor. Rating: 3.

Possible Explanations. Most of the abnormal features of the green flash can probably be explained as due to abnormal refraction; that is, local changes in the optical properties of the atmosphere in the vicinity of the setting sun. Gravity waves have been suggested as possible modulators of the index of refraction. The observations in X8 and X9 may, in fact, insist that all green flashes arise from anomalous refraction. Further, from X7, it seems that some green flashes have a psychological origin.

Similar and Related Phenomena. The Krakatoa sunsets (GEL5); Bailey's Beads (AL); unusual refraction at the horizon (GEL4); puzzling mirages (GEM).

Examples of Puzzling Features of the Green Flash

X0. Green flash observations that may be considered "normal" but nevertheless display some out-of-the-ordinary aspects.

No date given. The Southern Ocean (?). Multiple and long-duration flashes created by observer motion. "This 'green ray' is seen to best advantage at sunrise, owing I imagine to the eye not being wearied with watching the previous glare, as is apt to be the case at sunset. At the same time, I had many very satisfactory observations at sunset, one in particular, when we were running before a very heavy sea in the Southern ocean, and the 'green ray' was seen no less than three times in as many seconds, as the ship rose and fell on the huge waves, causing as it were two sunsets, with a sunrise in between them. The best displays took place when the refraction near the horizon was of such a character that the sun assumed a balloon, or vase, shape as he came close to the sea lines." (R1, R3)

December 4, 1962. South Pacific Ocean. "Near sunset at 0130 GMT, three brilliant green flashes seen as the sun's upper limb sank below three separate layers of Sc.(R33) In this example, each cloud layer constituted an effective horizon. Often in such situations a red flash will appear as the sun's lower limb emerges from behind a cloud layer. (WRC)

No date given. Egypt. "I have often in Egypt watched the sunrise light descending the western hills, and when the edge of the shadow reached me, turned to view the sun. The first appearance of the sun is a blue segment, changing to green, and then to white. The is exactly the converse of the colours of the sunset segment; as the rising light cannot be due to an after-image, no more is the setting light..... Further, I often have protracted the sunset blue by walking up a slope, and so keeping it in view, on and off, as long as I go upward. The least distance of horizon that is effective is about two miles, at which the change of colour is just visible. (R15) Two interesting points are made in this note: (1) Given a very clear atmosphere, one can observe the flash go through a portion of the spectrum; and (2) By changing one's altitude, the duration of the "flash" can be altered. (WRC)

Circa 1941. Portpatrick, United Kingdom. "I remember that, some twenty years ago, at Portpatrick, I happened to be on the east side of a breakwater as the Sun was setting. The whole disk of the Sun was still above the horizon but, by stooping, the top of the breakwater could be brought upwards across the Sun. At the instant when the disk was just covered the Green Flash was seen. It could be repeated as often as one wished by performing the necessary gymnastics." (R30)

August 21, 1952. North Atlantic Ocean. "As the sun's lower limb, altitude 35', appeared to emerge at 2208 G. M. T. from the base of a large Cu (estimated at 2,500 ft), a brilliant red flash was observed. As the sun's disc sank below the horizon (2215) a brilliant bluish-green flash was seen from its upper limb.... Note. This observation is of very great interest, observations of the red flash being extremely rare, though it might be seen more often if it were definitely looked for in suitable conditions. In conditions of normal refraction the sun's upper limb is bordered by a narrow rim of bluish-green or green colour and the lower limb by a similar rim of red colour, when the sun is near the horizon. These rims are wider and therefore more likely to be observed in conditions of abnormal refraction. The upper green rim

Three successive green flashes observed in the South Pacific as the sun sank behind three cloud layers (X0)

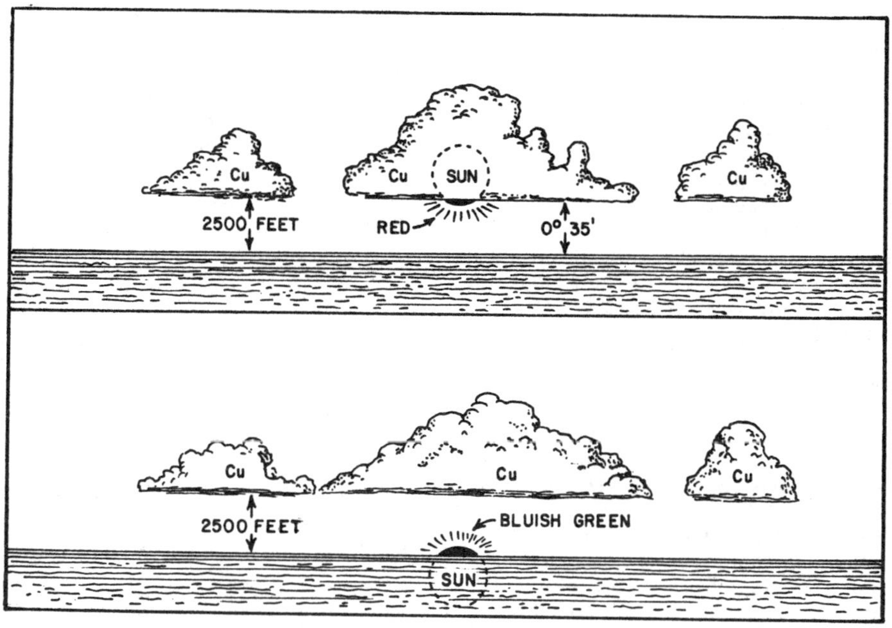

Red flash followed by green flash in the North Atlantic (X0)

gives the well-known green flash when the sun sinks below the horizon; almost all the sun is hidden before the green colouration can be seen. The green flash has also been seen where the sun goes down behind a cloud with a sufficiently well-defined upper edge. The red flash can never be seen at the horizon since the red rim would be the last part of the sun to rise and it would be lost in the sun's general light." (R25)

May 19, 1960. South Atlantic Ocean. "The moon on rising at 0339 GMT was seen, when examined through binoculars, to have assumed the distorted appearance shown in the sketch. As the lowest part of the limb cleared the horizon a brilliant red flash was observed for 1 1/2 sec. approx." (R29) An amusing exception to the above-mentionod rulo that tho red flash is never seen at sunrise (or moonrise by extension).

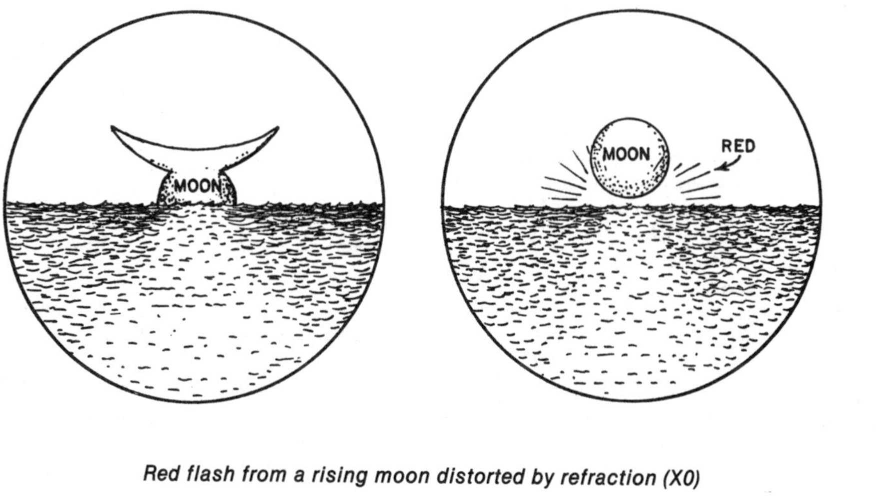

Red flash from a rising moon distorted by refraction (X0)

March 28, 1960. Washington, District of Columbia. "Just prior to sunset a high-flying airplane produced a contrail several miles in length, with almost perfectly oriented east-west direction, over the western Washington, D.C. area. The initial beauty of this orange-red path in the almost cloudless sky first attracted my attention. After the sun had set, with respect to my ground position, the color remained in the contrail. As the sun set on the contrail the orange-red coloration was erased along the trail and was replaced by a blue-gray, but interposed between the red and the gray was a thin strip of green which raced westward on the trail. The time of passage was on the order of 1-2 sec and must represent the 'green flash' as might be seen by a string of observers positioned along the contrail's path." (R31) A delightful instance of a moving green flash as seen by a stationary observer that contrasts with the foregoing examples of a moving observer and stationary green flash. To these must be added the observations of polar explorers of green "flashes" of up to 30 minutes duration occasioned by the upper rim of the sun skimming the horizon during the polar night. (R21)

April 12, 1957. South Pacific Ocean. "At 0053 G.M.T., the green flash was observed under almost perfect conditions. There was not evidence of any abnormal refraction, the sun sinking uniformly below the horizon without distortion. There was, however, a more striking series of colour changes than is normally seen. Just before the top edge of the sun's disc finally disappeared, the green colouration quickly changed to blue, from pale blue to intense blue and finally to a purple, almost violet as the last rays of light vanished." (R28) This and the following examples illustrate that the "green" flash can run through much of the visible spectrum. (WRC)

July 28, 1971. North Atlantic Ocean. "During the last few seconds of sunset, the remaining solar segment changed color through the whole spectrum: first more yellowish, then greenish, and finally blue and violet just before the last of the solar disk disappeared. Due to preferential atmospheric absorption, the blue and violet effects were faint, perhaps only 1/10 as bright as the green, but they were definitely visible. This complete range through the visible spectrum at setting of the sun's upper limb clearly indicates the phenomenon's general cause: the dispersion in astronomical refraction, which becomes greatest at the horizon. With a telescope at night, one can easily see that stars within 20 degrees or so of the horizon appear as vertical spectra with the violet at the top. Since the spectrum is approximately proportional to the mean astronomical refraction, I estimate that the dispersion at the horizon would be about 50 seconds of arc from the red to the violet." (R36)

Although some observations of the green flash may be psychological in origin, the physical reality of the phenomenon has been demonstrated by rare photographs (R34) and spectrographic analysis of the light from the setting sun. (R24)

X1. Multiple green flashes arising from mirages and other vagaries of the atmosphere.

June 24, 1975. Scotland. "On 24 June 1975, I witnessed a double occurrence of green flash as a superior mirage above the top of the setting sun. After a bright and warm afternoon, I was watching the setting sun through binoculars from a rocky vantage point at 875 ft (grid reference NG 82335925) on the west side of the pass between Alligin and Diabaig to the north of Upper Loch Torridon. From about 1025 pm (BST) through a medium haze, as its lower limb was sinking behind a low bank of cloud over the Butt of Lewis, the sun had taken the form of a bright orange irregular ellipse with a shimmering edge, when a detached saucer-shaped segment formed separately from, but close to and centrally above, the upper limb. This was clearly an inverted image of the top rim and having the same apparent brightness as the main image. The segment gradually diminished in depth and diameter towards its lower edge (over a period of about 10 secs) until the residual tip turned bright 'traffic light' green for perhaps 3 secs before dissolving completely. To my surprise, less than a minute later, the whole sequence was repeated almost exactly as before, but perhaps over a slightly shorter period, and by about 1030 pm the sun's disc had settled behind a low cloud." (R38)

March 8, 1981. South Atlantic Ocean. "At sunset the atmosphere was clear, with no clouds on the horizon, and it appeared to be a good day on which to observe the green flash, as indeed was the case. The sun, as it began to dip below the horizon, was large and orange and very distinct. During the 28

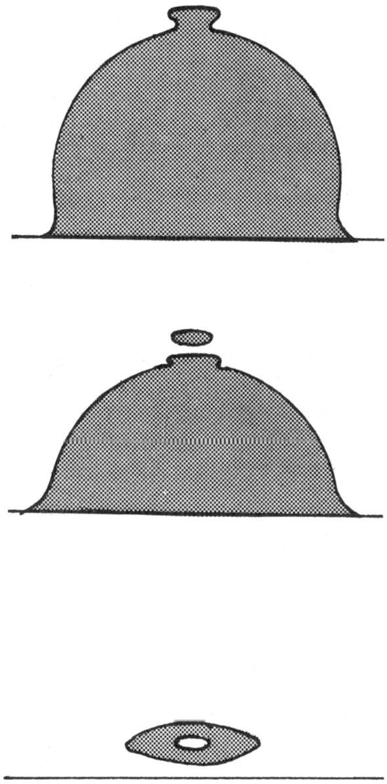

Complex green flash sequence. Top: first cigar shape breaks away. Middle: cigar shape turns green and disappears as subsequent cigar shape breaks away. Bottom: final green flash has a lighter circular area in center and lasts about 2 seconds. (X1)

seconds or so that it took the sun to disappear totally below the horizon, five separate green flashes were observed. These occurred when the upper limb of the sun broke away from the main body of the sun and formed a cigar shape approximately one-sixth of the diameter of the sun in length. This cigar shape almost immediately disappeared, turning a faint and rather watery green as it did so. This occurred on five separate occasions as the sun set, each new cigar shape breaking away as the previous one disappeared. The duration of these flashes was a split second each. The final and sixth flash, observed as the sun sank below the horizon, was very pronounced. It appeared very slightly above the horizon and lasted approximately two seconds. A feature which the observers had not witnessed before was that in the centre of the green colouring (which was a bright emerald green) was a small circular area of a much paler green colour; this only occurred, however, in about the last half-second of the duration of the flash." (R40)

General observations. Frequent observations of multiple green flashes may be due to gravity waves, with wavelengths about 0.2 to 2.0 kilometers, which perturb the atmospheric index of refraction. (R37)

X2. The appearance at sunset and sunrise of multiple green rays of light ascending from the point of the sun's disappearance or appearance. This phenomenon may be related to the crepuscular rays (GEL2), which themselves display mildly anomalous features.

General observations. "The appearance of a green light at sunset, like many other phenomena supposed to have only recently attracted attention, was noticed and commented upon by the ancient Egyptians, and more particularly so because in the clear air of Egypt the tints of sunset are peculiarly distinct. As the sun there descends nearer and nearer to the horizon, apparently hastening to disappear behind one of the Libyan hills, as if burying itself in the sand at their base, the immensely enlarged flaming disc suddenly becomes, for an instant, of a brilliant green colour, and immediately a series of green rays suffuses the sky in many directions, well-nigh to the zenith. The same phenomenon appears sometimes at sunrise, but to a smaller extent." (R6)

July 1947. Outer Hebrides, United Kingdom. "One evening in July, I was standing alone on the shore watching a glorious sunset---a red ball slowly falling into a glassy sea. When about three-fourths submerged, the red quite suddenly changed to bright green and a beautiful green aura spread outwards like a fan. Almost before I realized what I was seeing, the fan collapsed, the green arc of the sun disappeared and only a red glow remained. Total time, possibly 7 or 8 seconds." (R26)

October 10, 1915. Co. Down, Northern Ireland. "I was watching the sun go down behind a dense, almost black, bank of clouds over the Co. Louth mountains, when suddenly a remarkable series of flashes darted upwards and towards the north, with great rapidity. So, thought I, I've seen the green flashes at last; but alas! they were all blue, showing darkly on a very pale-blue sky above the black bank of clouds. A friend passing, I called his attention to them, and asked what colour he thought they were, and at once said, 'Blue, certainly.'" (R11) The sketch submitted with the report shows the flashes to be ray-like. They persisted some 10 minutes, a feature which suggests they may have been only crepuscular rays. (WRC)

General observations. "Knowledge of various phenomena related to the green flash, but not observed on the sun itself, is also being built up. Examples are green rays shooting up into the sky at the instant of sunset, a transient green mist above the sun and more persistent rays or cones of green or bluish light." (R26)

X3. Bead-like appearances of the green flash. There are certain resemblances here to the Bailey's Beads associated with total solar eclipses.

April 19, 1901. England. Watching for the green flash with field glasses. "The object of this communication is to give some details of an observation made here on April 19 of this year. The sun set in a manner exceptionally favourable for observing the green flash: sky quite clear, and without any trace of haze or mist even near the horizon. The green flash first appeared in the shape of a single brilliant emerald bead, then another, and a third, and in about a second a full display of Baily's (sic) beads had developed in much the same way as in a solar eclipse, but to a much smaller extent and for a much shorter time. The first bead I noticed was towards the S. point of the horizon, the others close toward the north., until my eye reached the fourth, when I found a string of about seven beads.

My idea is that they would form at either end of the segment and join on the middle of the row; but my surprise at the sight of Baily's beads without an eclipse was so great that I did not notice this point." (R2)

August 4, 1957. Cambridge, Massachusetts. "While searching for Comet Mrkos on August 4th, we observed a phenomenon apparently related to the green flash. The western horizon was cloudless but hazy, and the setting sun deep red. We turned a mounted 3-inch 21x terrestrial refractor on the sun as it neared the horizon. Owing to layering of the air, the sun's limb appeared slightly notched; as the sun set, pairs of corresponding notches moved upward across the disk, pinching off small 'bubbles' from the top. Each bubble gradually turned orange, yellow, and finally pale green before it faded out, the whole process taking up to three seconds. By comparison with the telescopic disks of Venus and Jupiter, we estimated that the bubbles were about half a minute of arc in diameter. The upper limb of the sun also showed a yellow-green fringe, considerably narrower than the bubbles. Neither bubbles nor fringe could be seen in hand-held 7-power binoculars." (R27)

October 26, 1927. Cairo, Egypt. "Watching the sunrise this morning across the Nile valley, under conditions of very good visibility and no local wind, using Goerz x8 binoculars, I saw what is, to the best of my knowledge, a novel form of the 'green flash.' From the top floor of Mena House Hotel the sunrise was behind Gebel Tura, with its skyline at about 300 metres elevation, and some 25 kilometres away. The dawn was clear, a light white mist lying low on the valley floor; the sky was cinnamon colour, changing to amber as the sun rose; the desert hills appeared deep blue to the naked eye, and brownish purple through the field glasses. The form of the actual flash was totally unexpected: it preceded the emergence of the limb of the sun itself by perhaps as much as two seconds, and I can only describe it as resembling a shallow and turbulent river of apple-green water hanging ready to break over the hill crest. But his river flowed with a surface which seemed to follow the minor prominences of the skyline, instead of first appearing between them; it appeared suddenly, not seeming to widen much after its first apparition, the greatest width attained being perhaps one-third of the diameter of the sun's disc. At the emergence of the orange-gold disc in the center of the pale-green band, the two colours seemed momentarily to co-exist; this was probably persistence of vision, but indicates that there was no eye fatigue." (R18) Perhaps this example is not precisely bead-like, but it does indicate a related horizon effect in any case. (WRC)

1909. Bombay, India. "Seven years ago from the roof of a high building in Bombay, looking seaward, at sunset, with a pair of field-glasses, at the setting sun, I saw distinct green flames of considerable height, nearly one-quarter the apparent diameter of the sun. I watched it for an appreciable time, at any rate from the first contact with the horizon until the final disappearance of the sun, after which there was still a floating wisp of green. The appearance was that of serrations, as in a very rough saw, the edges of which from time to time came together and sort of floated in parachute form, again and again." (R12)

X4. Unusual fine structure of the green flash as seen through binoculars and telescopes. These apparitions are closely related to the preceding multiple flashes and bead-like structures.

August 31, 1903. Scourie, United Kingdom. "At the last contact of the sun's upper limb, the 'blue line' appeared. When the sun's lower limb was about 15' above the horizon, it appeared intense red. The blue line began to show evidence of its coming about 30 seconds before the sun disappeared. At first the whole of the visible portion of the disc of the sun turned intense fiery red. Then when a little more than half the disc had disappeared it changed to yellow. Then when only about 1/6th was left, there were green flashes round the edge of the disc lasting perhaps a second each, and passing round a short arc beginning always at the horizon. Then similar blue flashes appeared. Then when the sun was almost down the whole visible part became greenish with violet flashes. Then the part of the disc visible became a line with deep violet at each end, blue next to the violet, and green in the middle parts of the line. Then the violet took the place of the blue, and blue was in the middle, while deeper, almost ultra-violet, colours the ends of the line. Then there was a lingering, rippling flash of deep violet, and the sun was gone. The flash (as a whole) was much clearer and lasted longer than I have ever seen it before." (R5)

No date or place given. Similar colored flickering observed through a telescope. (R7)

April 22, 1923. Seaton, England. "Special atmospheric conditions prevailing here on April 22 produced the apparition of a succession of green rays at sunset which it was very interesting to watch. The conditions were a very clear sky generally, but a very dense mass of cumulostratus cloud hung in the N.W. sky, terminating in a belt of apparently clear sky, 2° wide all along the horizon. The horizon was distant land, 0° 30' elevation. I found the Sun entering this clear belt, and watched with optical assistance of x13, no eye shade being necessary. Under telescopic scrutiny this zone was seen to be not clear sky, but a stratified form of cirrus in horizontal lines and bands, which increased in density rapidly near the horizon, so that the Sun's lower limb disappeared before reaching the land contour. These bands were so faint that they were scarcely to be seen on the Sun, but they produced remarkable results with the green ray effect. They formed faint chords across the disc, at the extremities of which, in the upper segment, shining prominences were projected from the Sun as elongations of the chords. These prominences rose upwards on the circumference as the disc sank. Approaching the top, they turned green, and meeting there they united and floated clear as a tiny cloudlet of brilliant green. I watched a succession of about half a dozen of these rays, until the disc became too faint to see them. The atmospheric bands thus formed separate horizons, each producing the effect that the actual horizon does in a clear sky." (R17)

August 27, 1981. Anglezarke Moor, England. An effect almost identical to the preceeding. (R41)

May 20, 1937. No place given. "I was looking at the setting Sun through a pair of binoculars just before its lower limb touched the sea horizon. There were some strips of horizontal cloud about, but the atmosphere must have been fairly clear as the Sun was setting yellow, not at all red. The obscuration was enough to allow me to look at the Sun quite comfortably through the glasses without any dark glass. I was admiring the beautiful streak of spots then on the Sun, and was trying to see what I could of it with the glasses, and observing it very carefully as it set below the horizon as a noticeable streak sloping down nearly vertically. Now as the lower end of the streak got to the horizon it turned pink, and the pink wandered up the streak (of course as effect of the almost vertical setting) until the last component spot had disappeared. Actually it looked rather like a cascade, black at the top, suddenly becoming pink at the bottom, and then as though the top of the cascade were cascading down too. I was much taken aback by this, and proceeded to see what would happen to the other spots visible on the disc. Again, as these neared the horizon they suddenly became pink, and with a slight acceleration, like water just lipping over a waterfall, they disappeared. The time between the first beginning to show colour and their disappearance, during which time they became progressively pinker, was of the order of a second; I saw this for at least three spots, I think five actually. When the spots had gone I waited for the customary 'green flash.' This was the finest I have ever seen, remarkably long (some four seconds at least) and very intense, a deep aquamarine blue. I always see the 'green flash' as blue. There was quite an appreciable segment of the yellow disk still showing when the blue made its appearance on either side. As far as the colour of the sunspots was concerned, what is termed pink above should perhaps more accurately be described as cherry red, used in the sense in which it is used when tempering tools; a red on the border of becoming white-hot." (R22)

March 8, 1981. South Atlantic Ocean. The last in a series of green flashes, caused when lens-shaped portions of the setting sun broke off due to refraction, was bright emerald green on the outside and pale green inside. See full description in X1. (R40)

X5. The apparent greening of local objects prior to the green flash proper.

September 20, 1922. Wadhurst, England. "The sun set over low hills in the Ashdown Forest region and at a distance estimated at eight miles from the point of observation and several isolated rounded trees were projected on the disc for some time before the sun's disappearance. The largest of these trees subtended an angle estimated at about two minutes of arc. The observation was made with a telescope of three inches aperture magnifying eighteen diameters. The phenomenon occurred in two stages: (a) Every tree projected on the disc became suddenly outlined with a narrow hazy border of green light, the rest of the sun retaining its normal colour; (b) after a perceptible interval of time the whole of the remaining segment of the sun turned a brilliant clear emerald green and almost simultaneously disappeared below the horizon." (R14) Which optical phenomenon fringed the trees with green? Diffraction? See also landscape fluorescence (GEL8). (WRC)

X6. The appearance of appreciable green well before actual setting of the sun.

General observations. "Green coloration is not infrequently seen in connection with the upper limb, when a considerable part of the disc, half or more, still appears above the horizon. One example, which I have seen with binoculars, is the appearance of vapour coming off the limb, turning green as it leaves the sun. Other forms consist of parts of the sun detached from the main body by refractive distortion and assuming a green colour." (R26)

X7. The possible psychological basis of the green flash.

No date given. North Atlantic Ocean. "I have observed (the green flash) not less than fifteen or twenty times. I have usually seen it from ships at sea, mostly in tropical

American waters. However, I also saw it very clearly from the window of a Philadelphia apartment building, and once saw it over a distant mountain range in the Arizona desert. While the phenomenon has varied from the strikingly clear to the dimmest, all the observations I have made have certain factors in common. In each case the green light has displaced the last lingering fragment of the sun's limb at the instant of disappearance of the latter. It has never lasted more than 1.5 (estimated) or less than 0.5 seconds. There has never been any gap between the disappearance of the sun and the appearance of the green light. The shape of the green light (which has only rarely been definable) gives the impression of a narrow ellipse whose longest axis is horizontal. From these many observations I have inclined strongly to the belief that the phenomenon is an entirely subjective one. When attempting to see it, one stares fixedly at the vanishing limb of the sun, which, under clear conditions, emits an intense red light. Upon its disappearance, the optic nerve, temporarily dulled to red by this fixity of observation, immediately responds with the complementary color green. An experiment which I tried tends to corroborate this. My brother, Brandon Barringer, and I stood at the rail of a ship watching for the flash in company of at least twenty other people, all of whom had been warned to look for the phenomenon. The location was north of Puerto Rico and the evening was exceptionally clear. I observed with my left eye and at the same time kept my right eye glued to the telescopic finder of an 8 mm. motion-picture camera. I was thus able not only to make a good observation myself but to record it on color motion picture film. The resulting flash was so brilliant that a universal gasp of astonishment went up from the passengers at the rail. It was perhaps the finest example I have seen and looked like nothing so much as a tiny green traffic light, which glowed for perhaps a second. The film, when developed, was very sharp and clear and gave a beautiful picture of the sunset, but the flash did not appear. The flash was so bright that I feel confident the film would have picked it up if it had been anything but a phenomenon confined to the optic nerve of the observer. It has been previously suggested by Russell and others that the flash was caused by differential refraction of the spectral colors of the sun's light in the atmosphere, and that as the red, or straightest rays disappeared one became momentarily conscious of the other rays, which were more bent and hence could be seen a little later. I do not think that this hypothesis would account for the complete absence of any other color but red in the movie film. Another reason that I suspect the 'spectrum' hypothesis is that no color but green has ever been reported. If the spectrum flashed past the observer's eye immediately after the disappearance of the sun, one would expect some sensation of the other colors, such as yellow and blue. No mention of these has ever been made in the literature, to my knowledge." (R23) Of course the other colors <u>have</u> been reported and the green flash <u>has</u> been preserved on film (X0). (WRC)

Summer 1914. South Pacific Ocean. "So much has been written about the green-ray at sunset that I am somewhat diffident about adding anything. But as I find myself unable to accept the orthodox explanation of the phenomenon usually seen I write this note. This phenomenon, as seen by me on several occasions during the last summer on my way to Australia, always consisted in the last segment of the red sun before disappearance becoming a bright green (without any transition through intermediate tints); this green was as nearly as could be judged the complementary to the red of the sun itself. On one occasion I shut my eyes immediately after the green tint appeared, and it <u>remained visible</u>. There can be no doubt that what I saw was the purely subjective after-image of the disappearing segment of the sun." (R9)

No date given. Jaffa, Israel. "One evening when expecting to see this flash, a party of four of us, including two doctors, had discussed the possibility of this phenomenon being due to some effect in one's eyes and not to refraction: to try this out two of us watched the sun carefully until it had set, while the other two looked eastwards and on signal from the sun watchers turned quickly just as the sun was disappearing. The two who had watched continuously saw the green flash, while to those who did not look till the very last moment, the sun went down orange-yellow, and no green was seen." (R20)

Circa 1908. England. A bluish-green color overspread the entire upper portion of the sun. Could not have been due to refraction and must have been a contrast phenomenon. (R8)

X8. The capriciousness of the green flash.

General observations. "The appearance of the 'green flash' is capricious, as is agreed by all reasonably constant observers. There is no series of observations available for land horizons, but in Table 4 (not reproduced) I have collected the data of four observers who have made reasonably long series with sea horizons, and in varied localities. From these it seems that in about half the cases of marine sunrise or sunset observed, no 'green flash' has been visible, though no apparent reason existed for its non-appearance. In Egypt, where, according to W. Groff, it is very common, he also says that 'even sometimes, when, in appearance, there are neither clouds nor haze, the phenomenon does not take place.'" (R13) Several other students of the green flash term this non-appearance under what seem ideal conditions to be the greatest mystery connected with the phenomenon. (R4, R10, R19, R37)

March 11, 1962. Gulf of Aden, Indian Ocean. "At 1947 GMT, the moon, which was crescent-shaped was seen to be a dark red colour as it sank towards the horizon. When the left-hand cusp was disappearing below the horizon a pale green flash was seen through 7x50 binoculars. There was no flash when the right-hand cusp disappeared." (R32) The phenomenon displays variability over a short section of horizon and a time period of minutes. (WRC)

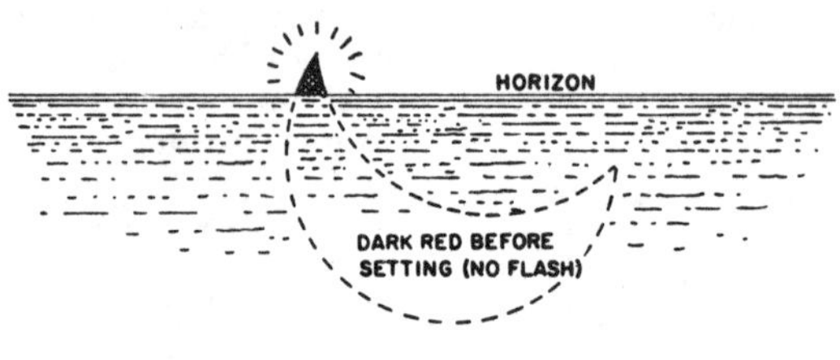

A crescent moon setting in the Gulf of Aden provided a green flash from the left cusp; nothing from the right cusp. (X8)

September 1923. South Pacific Ocean. "On the outward voyage to Australia to observe the solar eclipse of September last, I was struck with the faintness of the green flash at sunset, although the sky was clear down to the horizon. It was visible in binoculars (x8), but scarcely, if at all, to unaided vision. On the return journey, on the other hand, the phenomenon was brilliant every evening on the run between the northwest coast of Australia and Java, and I was able to observe also what happened when Venus set in the sea." Author concludes that the existence of mirage conditions enhances the green flash. (R16)

X9. Evidence that normal refraction of sunlight near the horizon is insufficient to cause the green flash.

General observations. "'With normal refraction, the width of the colored rim, usually green, is 10 seconds of arc, only half of that perceptible by the normal human eye. This fact seems to dispose of the view of Rayleigh that the flash is due to the usual refraction and dispersion of the atmosphere." (R35, R42) In other words, normal refraction at the horizon, the accepted explanation, seems inadequate. (WRC)

Mathematical analysis. "It has been shown by O'Connell that the green flash occurs at the vanishing of apparently-detached portions of the solar disc. It does not seem to have been previously noted that such detached portions cannot be produced above the astronomical horizon if the terrestrial atmosphere is modelled by a spherically symmetric or plane-parallel system. However, a system in which the refractive index varies in the horizontal is capable of producing these effects, especially if this variation causes the surfaces of constant refractive index to be of wave-like form." (R39) A gravity-wave origin for such special refractive structures has been proposed. (R37)

References

R1. "The Green Ray," Scientific American,

65:168, 1891. (X0)
R2. Franklin-Adams, J.; "The Green Flash at Sunset," English Mechanic, 73:422, 1901. (X3)
R3. Treadle; "The 'Green Flash' at Sunset," English Mechanic, 73:450, 1901. (X0)
R4. Childe, E. A.; "The Green Flash," English Mechanic, 80:299, 1904. (X8)
R5. Worthington, James H.; "The Green Flash at Sunset," Knowledge, 3:376, 1906. (X4)
R6. Offord, Joseph; "The Green Tints of Sunset," Nature, 75:342, 1907. (X2)
R7. Whitmell, C. T.; "The Green Flash and the Red," English Mechanic, 86:264, 1907. (X4)
R8. Lippincott, R. Claud; "The Green Flash," English Mechanic, 90:188, 1909. (X7)
R9. Porter, Alfred W.; "The Green Flash," Nature, 94:672, 1915. (X7)
R10. Whitmell, C. T.; "The Green Flash," Nature, 95:35, 1915. (X8)
R11. Mansergh, A. W.; "The Green Flash," English Mechanic, 102:346, 1915. (X2)
R12. Gieve, Percy P.; "The Green Flash," English Mechanic, 104:48, 1916. (X3)
R13. Fisher, Willard J.; "Low Sun Phenomena--IV. The 'Green Flash,'" Popular Astronomy, 29:382, 1921. (X8)
R14. Barlow, E. W.; "The Green Flash---An Unusual Feature," Meteorological Magazine, 57:246, 1922. (X5)
R15. Petrie, W. M. Flinders; "The Green Ray at Sunset and Sunrise," Nature, 110:604, 1922. (X0)
R16. Evershed, J.; "The Green Flash at Sunset," Nature, 111:13, 1923. (X8)
R17. Housman, W. B,; "Multiple Green Rays," English Mechanic, 117:192, 1923. (X4)
R18. Balls, W. Lawrence; "The 'Green Flash' at Sunrise," Nature, 120:728, 1927. (X3)
R19. Wood, R. W.; "Factors Which Determine the Occurrence of the Green Ray," Nature, 121:501, 1928. (X8)
R20. Starr, C. Vaughan; "The Green Flash," Meteorological Magazine, 64:290, 1930. (X7)
R21. "The Green Flash," Nature, 135:992, 1935. (X0)
R22. Lindley, W. M.; "Sunset Phenomena," British Astronomical Association, Journal, 47:297, 1937. (X4)
R23. Barringer, D. Moreau; "Note on the 'Green Flash'," Popular Astronomy, 57:252, 1949. (X7)
R24. Jacobsen, T. S.; "On the Spectrum of the Green Flash at Sunset," Royal Astronomical Society of Canada, Journal, 46:93, 1952. (X0)
R25. Thompson, H. M.; "Green and Red Flashes," Marine Observer, 23:144, 1953. (X0)
R26. Barlow, E. W.; "Green Flash," Weather, 12:199, 1957. (X6)
R27. Howard, W. E., III, and Young, A. T.; Sky and Telescope, 17:15, 1957. (X3)
R28. Bennett, J. D.; "Green Flash," Marine Observer, 28:76, 1958. (X0)
R29. Thomas, W. S.; "Red Flash at Moonrise," Marine Observer, 31:68, 1961. (X0)
R30. "The Green Flash," British Astronomical Association, Journal, 71:35, 1961. (X0)
R31. Seebold, Robert E.; "Green Flash," Optical Society of America, Journal, 51:237, 1961. (X0)
R32. Longster, R. F.; "Lunar Green Flash," Marine Observer, 33:20, 1963. (X8)
R33. Milne, N. S.; "Green Flashes," Marine Observer, 33:188, 1963. (X0)
R34. Taylor, John H., and Matthias, Bernd T.; "Green Flash from High Altitude," Nature, 222:157, 1969. (X0)
R35. Botley, C. M.; "The Green Ray," Weather, 26:354, 1971. (X9)
R36. Wilson, Raymond H., Jr.; Sky and Telescope, 42:327, 1971. (X0)
R37. Fraser, Alistair B.; "Green Flash and Clear Air Turbulence," Atmosphere, 13:1, 1975. (X1, X8)
R38. Bartholomew, J. C.; "Double Green Flash over Western Isles," Weather, 31:30, 1976. (X1)
R39. White, R.; "A New Theory of the Green Flash," Journal of Meteorology, U.K., 4:270, 1979. (X9)
R40. Dodkins, T. P.; "Green Flash," Marine Observer, 52:28, 1982. (X1, X4)
R41. Missira, A.; "An Unusual Observation of the Green Flash," British Astronomical Association, Journal, 92:147, 1982. (X4)
R42. O'Connell, D. J. K.; "The Green Flash and Other Low Sun Phenomena, New York, 1958. (R42)

GEL2 Anomalous Diverging Rays at Sunset and Sunrise

Description. Sharply defined, fan-like rays, usually pink in color, diverging from the vicinity of the sun and/or antisolar point. In favorable climes, such as Africa, these rays may span the entire sky, connecting solar and antisolar points. Termed crepuscular rays when seen close to the sun and anticrepuscular rays when located near the antisolar point, they are conventionally explained as shadows cast upon the sky by clouds or landforms obstructing the low sun. While real crepuscular rays certainly exist, the following features of the phenomenon suggest that the situation may be more complex than supposed: (1) The rays are sometimes uniform and evenly spaced; (2) A few displays remain static for "long" periods of time; (3) A few displays with fixed geometries cannot be associated with fixed topographic features; (4) Rays are frequently seen in some localities where the skies are cloudless over immense areas; (5) A few ray displays diverge from points other than the solar and/or antisolar points; (6) More than one focus may occur; (7) Rays may persist well after sunset; (8) Crepuscular rays may appear in a cloudless sky before sunset; (9) Antisolar rays sometimes appear with no corresponding display around the sun; (10) Rarely the rays may wave like auroral streamers; (11) One variety consists of one or two long white arcs and has few of the standard features of the crepuscular rays; (12) Luminescent filaments and whiskers have been seen on the horizon near the solar point from spacecraft; (13) Some crepuscular rays may be highly colored---something like rainbow arcs; and (14) Single, very broad rays rarely cover much of the sky in one rare type of display.

Background. Virtually everyone has seen beams of sunlight shining down through broken clouds. As the old phrase goes, the sun is drawing water! When the sun is on the horizon, it is perfectly reasonable for the same effect to occur, with the sun shining tangentially through cloud layers. The situation seems so simple that scientists have rarely looked deeper.

Data Evaluation. The entire body of anomalous observations is rather impressive. One must admit that there is strong evidence contradicting the simplistic cloud/mountain explanation. Many of these observations, however, lack scientific precision. In addition, some cases may be confused with manifestations of the aurora or zodiacal light. Rating: 2.

Anomaly Evaluation. If clouds and mountains cannot account for all observations of diverging rays, we must look at other atmospheric structures for the shadowing effects (see below). It is generally recognized that waves and other disturbances propagate through the atmosphere, so it would not be stretching the current picture of the atmosphere too much to try and adapt such disturbances to the crepuscular ray problem. But, since few atmospheric scientists will acknowledge that a problem exists, we must assign a rather high rating to the anomalies of crepuscular rays. Rating: 2.

Possible Explanations. Fan-like rays (at least some of them) may be just one more indicator of regular structures high in the atmosphere. If such structures can segregate windrows of particulate matter (ice crystals, volcanic dust, etc.), we may have a basis for explaining some of the anomalies below.

Similar and Related Phenomena. Polar bands and other aligned cirrus clouds (GWC3); the banded sky phenomenon (GLA23); anomalous auroras in general (GLA); puzzling displays of the zodiacal light (AZ); eclipse shadow bands (GES1); spoke-like structures on Venus (AV).

Examples of Anomalous Diverging Rays at Sunset and Sunrise

X0. Crepuscular rays with unusual features but which can very likely be explained within the bounds of conventional theory.

Lunar crepuscular rays. September 4, 1876. Shanghai, China. "Last year (1876), on the morning of September 4, I enjoyed a most interesting sight. It was about 5 a.m., the moon, then on her nineteenth day, was above the western horizon, and just being partially eclipsed; now from her bright disc, as from a radiating centre, shot out a number of those bands or blue beams; they traversed the whole expanse of the sky, and seemed to converge towards a point whose situation in the east below the horizon corresponded with that of the moon in the west above the horizon." (R6) Rays at moonset are very rare, only one other example having been found so far. (R24)

Slow rotation of the system of crepuscular rays across the sky. March 13, 1906. Near Cuba, North Atlantic Ocean. "On Tuesday March 13, 1906, about 5.30 P.M., in lat. 23° N. and long. 79° 20' W., whilst navigating between the north coast of Cuba and the Bahama Bank, I witnessed a phenomenon which I thought to be unusual. The sun still had some degrees to fall before reaching the horizon, a light E.N.E. breeze was ruffling the sea; to the westward a few high cloud wisps presented a gorgeous spectacle, showing an inky blackness skirted with brilliant gold, which, as the sun set, turned to an intense crimson. Looking to the east, I was somewhat startled by seeing streamers emanating from an apparent centre below the eastern horizon. These streamers were of a whitish colour, and extended for about 60° towards the zenith, and widened a little as they approached it. I counted over 20 of them. Examining them, they appeared to present a similarity to a white film rather then light, with well-defined edges contrasting with the deep blue of the sky. I then was still more astonished by noting two distinct motions of the phenomenon. The rays revolved slowly round their centre to the extent of about 15° in as many, or possibly 20, minutes. Also the apparent centre was rising though still below the eastern horizon. I turned to look at the setting sun, and found that the apparent centre of the phenomenon was about as much below the eastern horizon as the sun was above the western. As the sun neared the horizon so the radiant point rose. The sun disappeared below the horizon, and the phenomenon vanished. For a minute or two I was wondering what could be the explanation of what I had seen, when the whole burst into view again, but with this addition: the centre was visible well above the eastern horizon. The centre was a dark disc of 3° or 4° in diameter, and from the circumference of this disc the straight white rays projected all round, still retaining the rotary motion, which, of course, was very slow. I tried to measure the arc of this rotary movement, but was unable to do so with any exactitude, but from the time of first sighting the rays at about 5.30 P.M. till they finally disappeared at about 6.20 P.M. A.T.P., I judged that each individual ray performed an arc or 40°. I noticed also that where a ray crossed an extremely high cloud it obliterated the portion it covered. This struck me as being peculiar, as I was inclined to think the phenomenon was exterior to the cloud region, and upon further scrutiny I could not be certain that what I had first taken to be cloud was really such; it was a dark formless shape of great altitude, similar in appearance, except shape, to that inky black semicircle which appears on the horizon beneath a display of the aurora." (R17)

This rotary motion has been remarked upon by others. October 4, 1933. The Azores. Referring to a preceding description of crepuscular rays. "Similar shafts of pale-rose light were observed for a few minutes beginning at 17h.30m., Azores Time, on the eastern horizon, radiating from a point diametrically opposite the sun just set. The point of radiation (possibly an anthelium) I could not see, as it was hidden by a big cumulus behind the line of mountains. I also observed that the shafts were moving slowly to the right around the supposed anthelium." (R25)

In the conventional view of sunset rays, this general rotary motion is the natural consequence of cloud motion. (WRC)

The joining of solar and antisolar rays overhead. September 1894. Turkey. "It was in the month of September of this year, on the arid plains that lie between Erivan and Mount Ararat. The sun was just below the horizon, and from the point where it had set there came a number of broad streamers. From the opposite direction similar rays were shooting forth, and joined the others in the sky above---one might say, huge fingers coming together." (R11) This apparent development on opposite horizons and consequent joining overhead is probably an effect of contrast. The shadows cast in the west would be seen easily against the light of the setting sun, while the rays of light would, at first, stand out against the dark eastern sky. Finally, as the sun sank further, both sets would join at the zenith. (WRC)

Crepuscular rays created by mountain ranges. July 1934. Ganderbal, Pakistan. "This year I had the good fortune to see the rays on eleven consecutive days, from July 14th to 24th inclusive, at Ganderbal in the vale of Kashmir. On each occasion the visibility at Ganderbal was good, and the sky was clear except for a little detached cumulus over very high peaks to the north. The rays appeared to occupy the same positions every evening but to vary in intensity. By climbing an adjacent hill it was possible, owing to the uneven spacing of the rays, to identify one or two of the most prominent peaks which produced the most conspicuous shadows. The rays extended well beyond the zenith and varied from purple to pink in colour." (R26) Shadows cast upon the evening sky by high peaks in the Rocky Mountains are not uncommon to the east of this range. (WRC)

X1. Uniform, evenly spaced rays that seem to contradict a cloud or mountain origin.

January 29, 1959. North Atlantic Ocean. "Before sunset, when the sun was 7° above the horizon, bearing 246°, several bands showing a darker colour than the sky, which was rose pink, were observed to be radiating from a point just below the horizon diametrically opposite to the sun (see sketch). As the sun set, the pink colouring gradually faded and the bands disappeared." (R27)

Evenly spaced rays over the North Atlantic (X1)

General observations. Ethiopia. "In Ethiopia, at sunset, the lower part of the eastern horizon is normally pink. The pink layer moves up, becomes better delineated as a dark segment develops underneath it: it soon becomes an anti-twilight arc with a horizontal extent of some 180 deg. Three to five minutes after the upper limb of the sun has disappeared below the western horizon, from a point 4 to 5 deg above the eastern horizon and in the centre of the anti-twilight arc, dark blue bands suddenly radiate in all directions through the pink-coloured sky, progressively cover the whole hemisphere and converge towards a focus apparently lower under the western horizon than the actual position of the sun. Then the blue bands slowly vanish from the east, covered by the earth's shadow advancing westward. The number of dark blue bands may vary from one to seven or nine. When unique, the preferred zenithal angle seems to be 45° north or south. By the time of their disappearance in the earth's cone of shadow at the upper limit, the twilight arc (on 16 December 1966, about 10 deg above the western horizon 33 min after apparent sunset) the height of the dark blue bands was estimated to be 30-35 km. The same phenomenon is also seen before sunrise, but I never observed one in the morning as geometrically perfect and complete as in the evening. Moreover, the time of disappearance before sunrise seems shorter (20 min, 30 December 1966) than at sunset. This difference in time of disappearance at dusk and dawn could probably be interpreted as daily variation in height of the same atmospheric layer, provided the elevation of the point of disappearance be the same in both cases. The pink or reddish colour of the evening may be attributed to volcanic dust in the stratosphere. The pattern is usually observed to be more or less geometrically perfect, from October to the end of December in the Addis Ababa region. Although it is more spectacular when the sky is clear, the clearness of the sky is not a guarantee that, on a particular day during that period, the phenomenon will take place. The azimuth of the solar point at sunset is not an indication either; the solid angle swept by the solar point on the horizon from January to 15 March is the same as that covered from October to January, and no complete display has been reported after

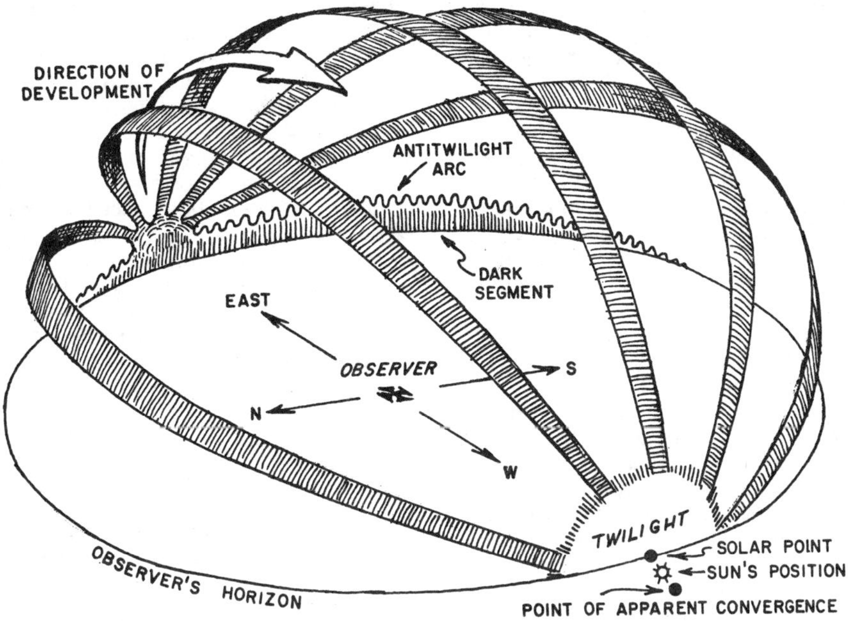

Sketch illustrating the development of twilight phenomena in Ethiopia (X1)

1 January." (R28) See also R2 for a similar appearance. What is most striking here is the geometric perfection of the apparition. Since the number of bands varies, mountains are ruled out. Further, the phenomenon is common and clouds in this region are not. (WRC)

X2. Displays of diverging rays that remain static for "long" periods of time.

October 21, 1880. Kentish Town, United Kingdom. "Last evening (October 21) at 5.45 p.m. I observed four huge radiating arms of faint white light, like the spokes of a gigantic wheel, rising from a centre apparently on the west-southwest horizon, and extending almost to the zenith. I say apparently on the west-southwest horizon because an intervening house prevented me from seeing the nucleus of the diverging rays. The aspect of the phenomenon was more suggestive of an aurora than anything else I know of, but the beams of light seemed to be quite stationary, and although I fancied their brilliancy increased at one time for a few moments, I cannot be sure. Other fainter rays appeared to me to divide the west-southwest sky with those I have mentioned; but on that point I am also not sure. The sun set at 4.53 p.m., and twilight ended about 6.43 p.m., at which time the appearance I have attempted to describe was no longer visible. The sky was heavily clouded." (R5) The rays lasted roughly one hour and did not change---true crepuscular rays do not behave this way. (WRC)

August 23, 1916. Port Arthur, Ontario, Canada. "On the night of August 23, in the east, almost directly opposite to where the sun was setting in the west, appeared four distinct streamers of light of a pinkish shade. These streamers must have lasted for more than an hour. I saw them first about eight o'clock and, at intervals, till after nine, though by that time they were getting rather faint. They were waving slightly, something like the Northern Lights, and extended from the horizon midway to the zenith." (R21) This and the preceding example have auroral overtones, but the locations of the displays were inappropriate for auroral activity. (WRC)

X3. Repeated displays with fixed geometries. No fixed terrestrial obstacles (mountain ranges) have been associated with these displays.

No date given. Southern India. "The usual appearance was a blue double segment of a circle, or middle section of a double convex lens starting from the sun's place very shortly after sunset, gradually widening out to a width of about 40° at the zenith, then gradually diminishing to a vanishing point a little above the horizon opposite to the sun's place. The blue was deepest at the zenith, and was flanked on each side by a whitish, hazy-looking atmosphere---brightest at the zenith, and shaded off to north and south to the ordinary light blue of a dry-season sky just after sunset. Sometimes the borders had a light-pinkish tinge, and sometimes the broad band was broken up into two or three by whitish rays similarly double convex; but the single band, as above, was the more general appearance. The phenomenon occurred in the dry season, and in a cloudless sky, when there was nothing visible to lead one to expect its appearance." (R9)

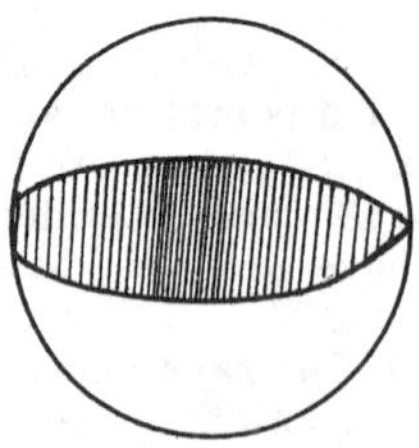

Lens-shaped blue segment reaches from horizon to horizon after sunset in India. (X3)

No date given. United Kingdom. "<u>The Five-Rayed Sunsets.</u> Correspondence in the papers has taken place about the rays observed after sunset. These rays are usually five in number. The third one, careful watching has apparently shown, has never quite died out, but has become, in twilight, what we understand by the Zodiacal Light. These rays stretch from west to east, and, on frosty evenings, are frequently brighter at first in the east than in the west. They are at no great height, not nearly so high, we may suppose, as the Martinique dust, because they appear almost immediately after the sun has disappeared, and have nearly vanished by the time the 'glow' is at its height. They have, of course, nothing to do with clouds passing over the sun when below the horizon, for they can, when visible at all, be found always at the same relative distance apart, and they do not revolve, as they would were they caused by any moving vapour. Are they a sign of atmospheric collapse? Taken in regard to the dun sky and large amount of volcanic dust, which has been sent into and beyond the atmosphere, they might be accepted as 'long distance' signals or messengers of a gradual change taking place in the atmosphere. Is that change doubtless due to astronomical forces of some kind?" (R15) This is one of the few surmises in the literature that some crepuscular rays might not be due to clouds. (WRC)

X4. Diverging rays observed during cloudless weather over a wide adjacent area.

General observations. India. "In the clear dry mornings of these regions (the Kymore Hills) a curious optical phenomenon may be observed of a sunrise in the west, and sunset in the east. In either case bright and well-defined beams rise to the zenith, often crossing to the opposite horizon. It is a beautiful feature in the firmament, and equally visible whether the horizon be cloudy or clear, the white beams being projected indifferently against a dark vapour or the blue (sky?). In the 'sunrise' I witnessed the rays crossed from west to east. It must be remembered that a glow in the East at sunset is a very common occurrence." (R32) Note the effect is independent of clouds on the horizon. (WRC)

> General observations. Shanghai, China. Sky-spanning bands often seen at both sunset and sunrise in cloudless weather. (R6) The sunrise here occurs over the open sea, so neither clouds nor mountains seem to be involved. (WRC)

General observations. Republic of South Africa. "In the dry uplands of South Africa, in the brilliant dry weather one gets in winter, I have night after night observed these lovely pink streamers stretching right across the sky, and meeting at a point opposite the sun's position; and I have often speculated as to their origin. They would never be quite the same two evenings running; sometimes there would be several of the rays, well-defined, all a beautiful pink, contrasting exquisitely with the pale blue of the twilight sky. Next evening there might perhaps be one huge fan-like one stretching laterally over many degrees of the sky, with one or two narrow ones. I quite admit that they are exactly analogous to the rays of the sun piercing an aperture in clouds which cut off his direct light from the space surrounding the cylinder of atmosphere illuminated directly. But the difficulty always occurred to me, where were the clouds that produced the streamers or rays, because we know for certain (in the Transvaal where I was then stationed) that in the winter season the whole sky, over an enormous tract of country, was absolutely free from cloud." (R8, R13, R14) Since the rays varied in number and position, mountains hardly seem a good explanation on cloudless days. (WRC)

General observations. Krugersdorp, Republic of South Africa. Rays often seen stretching across the entire sky in cloudless weather. Distant mountains considered a possible cause. (R16)

May 23, 1910. Cumberland, England. "Whilst staying at Rosthwaite, Borrowdale, Cumberland, I observed on Monday, May 23, a series of dark rays radiating from the sun as a centre to a distance of some 60°. They were of a ruddy-brown colour, and very distinctly marked. For the most part, the day was cloudless; but, in the afternoon, whilst a small thundery cloud occupied a space about 15° below the sun, I saw the brown rays with the greatest distinctness. I thought they might emanate from Halley's Comet, and so noted the phenomenon in my diary." Similar effect reported in America according to the author. (R19) This occurrence was before sunset with no visible clouds above the sun! (WRC)

General observations. South Africa. "With reference to the illustration of the anti-solar rays and the description by Mr. G. A. Clark, which was published in the February number of the Meteorological Magazine, this phenomenon has long been seen and commented upon by residents in the Buchuanaland Protectorate. It is a common sight north of latitude 25° in South Africa. It is sometimes seen quite clearly much south of this, even as far south as Kimberly, latitude 29°. For years past I have observed the rays at sunset. I was first of the opinion that they were the converging shadows of cirrus clouds, but on many occasions these rays have been observed quite clearly when there were no clouds in the sky whatever. I came to the conclusion that they were a contra-reflection of the rays of the setting sun inside the atmosphere, such as might be observed if seen from the inside of a crystal sphere. They would probably become visible under such circumstances if there were a certain amount of atmospheric dust." (R22) Mountains cannot be ruled out in this account. (WRC)

X5. Rays diverging from points other than the sun or antisolar point.

September 12, 1894. Juvisy, France. "I enclose a sketch of the optical phenomenon observed at Juvisy on September 12th, 1894, 5h. 45m., Paris mean time, which may be described as follows: The sun was nearly setting, when, all of a sudden, two very narrow lines, like telegraph wires, but white, originating from a point of the

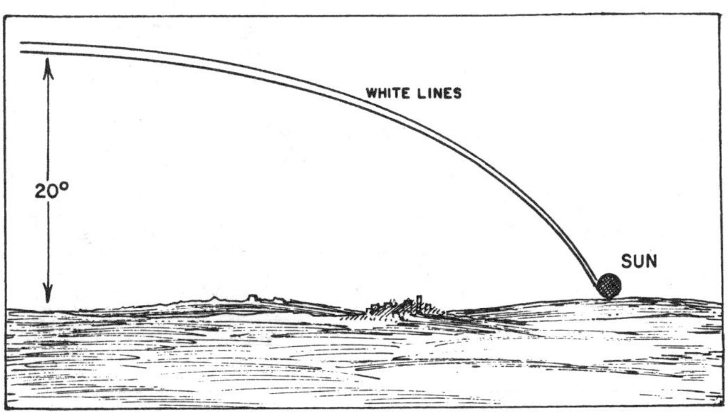

Two narrow white lines diverge from vicinity of sun (X5)

western horizon near the sun, rose in a gentle curve towards the east, attaining on the south their maximum height (20°) above the horizon. The lines were probably parallel, but owing (apparently) to perspective, their distance, which was over the south some 1° apart, was only half that amount in the vicinity of the sun. The apparition lasted more than half an hour, having gradually faded away at 6h. 22m. M. Flammarion, who happened to be present, observed also this remarkable phenomenon." (R12)

June 1894. Eastbourne, England. "On Monday evening last, whilst walking on the parade at Eastbourne, I was struck by an appearance such as might be expected directly after sunset, streams of light radiating from a point below the horizon; but as the sun was still above the horizon, and in the opposite quarter of the sky, I can only suppose the phenomenon was that due to the effects of celestial perspective mentioned in Sir John Herschel's 'Outlines." (R10) This instance is probably explicable as antisolar rays diverging from the antisolar point still below the horizon. (WRC)

January 7, 1928. North Atlantic Ocean. ".... observed rays of light emanating in the eastern section of the sky and extending to the southward bearing S 77° true, and to the northward bearing N 35° E. The extreme altitude was 10° and they appeared in white and grey contrasts. 1956 G.M.T., a broad ray appeared bearing N 50° E., true, and had the appearance of being light green in colour. At 2011 G.M.T., there were no rays visible. The computed time of moonrise U.L. was 6.26 A.T.S. and bearing N 67° E." (R23)

X6. Multiple foci. The following two examples are included for their curiosity value only, no anomalies are to be implied.

February 13, 1826. Berlin, Connecticut. "About half an hour before sun-set, I was walking on elevated ground, the sky was very clear, except a bank of light clouds to the N. of E., extending a few degrees above the horizon. I observed, directly opposite the sun, a perpendicular beam of light, shoot up on the clouds, and further S., near the horizon, a cluster of such beams, very oblique. Very soon, several beams, that were perpendicular, or nearly so, shot up around the first I saw, and some pale and scattered ones in the interval of the two clusters, and to the N. The whole was very exactly like the diverging beams that shoot out from the horizon just before sun-rise or after sun-set; its point of divergence, or centre, was below the horizon, apparently as many degrees below as the sun was above. The brightness, particularly of the two clusters, was very considerable---light soft and white." (R1) Though not so stated, it is possible that the two clusters had the same focus. (WRC)

July 7, 1914. Ipswich, England. Two sets of crepuscular rays; one from the

Confusion of diverging bands observed over a wide range of azimuth in the North Atlantic (X5)

setting sun, the other from the rising moon. (R20) This account is very confusing; and some of the rays may be auroral in origin. (WRC)

X7. Crepuscular rays may persist well after sunset.

July 7, 1914. Ipswich, England. An exceptionally long-lived display of diverging white rays from the spot where the sun had set. Total time: 1 hour 40 minutes. (R20) Auroral phenomena may have been involved here. (WRC)

X8. Crepuscular rays well before sunset.

May 23, 1910. Cumberland, England. Sun above horizon, no clouds. See full description in X4. (R19)

X9. Antisolar rays with no hint of normal crepuscular rays in the vicinity of the sun.

March 27, 1907. Bengal, India. An appearance of antisolar rays with the following unusual features: "(1) That they were without any pink colour; (2) That they appeared <u>before</u> sunset and disappeared at <u>or before</u> sunset; (3) That no corresponding rays were visible in the west." (R18) Note that antisolar rays are normally pink. (WRC)

X10. Waving crepuscular rays.

August 23, 1916. Port Arthur, Ontario, Canada. Antisolar rays waving slightly like northern lights. (R21) See full description in X2.

X11. One or two long white arcs, like searchlight beams.

April 10, 1871. No place given. "At 4.30 P.M., a whitish arc was seen, almost east, against a cloudless azure sky. On the previous night there had been a magnificent aurora borealis." (R3)

September 12, 1894. Juvisy, France. Two narrow white arcs rising from the west where the sun was setting. (R12) See X5 for the full account.

October 14, 1983. Near Lafayette, Louisiana. Rising from the antisolar point in the east. "At first it appeared very bright, like a searchlight beam. But I questioned the fact that a beam of light could be so bright and plain in broad daylight, with the sun in the opposite direction, shining brightly. There were no clouds, not even thin ones, anywhere in the azure blue sky. In time---after about 5 minutes---the light began to dim and faded after about 20 minutes or so. It changed during this time from a bright white to a dull grey white, until it disappeared completely..... The size of the light shaft was about 2° of arc at the base where it intersected the horizon, to about 5° where it gradually merged into the blue color of the sky. The shaft remained in the same position at all times---about 15° to 20° off from the vertical. The shaft was perfectly straight and the edges were well defined, but became more and more hazy as the sun moved closer to the time of setting." (R31)

X12. Luminescent filaments and whiskers observed from spacecraft.

General observations. "The optical effects consisting of luminescent filaments and whiskers, first observed by the astronauts of Soyuz-9, are discussed. During sighting of the nighttime horizon around the Earth there is observed a fine, weakly luminescent layer (braids) at an altitude of about $3^{o}40'$ (95 km). Before sunrise, in a symmetrical direction relative to the Sun, a stable, vertically radiating structure is observed in the form of individual luminescent filaments (columns) that extend to heights of 200-250 km. During the flight up to the terminator, the whiskers effect is also observed in the form of two luminescent zones symmetrical relative to the solar direction and, extending tangent to the horizon to heights of 200-300 km. These optical phenomena are explained qualitatively (fine structure, stability, localization, etc.) and quantitatively (brightness, absence of influence on the brightness of large stars) by a combination of optical phenomena in relatively dense atmospheric layers (braids) and the solar corona (filaments). The whiskers effect may be considered, also, as the effect of superposition of two light phenomena: the twilight aureole and the corona." (R29) It is not certain whether there is any connection between these phenomena and the diverging rays seen from the ground. (WRC)

X13. Spectral crepuscular rays.

April 17, 1974. North Pacific Ocean. A band of crepuscular rays (?) about 5° wide crossed the sky from west to east. The colors in the band were, from the two edges of the band, deep blue, pale green, pale orange, and, in the center, cream. (R30) For a full description plus the illustration submitted, see GEM9-X9.

X14. Single, very broad, sky-spanning bands.

July 3, 1877. Winona, Minnesota. "About eight o'clock in the evening of July 3d my attention was called to a peculiar appearance of the sky. The sun had been below the horizon about an hour. From the point in the horizon where it was last seen a broad band of pink or rose-colored light followed the ecliptic across the sky to the opposite horizon. Its south limit was sharply defined---its intensity was nearly the same from horizon to horizon. Its north limit was not determinable, the pinkish light extending nearly or quite to the horizon, filling the entire northern sky. The southern sky from the ecliptic was of the normal blue color, with the exception of a single streak of darker blue, extending from the point where the sun sank below the horizon about 90° into the southern sky, making an angle of say 30° with the southern limit of the rose-colored light, or the ecliptic. There were none of the auroral characteristics. The light was steady, and the entire exhibition as described lasted for twenty minutes, when it all faded away gradually, leaving a perfectly clear sky, with only a trace of pink in the west." (R4) The short life of this phenomenon makes the zodiacal light explanation unlikely. Still, some anomalous displays of the zodiacal light are rather similar in geometry (AZ).

References

R1. Percival, J.G.; "Curious Effect of Solar Light," American Journal of Science, 1:12:164, 1827. (X6)
R2. Sprung, A.; "A Rare Phenomenon," English Mechanic, 13:622, 1871. (X1)
R3. Jeremiah, John; "A Few More Words on Daylight Auroras," Nature, 4:47, 1871. (X11)
R4. Morey, Charles A.; "The Zodiacal Light," Popular Science Monthly, 11: 617, 1877. (X14)
R5. "Atmospheric Phenomenon," Nature, 22:607, 1880. (X2)
R6. Dechevrens, Marc; "A Curious Halo," Nature, 27:30, 1882. (X0, X4)
R7. Wambaugh, W.J.; "Sunset Phenomenon," American Meteorological Journal, 8:229, 1891. (X0)
R8. Markwick, E.E.; "Pink Streamers after Sunset," English Mechanic, 54:421, 1892. (X4)
R9. "Sun Streamers after Sunset," English Mechanic, 57:541, 1893. (X3)
R10. Gruning, H.H.; "Celestial Perspective," English Mechanic, 59:437, 1894. (X5)
R11. Rickmers, Willy Rickmer; "Phenomenon at Sunset," English Mechanic, 60: 366, 1894. (X0)
R12. Antoniadi, E.M.; "Optical Phenomenon," Knowledge, 18:205, 1895. (X5, X11)
R13. Markwick, E.E.; "Sunset Phenomenon," Knowledge, 24:88, 1901. (X4)
R14. Markwick, E.E.; "Zodiacal Light," English Mechanic, 76:105, 1902. (X4)
R15. Bridges, Guy J.; "Some Thoughts on the Zodiacal Light," British Astronomical Association, Journal, 13:242, 1903. (X3)
R16. "Red Rays at Sunset," English Mechanic, 81:199, 1905. (X4)
R17. Raisin, F.W.; "A Curious Phenomenon at Sunset," British Astronomical Association, Journal, 16:311, 1906. (X0)
R18. Hubback, J. Austen; "White Anti-solar Rays at Sunset," British Astronomical Association, Journal, 17:315, 1907. (X9)
R19. Haswell, John; "Dark Solar Rays," English Mechanic, 91:425, 1910. (X4, X8)
R20. Cook, A. Grace; "Aurora Borealis?" English Mechanic, 99:544, 1914. (X2, X6, X7)
R21. Fyfe, P.; "Streamers in the Eastern Sky," English Mechanic, 104:171, 1916. (X2, X10)
R22. Wallis, A.H.; "Anti-solar Rays," Meteorological Magazine, 58:105, 1923. (X4)

R23. Edwards, W.; "Light Rays," Marine Observer, 6:10, 1929. (X5)
R24. Vedy, L.G.; "Moonbeams Analogous to Crepuscular Rays," Meteorological Magazine, 64:115, 1929. (X0)
R25. Agostinho, J.; "Strange Sunset Effect," Meteorological Magazine, 69:165, 1934. (X0)
R26. Veryard, R.G.; "Anti-Solar Rays," Meteorological Magazine, 69:216, 1934. (X0)
R27. Thomas, W.S.; "Anti-Solar Rays," Marine Observer, 30:19, 1960. (X1)
R28. Gouin, Pierre; "Unusual Twilight Phenomena over Ethiopia," Weather, 23:70, 1968. (X1)
R29. Simonov, A.I.; "Nature of Optical Effects Observed from Spacecraft during Sunrise," Kosmicheskiye Issledovaniya, 11:490, 1973. (X12) As translated in Cosmic Research.
R30. Baker, J.; "Crepuscular Rays," Marine Observer, 45:68, 1975. (X13)
R31. Calais, Ron; "Ray from the Antisolar Point," Personal communication, October 1983. (X11)
R32. Paris, M.; "A Morning Sunset," English Mechanic, 15:465, 1872. (X4)

GEL3 Color Phenomena and the Earth's Shadow on the Sky

Description. Sudden color changes of the earth's shadow when cast upon the sky by a low sun. The change is usually from a purple or deep blue to a light blue. Also included here are unusual color fringes at the top of the deep-blue wedge of the shadow.

Data Evaluation. These phenomena are poorly studied; reports are scarce and of a testimonial nature. Rating: 3.

Anomaly Evaluation. Colored fringes on the upper portion of the earth's shadow seem to be due to the scattering of sunlight, as described below. But the sudden changes of the hue of the earth's shadow has no ready explanation. Rating: 2.

Possible Explanations. Fringe colors are probably the result of atmospheric scattering of the shorter wavelengths from the sunlight illuminating the sky just above the earth's shadow. Red light penetrates the atmosphere best and yellow somewhat less well; so we would expect to see red under yellow in the fringe structure. As for the hue of the shadow itself; it is normally purple or deep blue. There are no obvious reasons why this normal color would suddenly switch to light blue.

Similar and Related Phenomena. The purple light (GEL12); twilight bows in the east at sunset (GEL13).

Examples of Color Phenomena Associated with the Earth's Shadow Cast on the Sky

X1. April 6, 1956. South Pacific Ocean. "An interesting example of sky coloration was observed at sunset. A wedge-shaped piece of sky, in the E., was coloured deep blue, almost purple, and the sky on either side of the wedge was a very pale blue. The wedge was extremely well defined and covered an arc of from approximately 10° above the horizon, where it disappeared into the clouds, to the observers' zenith. The thick end of the wedge, directly overhead, was covering an arc in the sky of about 50°. A particularly heavy shower of rain was directly in line between the sun and the observed phenomenon. The outstanding feature of this peculiarity was the extremely fine definition of the sides of the wedge and the sudden change from purple to pale blue......
Note. This is an interesting observation since the colour effect described is not what is usually seen. The normal aspect of the clear eastern sky after sunset is as follows. At sunset a steely blue segment, darker than the rest of the sky, begins to rise up from the horizon opposite the place of sunset. This is the shadow of the earth thrown by the sun on the earth's atmosphere. The earth shadow has a narrow band of rose or purple colour, known as the counterglow, along its curved upper boundary. The whole rises

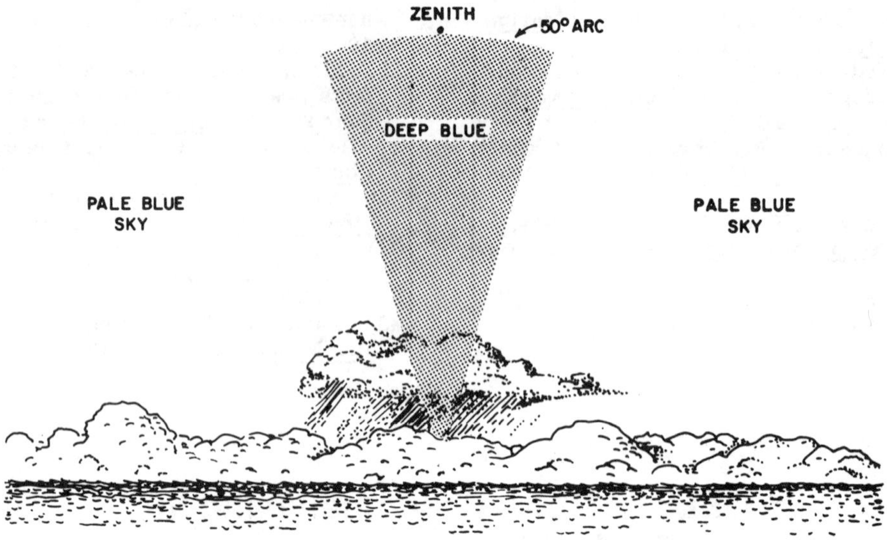

Earth's shadow seen in the east at sunset (X1)

fairly quickly in altitude with the shadow encroaching on and eventually obliterating the counterglow. The shadow rises to the zenith, but before this it generally becomes invisible owing to the progressive darkening of the whole sky." (R3)

X2. Frequent occurrence. Echo Mountain, California. "At every sunset, when the sky is clear, which it almost always is, about 15 minutes before he sinks below a range of hills, the entire surface of the valley south of an east and west line, assumes a dark blue, which moves east as the Sun descends, slowly ascending up to the horizon, causing the sky to assume the same dark blue, which extends 90° in azimuth. A few minutes after sunset the blue leaves the Earth and forms a belt some five degrees wide in the sky. A short time previous to this, a dark red band forms on the upper edge of the blue, but seems not to mingle with it. Its breadth is also about five degrees. As soon as the red belt disappears, or a little before, a broad yellow band appears above the red, and seems to slightly mingle with it. The yellow band is as broad as both the others. Of course the phenomenon is due to atmospheric refraction of sunlight. So much is certain, but what is unexplainable to me is, why the red band should be between the yellow and blue." (R1) The claim that refraction is the basic cause of the colors of the earth's shadow was challenged. The red section results because this portion of the atmosphere is illuminated by rays passing almost horizontally through the atmosphere and from which the shorter wavelengths have been scattered or absorbed. The yellow section is higher than the red and contains more of the shorter wavelengths. The blue color is a true shadow and has nothing to do with dispersion of the spectrum. (R2) The yellow section described above is not included in the standard descriptions of the earth's shadow. (WRC)

References

R1. Swift, Lewis; "Phenomenon Yet Unexplained," Popular Astronomy, 6:589, 1898. (X2)

R2. Keeler, James E.; "Color Phenomenon Attending the Earth's Shadow at Sunset," Popular Astronomy, 7:98, 1899. (X2)

R3. Jackson, I. W.; "Sky Coloration," Marine Observer, 27:86, 1957. (X1)

GEL4 Abnormal Refraction Phenomena as Astronomical Objects Approach the Horizon

Description. Image multiplication, image distortion, image motion, color changes, and other unexpected optical phenomena as the sun, moon, bright planets and stars approach the horizon.

Data Evaluation. Many excellent observations of a wide variety of abnormal refraction effects may be found in the literature, especially the marine literature. Rating: 1.

Anomaly Evaluation. Most of the phenomena described below are probably explicable by appealing to atmospheric convection along the line from the observer to the horizon. Complex shapes of shifting air masses at different temperatures can be employed to explain a great variety of image changes and color phenomena. Still these effects are fascinating and are not always easily accounted for. Rating: 3.

Possible Explanations. Refraction and spectral dispersion of the light from astronomical objects by the long section of turbulent atmosphere between the observer and the objects.

Similar and Related Phenomena. All mirages (GEM); the green flash (GEL1).

Examples of Weird Refraction Phenomena When Astronomical Objects Are Close to the Horizon

X0. October 28, 1961. North Atlantic Ocean. "Between 2321 and 2326 1/2 GMT, the rising moon, which was very red in colour, assumed the shapes shown in the sketches. The true time of moonrise was some 5 min. later, the moon then having an entirely normal appearance." (R21) This example is included to convey the fact that abnormal refraction near the horizon can do remarkable things to the shapes and colors of astronomical objects. (WRC)

Abnormal refraction distorts the rising moon in the North Atlantic (X0)

X1. Multiple suns and moons.

May 7, 1823. Cape of Good Hope, Republic of South Africa. "On the 7th May 1823, when the sun's lower limb had just dipped the water's edge, the Reverend Mr Fallows observed several parhelia, viz. 4 on the left, and 3 on the right hand of the sun, and all cut by the horizon like the real sun. They had the same shape as the real sun, and were as high, but not so long. When the upper limb of the sun came in contact with the horizon, it and the mock suns appeared as bright points upon the water's edge, and then one of them instantaneously vanished.....It deserves to be noticed, that, according to the diagram of Mr Fallows, all the mock suns were included in a space extending only one degree and a half on each side of the sun's centre." (R1) Real parhelia do not lie so close to the real sun! (WRC)

February 2, 1884. England. "The subjoined diagram will serve to show the apparent positions and relative sizes of the three suns when first seen, at about twelve minutes to five o'clock, from a window

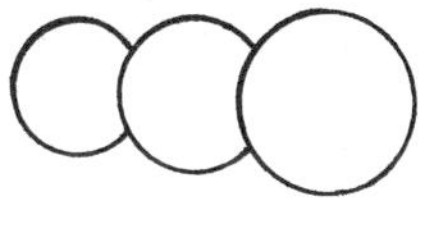

Multiple suns at sunset. True sun at right. (X1)

which commands an almost uninterrupted view to the horizon. I suppose No. 3 was the true sun, though all were of about the same blood-red colour and equally well-defined. On the horizon, and very slightly overlapping the lower edges of the three disks, was a heavy bank of clouds, behind which they rapidly disappeared---it may have been about three minutes from the time of my first observing them; but No. 1 had then become very indistinct, Nos. 2 and 3 remaining well defined till the last." (R3)

March 12, 1914. Observatoire du Jorat, Switzerland. "A rare optical phenomenon was observed by G. Isely at the Observatoire du Jorat, Switzerland, on March 12th, just after sunrise, when the sun was seen to be accompanied by two mock suns, one above and one below it, each at a distance of about 2 degrees from the luminary. This phenomenon is explained by Pernter as a form of mirage." (R9) Mock suns associated with ice crystals in the atmosphere do not occur this close to the sun, or in these positions. (WRC)

1925. England. "I was hurrying to get some work done before sunset, and in my garden I looked West. The moon presented the appearance of having a section cut off on the south upper side, just as a Dutch cheese would look if a small slice had been cut off the whole globe of it with a straight knife. The peculiar thing was that there appeared, as it were, straight in front of it a second image transparent and with the frustrum of the sphere on the opposite side to that of the Moon itself. Now what could be the optical reason for this image of the Moon reversed?" (R12)

December 6, 1937. North Pacific Ocean. "Mr. Lewin observed the strange phenomenon during a run from Balboa, C. Z., to San Pedro, Calif., while his ship was off the tip of Lower California. The crescent moon was going down through a thin haze, when it appeared double, the lower crescent scarcely touching the upper. Twenty minutes after he first noticed it, the moon set, still appearing double." (R13)

January 2, 1954. South Pacific Ocean. "When the moon had attained an altitude of 18 1/2° an image of it was observed resting on the tip of the right-hand crescent of the true moon. The edges of the image showed four colours of the spectrum, red, orange yellow and green. Sky completely cloudless, but previous to this phenomenon heavy rain showers had occurred..... Note. This is fifth observation of a double moon or sun received since the war, all observed at fairly high altitudes, 10° to 20°. As stated in the note to the observation of S. S. Rialto, Vol. 24, page 147, of this journal, no explanation can yet be given. The special point of interest in the present observation is the colour fringes, which clearly point to some abnormality of refraction." (R25)

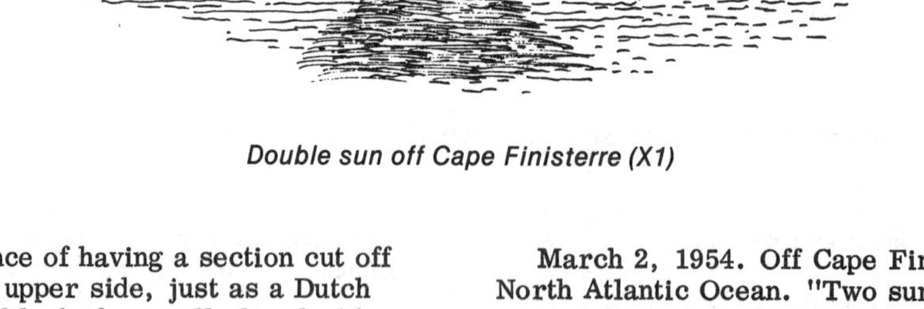

Double sun off Cape Finisterre (X1)

March 2, 1954. Off Cape Finisterre, North Atlantic Ocean. "Two suns were observed at sunset, the image being above the true sun. The observation was not very good because the true sun was obscured by clouds until a few minutes before sunset, and consequently the period of this phenomenon is not known." (R15)

March 28, 1955. Indian Ocean. "1300-1430 G. M. T. The crescent moon, with a faint hazy image close above it, was observed through patchy Cs through which the brighter stars were also visible. The phenomenon, which disappeared when viewed through a telescope, was unchanged in form during the whole of the period in which it was observed." (R26)

July 19, 1958. South Atlantic Ocean. "1800 G.M.T. The moon, which was 3 days old, presented the appearance shown in the accompanying sketch. The disc was feebly illuminated by 'earth shine' and superimposed upon it was a reflected image of the crescent (see sketch). The night was fine and there was no mist or haze." (R27)

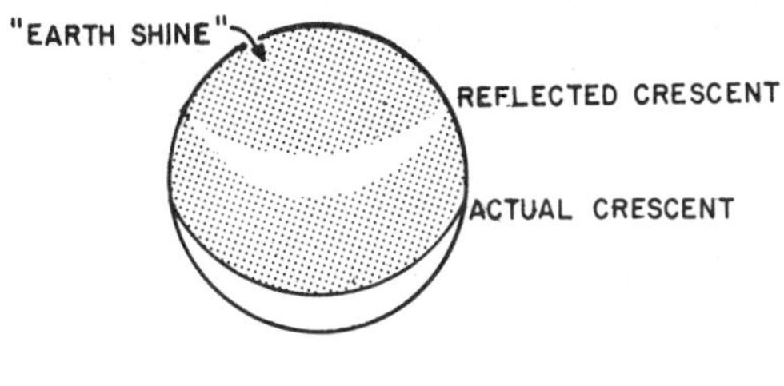

April 13, 1967. Off Guinea, North Atlantic Ocean. "At 2130 GMT, the moon (age 4 days) appeared to have two crescents. The real crescent was bright while the false one was dull and lay just above it. This abnormal appearance persisted for 7 min. No cloud was visible but there was a halo round the moon." (R24)

No date or place given. Two crescent moons seen side by side. (R2)

X2. Apparent jumping or moving of astronomical objects near the horizon.

May 1888. Maryborough, Australia. "We saw such a curious phenomenon on Sunday night, about 10.30. Miss C___, Miss H___, and I were sitting in the balcony, when we noticed the moon apparently dancing up and down. It is on the wane, so looked extraordinary. The motion was visible only when she was behind a narrow stratum of cloud and continued at intervals for thirty minutes. I felt quite seasick watching it." (R6)

No date of place given. A comment in response to a letter on "jumping stars." "On an autumnal evening, after twilight had ceased, I observed Jupiter about 15° above the horizon, towards the east, oscillating to the extent of half a degree in all directions. I did not observe the same movement in any of the stars, none of any magnitude being visible in that part of the heavens." (R7) Other letters in this reference describe experiences with jumping or oscillating landmarks and astronomical objects. (WRC)

X3. Color phenomena and mirage effects with setting planets.

November 11, 1965. Gulf of Aden, Indian Ocean. "At 1744 GMT when Venus was about to set, bearing 237°, it changed from yellowish white to red. Shortly afterwards the planet appeared double, i.e. with a red image a short distance between the true disc. The image was below the horizon line and it gradually rose to meet the horizon as Venus decreased in altitude. At the moment of disappearing the two red discs coalesced and for an instant changed to a light green colour." Probably a combination of a mirage and the green flash phenomenon." (R23) A similar situation was reported in which the two images of Venus were green throughout. (R11)

No date or place given. Excerpt from a book by Savage Landor. "At night, while back in camp, we saw to the W.N.W. quite

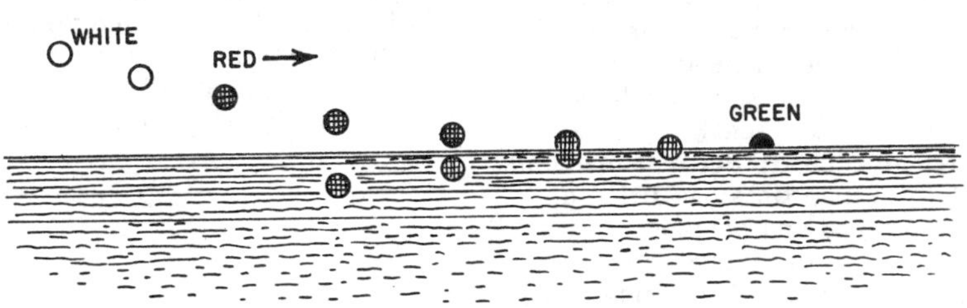

Venus setting over the Gulf of Aden, ending with a green flash (X3)

low on the horizon, a brilliant planet, possibly Venus. The stars and planets appeared always wonderfully bright and extraordinarily large on fine nights. Whether it was an optical illusion or not I do not know, but the phenomenon, which lasted some hours, was seen by all my men, and appeared also when the planet was seen through a powerful hand telescope. It seemed to discharge powerful intermittent flashes, red and greenish, only towards the Earth. Those flashes were similar to, and more luminous than the tail of a small comet, and, of course, much shorter---perhaps four to five times the diameter of the planet in their entire length..... At moments the planet seemed perfectly spherical, with a marvelous outline, and then the flashes shot out, especially to the right, as one looked at the planet, and downward slightly at an angle, not quite perpendicularly." (R8)

> December 6, 1957. Indian Ocean. The planet Venus was setting orange, when it changed to a double green image, one image above the other. The two images united and the planet set green. (R17)

July 4, 1962. Indian Ocean. "At 1415 GMT Jupiter suddenly appeared above the horizon with a bright red flash on a bearing of 105°. One image of the planet then rose slowly while the other maintained a steady altitude where it had first appeared. Both changed gradually from a bright red to an orange colour. At 1422 the images suddenly merged with a bright white flash and the planet continued to rise, slowly changing from orange to white." (R22)

X4. Mirage of a green flash at sunrise.

January 5, 1914. The Riviera, France. While watching the sun rise in the direction of Corsica (about 140 miles away), a mirage of the island's peaks appeared. "On the 5th January the island peaks again lay beautifully silhouetted on the bright orange sky; and they appeared slightly raised above the clear-cut horizon. As the sun rose between the two peaks there was a sparkling blue-green flash for one second. But just previous to this the bright orange tints had filled up the hiatus between land and horizon. As soon as the Sun was fully up the island vanished." (R10) This is a most unusual instance of the green flash appearing in a mirage. (WRC)

X5. Remarkable color phenomena during moonrise.

June 28, 1956. South Pacific Ocean. "At 1415 G.M.T., just prior to moonrise, a bright red glow was observed on the horizon on the bearing on which the moon was expected to rise. A red light suddenly appeared and at first was taken to be a hurricane lamp, lit by a small boat about 2 miles away, on our approach. The light spread to 2° in width, and was obviously in, not on, the water, as it did not lift to the swell; it had the appearance of a flaming log, but the "flames" were motionless. Three minutes later a line of white light appeared over an arc of about 10° of the horizon. The moon then rose and the red, flame-like light disappeared, to be replaced by two half-moon shaped glows in the water. As the moon rose, the irregularities on its upper limb were plainly visible to the naked eye, and when just clear of the horizon the lower limb appeared to have a dent in the edge at about five o'clock." (R16) It is difficult to explain how any mirage or refraction effect could produce an image of the moon below the horizon when the moon itself has not risen. (WRC)

December 5, 1884. Red House Observatory, Phelps, New York. "On the date mentioned, about fifteen minutes before moonrise, upon a cloud in the eastern sky at an altitude of about twenty degrees, I noticed a pinkish tint which, in a few minutes, became intensified to a brilliant reddish glow, nearly equalling any display I have witnessed in connection with the Sun. The red light extended about thirty degrees in azimuth and was central over the Moon, which, although it had not risen, was plainly marked by a peculiar cone-shaped light extending from the horizon to the cloud upon which the red light was displayed. I gave the whole phenomenon a most careful study, and noticed a singular flashing of the red light similar to the aurora. The maximum of intensity was reached the instant before the upper limb of the Moon became visible. When the Moon was half way up the glow was scarcely visible, and when fully risen all trace of the wonderful light had disappeared." (R4, R5)

X6. Triangular rays or patches beneath astronomical objects low in the sky.

September 18, 1961. North Atlantic Ocean. "The planet Venus was observed rising at 0645 GMT and while low on the horizon it changed at regular intervals of 1 sec. from blood red to white; it actually resembled a shore flashing light. The colour changes continued until an elevation of about 4° was reached. When the planet attained this altitude, a fairly bright triangular ray of light was observed to move from it down to the horizon. This persisted for about 2 min. and then faded out; no more flashing was seen, the planet shining with a white light which flickered slightly for a short time." (R20)

General observations. Sometimes, when the moon is low, a dark triangle will appear between in and the horizon, with the point down. This dark triangle may meet the apex or a bright triangle at the horizon, the latter being the reflection of the moon on the waves. The dark triangle may be a contrast phenomenon. (R11)

X7. Scintillation of the moon's lower limb.

August 24, 1959. South Pacific Ocean. "An unusual degree of scintillation was observed in the case of the stars, which became visible at 1/2° altitude. In these conditions of abnormal refraction, it was hoped to see the green flash at moonrise, but a cloud bank low on the horizon prevented an observation being made. However, when the moon had risen to 15 min altitude and was clear of the cloud, its lower limb was seen to scintillate with a brilliant scarlet colour. The coloured rim gradually contracted in length and the last of the scarlet disappeared when the moon was at an altitude of about 1°." (R19)

References

R1. "Unusual Parhelia Seen at the Cape," Edinburgh Philosophical Journal, 10: 363, 1824. (X1)
R2. "Curious Lunar Refraction," American Journal of Science, 1:10:187, 1826. (X1)
R3. Amyot, Thos. H.; "False Suns," Knowledge, 5:133, 1884. (X1)
R4. English Mechanic, 40:382, 1885. (X5)
R5. Brooks, William R.; "A Lunar Red Light," Sidereal Messenger, 4:26, 1885. (X5)
R6. Reade, T. Mellard; "Curious Apparent Motion of the Moon Seen in Australia," Nature, 38:102, 1888. (X2)
R7. Tennant, J. F., et al; "Jumping Stars," Observatory, 11:433, 1888. (X2)
R8. Rose, Ernest George; "A Planetary Phenomenon," Knowledge, 11:50, 1914. (X3)
R9. "A Rare Optical Phenomenon," Scientific American, 111:55, 1914. (X1)
R10. Carpenter, A.; "The Green Flash," British Astronomical Association, Jour-

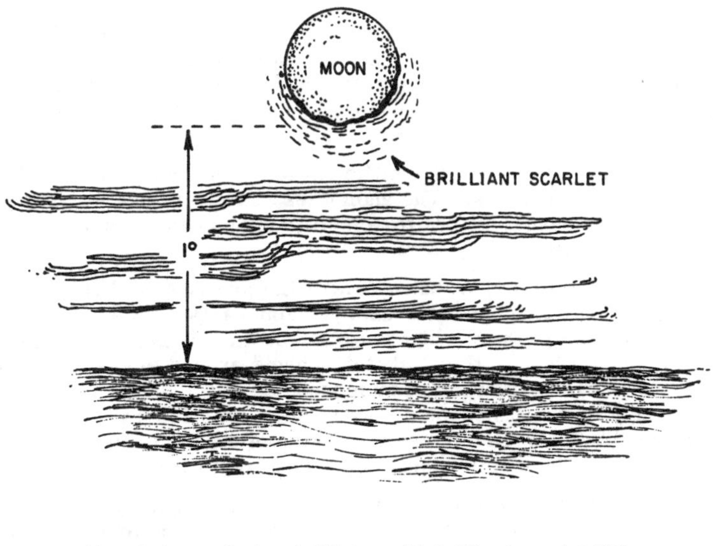

Moon's lower limb scintillates with brilliant scarlet (X7)

nal, 24:216, 1914. (X4)
R11. Hollis, H. P.; "The Green Flash," English Mechanic, 101:380, 1915. (X3)
R12. Toomer, E. Sloper; "A Lunar Phenomenon," English Mechanics, 122:311, 1925. (X1)
R13. "Ship's Officer Saw Two Setting Moons," Science News Letter, 33:62, 1938. (X1)
R14. Minnaert, M.; "Unexplained Contrast-phenomena," The Nature of Light and Color in the Open Air, New York, 1954, p. 138. (X6)
R15. Atkinson, W. S.; "Double Sun," Marine Observer, 25:31, 1955. (X1)
R16. Fraser, N.; "Abnormal Refraction," Marine Observer, 27:83, 1957. (X5)
R17. Sinclair, I. M.; "Setting of the Planet Venus," Marine Observer, 28:194, 1958. (X3)
R18. Townshend, C. R.; "Unusual Refraction of a Crescent Moon," Marine Observer, 29:178, 1959.
R19. Brittain, D. B.; "Scintillation of the Moon," Marine Observer, 30:135, 1960. (X7)
R20. Gray, J. H.; "Scintillation," Marine Observer, 32:122, 1962. (X3, X6)
R21. Johnston, L. H.; "Abnormal Refraction," Marine Observer, 32:182, 1962. (X0)
R22. Vizer, V. H.; "Abnormal Refraction," Marine Observer, 33:124, 1963. (X3)
R23. Borthwick, C. A.; "Venus---Abnormal Refraction," Marine Observer, 36:183, 1966. (X3)
R24. McLeod, A.; "Moon---Abnormal Refraction," Marine Observer, 38:71 1968. (X1)
R25. Harrison, P. P.; "Double Moon," Marine Observer, 25:34, 1955. (X1)
R26. Green, V. W.; "Double Moon," Marine Observer, 26:9, 1956. (X1)
R27. Lloyd, G. W. B.; "Abnormal Refraction," Marine Observer, 29:177, 1959. (X1)

GEL5 Anomalous Aspects of the Krakatoa Sunsets

Descriptions. Brilliantly colored sunsets prolonged well after normal twilight displays. The so-called Krakatoa sunsets officially began in late August 1883 and continued into 1885. The forms and intensities of the displays varied from place to place and time to time. A rather typical sequence of events is detailed in X0. The focus in this Catalog, however, is on hard-to-explain features, such as intermittence and recurrence well after 1885. Of even stronger interest are the apparent displays of the fully developed phenomenon in Africa several months prior to the actual eruption of Krakatoa.

Background. Spectacular sunsets have frequently followed major volcanic eruptions. Those that appeared around the world after the Krakatoa cataclysm on August 27, 1883 were so remarkable that some special scientific response seemed required. The Royal Society's Krakatoa Committee investigated the phenomenon and concluded, as most scientists suspected, that the unusual optical effects were caused by dust from Krakatoa circulated around the globe at high altitudes.

Data Evaluation. The newspapers and scientific literature of 1884-1885 are full of accounts of the remarkable Krakatoa sunsets. The anomalies, however, are not as well documented, being primarily in almost casual letters to scientific publications rather than serious descriptions and analysis. Rating: 2.

Anomaly Evaluation. Some of the Krakatoa data---the pre-August displays and very late recurrences---suggest that some of the sunset phenomena associated with Krakatoa may in fact have derived from either other volcanos or astronomical phenomena, as mentioned below. The irregular occurrence of the Krakatoa sunsets around the earth and subsequent intermittence even during the peak of the phenomenon seem to indicate that the Krakatoa dust was dispersed unevenly by the atmospheric processes. Our general ignorance of this process of dispersal constitutes a mild anomaly. On the other hand, an astronomical origin of some of the displays would be highly anomalous. Composite rating: 2.

Possible Explanations. Non-Krakatoa origins, as suggested at the time, might include unnoticed volcanos elsewhere in the world and extraterrestrial streams of meteoric dust that the earth penetrated in the 1883-1885 period and occasionally thereafter. The strange intermittence of the phenomenon could be ascribed to the segregation of the volcanic dust into long lanes (windrows) and other patterns.

Similar and Related Phenomena. Dry fogs (GWD4); dark days (GWD1); diverging rays at sunset and sunrise (GEL2); the banded sky phenomenon (GLA23); the Tunguska sky glows (GEL14).

Examples of the Anomalous Aspects of the Krakatoa Sunsets

X0. General observations. Hawaii. "These brilliant phenomena first began to be observed on the 28th day of August, 1883. They have continued with varying but diminishing intensity for more than two years. They first appeared in great splendor along an equatorial belt of 18,000 miles or more. They gradually extended with reduced brilliancy to the temperate zones, exciting the wonder of Europe and the United States in November, 1883. The most conspicuous of these phenomena take place during one hour or more before sunrise and after sunset. They may be considered as a great intensifying and prolongation of common twilight sky reflections, in consequence of a recent introduction into the higher regions of the atmosphere of some kind of finely divided matter which powerfully reflects the sun's rays, especially the red. The usual order of changes is as follows: Clouds not obscuring the view, the horizon where the sun has set is occupied by a bright silvery luster. Above this to a height of 30° or 40° a yellowish haze fills the Western sky. Although seemingly opaque and dense, the presence in it of Venus or the crescent moon shows it to be entirely transparent. This haze rapidly changes color and extent, ranging through greenish yellow and olive to orange and deep scarlet. As the dusk advances, orange and olive tints flush out on all sides of the sky, especially in the east. The chief body of color gathers and deepens over the sunset, rapidly developing the red. In from 20 to 30 minutes after sunset, deep scarlet has overpowered all other hues, flaming along 60° of horizon and 10° of altitude. This rapidly sinks and intensifies. There is a dark interval above the red. The stars begin to appear. While yet the color flames low, above the dark space appears a repetition of the orange and olive hues. Seen against the night sky, these secondary reflections or after-glows are seemingly more brilliant than the primary ones. Again the colors change and deepen into red, and after the stars are all out, and the earlier flame has sunk below the horizon, and far later than any common twilight, a vast blood-red sheet covers the west. It has been seen rising as high as 20°. As it sinks and rests low on the horizon, in the dark night sky, it precisely simulates the appearance of a remote and immense conflagration, for which it has in many places been mistaken. I have known our usual 30 minutes of twilight to be prolonged to 90, before the last glow disappeared." (R11)

Historical precedents. 1831. Malta. Red sunsets similar to those observed after Krakatoa appeared in the Mediterranean. Associated with the appearance of Graham's Island. (R4) 1831 was a year of "dry fog" and nights so bright that the smallest print could be read at midnight. See GWD4. (WRC)

1783. Europe. A year noted for Krakatoa-like sunsets, smoky fog, peculiar haze, alarming thunderstorms, and other unusual meteorological phenomena. Associated with the eruption of Skaptar, on Iceland. (R4) Once again see GWD4. (WRC)

Recent occurrences of Krakatoa-like sunsets. 1963. Africa, especially the southern tip. Attributed to volcanic activity on Bali. (R22)

1980. England. Spectacular sunsets consisting of a series of bands blending into one another. Blamed on the eruption of Mt. St. Helens in the United States. (R23)

X1. The appearance of Krakatoa-like sunsets several months prior to the actual eruption.

February 1883. Natal, South Africa. A letter from E. Neison, Director of the Government Observatory at Natal. "In England, you seem all busy over discussing the extraordinary sunsets. They began in Natal in February, 1883, but on a less grand scale, but gradually became more marked until June. Then for two months nothing was noticed. In the latter end of August they became most vivid. On the 21st and 22nd they were noticeable, but not vivid. The next five days were stormy, with much rain and lightning. On the 28th and 29th the sunsets were most vivid..... Now for a remarkable point. In the Transvaal they were first noticed in the beginning of September---the 2nd, I think---and were most vivid until the end of January, though, here, only some 250 miles off, nothing was seen." (R1) Several scientists remarked on the report from South Africa and wondered if Krakatoa could properly be blamed for the remarkable sunsets. (R8, R10, R15) It is possible, of course, that some other volcano erupted in a remote area prior to Krakatoa's cataclysm. (WRC)

X2. Prismatic colors during the Krakatoa sunsets.

1884. St. Andrews, New Brunswick, Canada. "This evening about 4 p.m., walking along the East Cliffs towards the town, I witnessed a remarkable and somewhat unusual sunset. The western sky above the sun (which was hidden, as also a great part of the western horizon by the ridges of Kinkell Braes), seemed to be covered by a light grey vapour---almost white, which shaded into the clear sky above without perceptible edge or break, as in ordinary clouds. Through rifts in this vapour, were clearly discerned bands of prismatic colours, as distinct as in the rainbow, but perhaps softer in tint, also the bands were broader. From the shape of the visible bands of colour it seemed as if they radiated from the setting sun in the same way as the fan-shaped pencils of light often seen on grey clouds in the west. They were all about the same height---rather less, I should think, than 45° above the horizon. " (R2) From this description it is difficult to assign a name or cause to the prismatic bands. This is the only example of spectral colors associated with the Krakatoa sunsets. (WRC)

X3. Unusual geometries of the Krakatoa sunsets.

November 17, 1884. Cyprus. "At 5.30 p.m. (local mean time) this evening, my attention was drawn to the abnormal appearance of the sea-horizon. From the south to the west a lurid red glow extended at intervals along the horizon for the complete quadrant. In three places the glow was similar to that caused by large conflagrations viewed from a distance of some miles. At the margins of these places what seemed to be minute tongues of red flame flickered above the sea-line. " (R3) The flickering might be caused by refraction close to the horizon. (WRC)

April 7, 1885. The Transvaal, South Africa. "On the 7th and 8th inst the light was very strong, and huge ribs or streams stretched right across the sky to the eastern horizon, terminating there over a sort of dark circle. These streamers are seen after the arc of light in the east has faded. The first, or rather red glow, lasts for about 25 minutes, when, as before, it settles on the horizon in a deep orange band; the sky now again brightens up, and a third glow goes through the same stages, lasting for another 30 minutes. On the 7th the sky seemed clear and the stars bright; but, on using the 6 in. on Jupiter, the planet presented, most of the time, just a luminous blurr (sic), with spurious images flapping on its sides, the same as I have before noted of Venus, and which seemed to me to be caused by the glow-producing medium. " (R9) South Africa is noted for its fine displays of sky-spanning crepuscular rays (GEL2), but this is the only report linking them to the Krakatoa phenomenon. (WRC)

X4. The intermittence of the Krakatoa sunsets.

General observations. The Transvaal, South Africa. "As before mentioned, I have since frequently observed the variability or irregularity of glow; one evening it would be very brilliant and full-coloured, while the following only a very dull display is seen. This peculiarity is very notable with the third glow. Trying to account for this, a thought occurred to me that maybe the dust or glow-producing dust travels about in partly detached cloud matter, driven before the air currents of the upper regions. This seems supported by the fact that whenever the glow is very bright, and that it fills that portion of the sky surrounding Venus and Jupiter, these two planets, especially the former, do hardly present a sharp outline, but are much blurred. " (R6)

June 1884. Glasgow, Scotland. The Krakatoa afterglow has displayed a strangely intermittent character. It would be absent for weeks and then return with brilliant displays. (R5)

General observations. "Allowing for cloudiness, on certain days, it will be seen that before September 8th (1883) a belt of the earth's surface 15° on either side of the equator was suddenly visited by the phenomenon. At first sight it might appear that there was a regular progression from the Indian Ocean westward, but on that supposition it would be hardly possible to explain why Venezuela should have seen the glow six days earlier than Ceylon, though somewhat farther north. The phenomenon might have been seen at Hawaii or even at Ceylon the last of August had it not been for cloudiness; and granting this, it is plain that appearances might have well nigh been instantaneous over the regions near the equator. The glow was marked at Yuma on October 19, 20 and 21, after which it ceased for a month. It was first seen in the eastern United States on October 30th, when the appearance was very brilliant; the same sight presented itself the next night; but after that it did not appear as bright, though carefully looked for, until November 27th. On this night the spectacle in the south-west was grand, and acknowledged by all as the finest even to the present time. The fire engines at Poughkeepsie, N.Y., and at New Haven, Conn., were summoned to 'quench the burning skies.' On the succeeding night the scene was nearly the same. Since November 28th the phenomenon has continued more or less brilliant, and with a few complete absences it has appeared down to the present (Feb. 3rd, 1884.) A very remarkable fact is to be noted in connection with the display of November 26th and 27th, and that is, the sudden brightening over an immense region extending over half of Europe, over nearly the whole of the United States and British Columbia, though it had not been specially noted for about a month previous." (R7)

General observation. The intermittence of the phenomenon makes it likely that the cause lies outside the earth's atmosphere. Perhaps the glows are due to a stream of meteoric matter than the earth regularly encounters as it orbits the sun. (R8) Other observers claimed a cyclic behavior of the Krakatoa sunsets and also associated them with the earth's orbital motion. (R16)

X5. Supposed recurrences of the Krakatoa sunsets at more distant times.

Summer-Fall 1902. England. More brilliant sunsets and a reappearance of Bishop's Ring. (R17)

November 21, 1903. Austwick Hall, England. "...one of the grandest displays of this curious phenomenon since its first appearance about 1884." (R18)

Throughout 1903. United States. More Krakatoa sunsets and Bishop's Ring is seen once more. (R19)

November 18, 1912. England. "Nothing has been seen like it since the Krakatoa glows." (R20)

Autumn 1917. Bright sunsets may be due to the smoke and dust injected into the atmosphere by the war. (R21)

References

R1. Noble, William; "The Recent Extraordinary Sunrises and Sunsets," Knowledge, 5:418, 1884. (X1)

R2. Werner, A.; "Curious Sunset," Knowledge, 6:532, 1884. (X2)

R3. "A Strange Form of the Afterglow," Knowledge, 6:490, 1884. (X3)

R4. Larrabee, W. H.; "Green Suns and Red Sunsets," Popular Science Monthly, 24:598, 1884. (X0)

R5. Gemmill, S. Maitland Baird; "After-Glow," English Mechanic, 39:369, 1884. (X4)

R6. Ballot, John; "The Afterglow in South Africa," English Mechanic, 39:510, 1884. (X4)

R7. Hazen, Henry A.; "The Sun Glows," Symons's Monthly Meteorological Magazine, 19:36, 1884. (X4)

R8. Abbe, Cleveland; Smithsonian Institution Annual Report, 1884, p. 381. (X1, X4)

R9. Ballot, John; "The Afterglow," English Mechanic, 41:284, 1885. (X3, X4)

R10. "The Fore- and After-Glow," English Mechanic, 43:29, 1886. (X1)

R11. Bishop, Sereno E.; "The Origin of the Red Glows," Scientific American Supplement, 21:8595, 1886. (X0)

R12. Rous-Marten, Charles; "The After-glow," English Mechanic, 47:303, 1888. (X5)

R13. Clapham, T.R.; "Green Ray, &c.," English Mechanic, 54:40, 1891. (X5)

R14. Johnson, S.J.; "Sun-Glows," British Astronomical Association, Journal, 1:317, 1891. (X5)

R15. "The Afterglow," English Mechanic, 58:36, 1893. (X1)

R16. "The 'Green Flash' at Sunset," Eng-

lish Mechanic, 73:450, 1901. (X5)
R17. Backhouse, T. W.; "Volcanic Dust Phenomena," Nature, 67:174, 1902. (X5)
R18. Clapham, T. R.; "Solar Glow," English Mechanic, 78:356, 1903. (X5)
R19. Clayton, Henry Helm; "A Second Bishop's Ring Around the Sun....," Science, 17:150, 1903. (X5)
R20. Nelson, Edward M.; "Remarkable Sunset," English Mechanic, 96:372, 1912. (X5)
R21. Hunter, A. F.; "Sunset Phenomena and the War," English Mechanic, 105:119, 1917. (X5)
R22. Keen, Gustav A.; "Spectacular Sequence in the South," New Scientist, 18:740, 1963. (X0)
R23. "Red Sky at Night," New Scientist, 88: 318, 1980. (X0)

GEL6 The Alpine Glow or Alpengluhn

Description. A beautiful second kindling of the rosy glow on high Alpine peaks some 15 minutes after the sun has sunk below the horizon. The first glow is, of course, the consequence of direct illumination by the setting sun, but a second (occasionally a third) glow seems to be the result of reflected light from somewhere. Some observers believe the Alpine Glow is synchronous with the so-called purple light (GEL11). The Alpine Glow seems to be enhanced by massive eruptions of volcanos. Strangely, similar glows are rarely reported from other high mountain ranges.

Data Evaluation. The Alpine Glow is a well-known phenomenon, but it is met with only infrequently in the scientific literature. No studies-in-depth have yet been uncovered in the English-language journals. Nevertheless, it is a permanent fixture of Alpine lore. Rating: 1.

Anomaly Evaluation. Although the Alpine Glow is commonly thought to be merely the result of light scattered down on the peaks by dust layers in the upper air, even the most recent investigations conclude that the effect is not well-understood. Just where is this vaunted dust layer? Why does the effect not occur in the Rockies, the Andes, and other high ranges as it does in the Alps? Rating: 2.

Possible Explanations. As indicated above, the conventional explanation states that the Alpine Glow originates in light reflected on the peaks by a layer of dust, or perhaps ice particles, in the upper atmosphere.

Similar and Related Phenomena. The Krakatoa sunsets (GEL5); the purple light (GEL11); mountain-top glows (GLD1).

Examples of the Alpine Glow

X1. General observations. An early report. "When the sun has set at Geneva, it is observed that Mont Blanc remains illuminated by its direct rays for a much longer time than the surrounding mountains. This phenomenon is owing to the great height of Mont Blanc. But, after it has ceased to be illuminated, the summit of Mont Blanc sometimes reappears at the end of ten or fifteen minutes, less intensely enlightened than at first, but nevertheless in a manner very decided, and often very brilliant. This phenomenon takes place especially when the atmosphere is very pure---highly charged with aqueous vapour in an invisible state---and consequently very transparent. The author has satisfied himself (by exact observation of the time which elapses between the two successive illuminations of the mountain, combined with the calculation of the sun's progress) that the phenomenon is due to the rays of the sun which traverse the atmosphere at a distance from the earth less than the height of Mont Blanc, but greater than half that height, and which arrive at rarer regions of the atmosphere, under an incidence so great that they are reflected instead of refracted. This interior reflection is facilitated by the humidity of that part of the atmosphere which the rays traverse until they reach the point of incidence. The reflected rays falling on the snowy summit of Mont Blanc, produce this second illumination; and the humidity (by augmenting the transparency of the air) renders the illumination more brilliant." (R1, R2)

A more poetic general observation. "And now, while evening is drawing on apace, the ruddy warmth that suffused the Alpine realms is no longer seen; each mountain outline grows less and less distinct, and the whole range is rapidly disappearing. Another minute, and night, that has already claimed the valleys as her own, will assert her dominion over even the towering monarchs of the land. But no. The wondrous effects of the second illumination descend upon the ice-world above; subdued yet still glowing hues tint once more the snowy summits, and the western light, with unwonted potency, throws from the mountains a shadow, soft yet distinct, upon the undulating snow-field beyond." (R3)

General observations of Alpine sunsets. "In fine weather the following sequence of events may be noted: When the sun has sunk nearly to the observer's horizon, the peaks to the eastward begin to have a reddish or golden hue. This fades gradually, but in a few minutes, when the sun is a little below the observer's horizon, but the peaks are still bathed in direct sunlight, an intense red glow begins down the slopes and moves upward to the summits. This is identical with the rosy 'twilight arch' that in clear weather rises from the eastern horizon as the sun sinks below the western; and it is bordered below by the blue shadow of the earth. Now, for a few minutes, the peaks are in the earth-shadow; their rocks and snows assume a livid appearance, aptly described by the inhabitants of the vale of Chamounix, whence the phenomena in question are well seen on the summits of Mont Blanc, as the 'teinte cadavéreuse.' Presently occurs the remarkable phenomenon known as 'recoloration' or 'afterglow.' In Chamounix it is called 'the resurrection of Mont Blanc.' The peaks, from which all color had faded, again assume a rosy tiny; but this time gradually, and without any sharp line of demarcation between the glow and the shadow beneath. The recoloration is by no means a daily occurrence---in fact, it rather uncommon---and it varies greatly in appearance and duration. Sometimes it lasts until an hour after sunset, and it passes away from below upward. Very rarely a faint second recoloration may be seen. All these phenomena may be seen, in reverse order, at sunrise, though they are less common then than at sunset. They are known, collectively, as the 'Alpenglow.' The recoloration of the Alps has been variously explained, but there seems to be little doubt that it is due, at least principally, to the reflection from the peaks of the rosy glow which forms in the western sky after sunset, known as the 'purple light,' and which is sometimes intense enough to constitute a secondary luminous source. The diffuse appearance of the recoloration, as compared with the well-defined zone of rosy light that rests upon the peaks while the latter are still in direct sunlight, is explained by the broad expanse of the 'purple light' as compared with the small disk of the sun. Sometimes the sky itself (i.e., the fine dust in the upper atmosphere) reflects the 'purple light,' giving rise to a second 'purple light' for an observer situated at a suitable angle of vision. This, in turn, may be reflected by the mountains, giving the rare phenomenon of a second recoloration." (R4)

General observations. The Alpine glow has been seen in the Apennines and, rarely, in a few other places. (R5, R6)

General observations. The discovery of low-sun color phenomena on the moon, stimulated a review of the literature on the Alpine Glow. Conclusion: terrestrial twilight phenomena are not really well-understood. It seems possible that the mechanism producing the Alpine Glow on earth (reflection from dust particles) might also contribute to the observed lunar phenomena. (R7)

References

R1. De la Rive, M.; "On an Optical Phenomenon Observed at Mont Blanc," Report of the British Association, 1837, p. 10. (X1)
R2. "Atmospherical Phenomena," Franklin Institute, Journal, 25:68, 1838. (X1)
R3. "Alpine Peaks and Passes," Eclectic Magazine, 56:160, 1862. (X1)
R4. "The Alpenglow," Scientific American, 108:482, 1913. (X1)
R5. "'L'Alpengluhen' Observé dans les Apennins," Ciel et Terre, 37:176, 1921. (X1)
R6. "Twilight Phenomena," Nature, 118: 711, 1926. (X1)
R7. Rawlings, G. C. I.; "Lunar Transient Phenomena: Quasi-Alpengluhn or Twilight-Colour Theory," British Astronomical Association, Journal, 77:309, 1967. (X1)

GEL7 Spectral Dispersion near the Sea's Surface

Description. The dispersion of sunlight into its spectrum near the surface of the sea when the sun is low. Bright and dark lines may appear in the spectrum.

Data Evaluation. A single account of fair quality. Rating: 3.

Anomaly Evaluation. This is a fascinating observation for which there is no obvious explanation, despite a great variety of optical possibilities. Rating: 2.

Possible Explanations. The observers suggested that ripples on the water acted as a diffraction grating, thus dispersing the spectrum. With no knowledge of ripple separation or the dimensions of the "spectroscope", it is hard to check this possibility. Halo and rainbow phenomena seemed to be ruled out because their colors are at fixed angles from the sun, and usually not aligned with it, although a patch of a halo is not an impossibility. Spectral dispersion by impurities floating on the water (like oil on a puddle) offers another possibility.

Similar and Related Phenomena. Horizontal rainbows (GEB13); the various types of halo displays (GEH).

Examples of Spectral Dispersion near the Sea's Surface

X1. February 11, 1911. Goat Fell, England. "At five o'clock on February 11, from the summit of Goat Fell, some 1,500 ft. above sea-level, we saw below the sun, which was then about three or four degrees above the horizon, what appeared to be a portion of the Solar Spectrum, crossed by bright and dark lines. So far as we could judge, the lower end of the spectrum appeared to rest on the sea, which at the time was obscured by a band of fog. Two bright and two dark lines were prominently visible, and several others were suspected. A vertical bright ray was also visible running down the middle of the spectrum. The colour was orange-yellow, and the dispersion apparently very great. The width of the spectrum corresponded exactly with the diameter of the sun. Wisps of cloud or fog were seen drifting across the upper portion of the spectrum, and ultimately obscured it. The phenomenon disappeared at 5h. 4m., reappear-

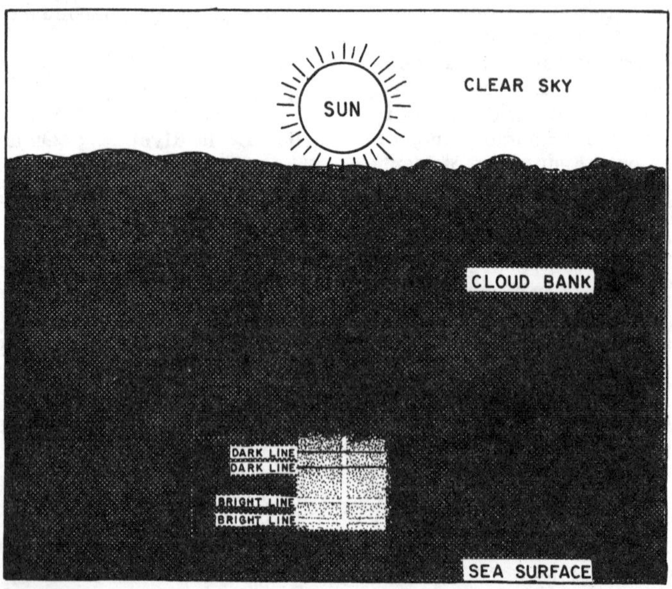

Spectral band seen on sea surface (X1)

ed for a minute or so about 5h. 8m., and then finally vanished. It was as sharply defined as if seen through a spectroscope. The sun was at the time shining brightly in a clear sky, immediately above a dense bank of cloud. Our theory is, that a rift in the cloud-bank immediately under the sun allowed a ray of sunlight to fall on the sea below, and the sea, being covered by ripples, acted as a huge diffraction grating." (R1)

References

R1. Walmesly, Cyril; "A Remarkable Phenomenon," English Mechanic, 93:84, 1911. (X1)

GEL8 Low-Sun Landscape Fluorescence

Description. The outlining of trees, bushes, birds, and other objects with silver, white, green, or possibly other colors when the sun is low. The remarkable silver (or white) fluorescence occurs primarily in mountainous districts when the objects are backlighted by the sun and observed against a deep-blue sky. Green fluorescence is more rare and seems to require a low sun but not mountains.

Background. When the word "fluorescence" is used here, it refers only to a brightening of objects, mostly around their edges, due to external light changes, various optical effects, and psychological contrast effects---not to any intrinsic brightening.

Data Evaluation. Silvery fluorescence of woods, their leaves, and birds is apparently quite common in the Alps but is seldom described scientifically. The few papers that do exist dwell mostly on the rather beautiful subjective effects rather than positions of light sources, angles involved, etc. The same situation applies to green fluorescence of the landscape, which has barely been touched upon the the literature studied so far. Rating: 2.

Anomaly Evaluation. In the cases of silvery fluorescence reported below, only the edges of the objects seem to be involved and the observer is shaded from the direct rays of the sun. The silver-edged objects are in full sunlight and perhaps also illuminated by reflections from snow. The phenomenon of diffraction could well make objects seem to have silver or white edges when they are viewed against the deep-blue sky. Green fluorescence, on the other hand, may be only a heightening of the green hues in the landscape due to the fact that the other wavelengths in sunlight are reduced in intensity due to absorption, scattering, etc. In summary, landscape fluroescence may be explained in at least two ways and does not seem particularly anomalous. Rating: 3.

Possible Explanations. See above.

Similar and Related Phenomena. Kaleidoscopic suns (GEH8); mountain-top glows (GLD1); large-scale St. Elmo's Fire (GLD3).

Examples of Low-Sun Landscape Fluorescence

X1. General observations. Switzerland. "I now come to the point which you particularly wished me to describe to you: I mean the luminous appearance of trees, shrubs and birds when seen from the foot of a mountain, a little before sunrise. The wish I had to see again the phenomenon before attempting to describe it, made me detain this letter a few days, till I had a fine day to go to see it at the Mont Saleve; so yesterday I went there and studied the fact, in elucidation of which I made a little drawing (not reproduced), of which I give you here a copy; it will, with the explanation and the annexed diagram, impart to you, I hope, the correct idea of the phenomenon. You must conceive the observer placed at the foot of a hill interposed between him and the place where the sun is rising, and thus entirely in the shade; the upper margin of the mountain is covered with woods, or detached trees and shrubs, which are projected as dark objects upon a very bright and clear sky, except at the very place where the sun is going to rise; for there all the trees and shrubs bordering the margin are entirely, branches, leaves, stems, and all, of a pure and brilliant white, appearing extremely bright and luminous, although projected on a most brilliant and luminous sky, as that part of it which surrounds the sun always is. All the minutest details, leaves, twigs, &c. are most delicately preserved, and you would fancy you saw these trees and forests made of the purest silver, will all the skill of the most expert workman. The swallows and other birds flying in those particular spots appear like sparks of the most brilliant white." The same effect is also seen at sunset. (R1, R5)

X2. General observations. Necker's observations (X1) confirmed by personal experience. The same phenomenon may be seen elsewhere to a lesser degree. Unquestionably due to diffraction. (R2)

X3. September 1881. Switzerland. A typical description of the effect. (R3)

X4. August 13, 1893. Zermatt, Switzerland. "On our right, towards the east, on the steep flanks of the mountains which enclose the valley of the Viege, rose a group of fir trees, the highest of which projected themselves against the azure of the sky, at a height of 500 m. above the road. Whilst I was botanising my son exclaimed: 'Come and look: the firs are as if covered with hoar-frost!' We paid the most scrupulous attention to the phenomenon. To make sure that we were not misled by an illusion we made various observations, both with the naked eye and with an excellent opera-glass. It was observed that not only the distant trees, but those lining the road, glittered in a silvery light, which seemed to belong to the trees themselves, and that the insects and birds playing round the branches were bathed in the same light, forming an aureole round the tops of the trees, somewhat resembling the light effects observed in the Blue Grotto. It is suggested that the light was reflected from the snow. Since it disappeared as soon as the sun rose above the hill, and has never been seen except in the presence of snow, this explanation seems plausible, but it is highly desirable that further and more detailed observations should be made of this <u>spectacle féerique</u>." (R4) Note that other observers <u>have</u> seen the phenomenon where there was no snow. (WRC)

X5. September 20, 1922. Wadhurst, England. At sunset all trees projected against the sun's disk were suddenly outlined with a hazy border of green light, although the sun retained its normal color. A green flash did occur, however, at the moment of sunset. (R6) See full account in GEL1-X5.

X6. July 8, 1924. Near Monmouth, England. "On Tuesday, July 8, when the weather was just picking up from wet into the present dry period, I was looking northwards from Bigsweir Bridge towards Monmouth, up the valley of the Wye, when after the Sun had gone down below the hilltops, one of the smaller branch valleys (at Whitebrook to be correct) was suddenly flooded with a light of intense green and of superb lustre, dazzling to the sight; it illuminated the foliage until this particular valley looked for all the world like fairyland---no other description could apply; the green rays did not sink far into the valley as they came over the top of the background of hills, the Sun having sunk down. This glorious transformation lasted fully five minutes, and then faded away again with the usual summer setting." (R7) This rather resembles X5, but seems quite different from the silvery effect of the Alps. (WRC)

X7. September 26, 1925. Hawick, England. "There was a shower that evening between 5 and 6 p.m., and shortly after 6 a rainbow appeared in the south-east, opposite my house. We were watching the rainbow---which appeared to rest on the hill top---along with two little boys of 5 years of age. Suddenly between us and the hill there appeared a number of bright, pure, white lights, with much sparkle. They seemed as brilliant and looked as big as motor headlights, and were all moving about in a cluster; one of the boys cried out: 'Look at the aeroplanes painted silver!' The sky was dull and birds flying could not be seen distinctly from our position unless they rose above the hill top. The lights were midway up. They disappeared for a moment and then came into being again, when it was noticed that they were the reflection of light on a considerable number of black-headed seagulls all wheeling together, only the white undersides being illuminated as they wheeled in and out of a narrow belt, beyond which, on each side, they could hardly be picked up against the hill. I have said the light was reflected---and evidently from below---as there was no sunshine visible; and the light was confined to a certain space. On saying that what we saw were not aeroplanes but birds, the other little boy said: 'They have sparks inside.' The remarks of the children really gave a good description of the appearance and twinkling of the light. I thought the twinkle would be caused by the movement of the wings. If seen from a distance which made it impossible to see the birds, any spectator must have beheld it with great wonder." (R8) Although this phenomenon resembles the others in this section, it may have a completely different origin---possibly contrast effects as the birds flew in and out of a shaft of sunlight. (WRC)

X8. October 1, 1953. Alston, England. "The hills near Alston, Cumberland, were capped with low cloud and this extended over the entire sky. At 1810 the scene was that of a typically overcast day with the valley dimmed with the drabness of an autumn evening. Quite suddenly this same continuous cloud sheet glowed a coppery-red, presumably due to the fact that it was of no great thickness and above it a brilliant red sunset was taking place. However, the extraordinary consequence of this reddening of the sky was to be seen in the landscape, for the fading greens of the trees and meadows had been transformed to a brilliance which might be described as 'fluorescent'. So brilliant were these greens that I developed a mild headache from the strain of observing. A normal street light at some distance from my home took up the same appearance as a fluorescent lamp." (R9)

References

R1. Necker, L. A.; "Observations on Some Remarkable Optical Phenomena Seen in Switzerland," Philosophical Magazine, 3:1:329, 1832. (X1)

R2. Talbot, H. F.; "Remarks upon an Optical Phenomenon, Seen in Switzerland," Philosophical Magazine, 3:2:452, 1833. (X2)

R3. "A Phenomenon," Knowledge, 2:439, 1882. (X3)

R4. Nature, 47:303, 1893. (X4)

R5. Murphy, Joseph John; "An Optical Phenomenon," Nature, 47:365, 1893. (X1)

R6. Barlow, E. W.; "The Green Flash---An Unusual Feature," Meteorological Magazine, 57:246, 1922. (X5)

R7. "Remarkable Sunset Illumination," English Mechanic, 120:24, 1924. (X6)

R8. Winning, John G.; "A Curious Optical Phenomenon," Meteorological Magazine, 60:241, 1925. (X7)

R9. Richardson, W. E.; "Optical Effect at Alston," Weather, 10:29, 1955. (X8)

GEL9 Low-Sun Spectral Bows

Description. Rainbow-like displays associated with a low sun which are neither rainbows nor halos. Most such bows are observed in the west near where the sun has set but not necessarily centered on the sun. The color sequences vary.

Background. Rainbows can usually be distinguished from low-sun bows by their angular dimensions, color sequences, and angular relationship with the observer and the sun. Halos, too, have similar distinguishing characteristics. See GEB and GEH introductory material.

Data Evaluation. The modern observations of low-sun bows are quite detailed but usually one-of-a-kind. The phenomenon is rare and complex, suggesting more than one optical effect at work. Rating: 3.

Anomaly Evaluation. Low-sun bows, especially those not centered on the sun, are most perplexing. Those displays that remain after all rainbow and halo phenomena have been eliminated have no obvious origin. Rating: 2.

Possible Explanations. One avenue to take in looking for an explanation involves selective scattering by different layers of air.

Similar and Related Phenomena. Offset rainbows (GEB3 and GEB6); the pink bows seen at sun rise and sunset (GEL9-X0); offset halos and anomalous arcs (GEH1).

Examples of Low-Sun Spectral Bows

X0. General observations of the normal pink twilight bow and dawn bow. "The bow in the morning begins to be defined in the west about half an hour before sunrise. The height of the arch is about 15° above the horizon and spans nearly, perhaps quite, 180°. Its first aspect is that of a blue belt, the red next appears like a faint blush above the blue, producing the purple as it mingles with the blue. Then appears the yellow above the red, producing the orange as it mingles with the red. As the sun advances towards his rising, the bow descends to coincide with the horizon, and at an angle of 8° above the horizon, or about 15 minutes before sunrise, the colors are most distinct and concentrated. At sunrise, the bow coincides with the horizon. In the appearance of the evening bow, the whole process of appearance and disappearance is in a reverse order: the bow is in the east; it rises at sunset, and disappears in about half an hour." (R1)

1884. Enhancement of the twilight bows during the Krakatoa sunsets. "There is one striking phenomenon of the remarkable sunrises and sunsets which I have been very much surprised not to see mentioned, and yet it is a phenomenon I never saw before they began. I allude to a great pink semicircle opposite the sun, which is at its best about twenty minutes before sunrise and after sunset. It does not last many minutes at each time, which may be the reason it has not attracted so much notice as the wide pink ring round the sun or moon, which is much less conspicuous, and doubtless arises from the same cause, for it, also, I never saw before. It has been more or less visible each morning that the sky has been clear enough, ever since the strange sunsets began. I have not seen it so often in the evening. I first noticed it on the morning of November 27th. It consisted of the upper half of a very broad bright ring, whose outer radius is perhaps 25°, and inner about half that; it is more or less pink, sometimes a deep pink, but sometimes the inner half is white. Its edges are indefinite. The space within is darker, and always more or less blue. The centre seems just opposite the sun." (R3) These twilight and sunrise arcs are on the periphery of the earth's shadow (GEL3).

X1. May 2, 1869. No place given. Shortly after 6:00 P.M. "The disc of the sun at the hour named was covered by thin clouds, but not sufficiently to obscure it, and, what I think is very singular, spanned at a few degrees distant by a faint bow, composed of the prismatic colours." (R2) This is manifestly a poor example, for we cannot decide whether this is a corona, a small-radius halo, or an example of the phenomenon at issue. (WRC)

X2. July 11, 1920. Folkestone, England. "On July 11th, shortly after sunset (20h. 13m to 20h. 35m. G.M.T.), a remarkable colouration of the sky occurred. The clouds at the time, which covered about nine-tenths of the sky, were all Fr. St. Cu., and they did not change materially for two hours. Just as the green light---the greenish sky often seen at sunset, but more brilliant---was disappearing, and image of the sun, yellow surrounded by red, appeared on the clouds in the eastern horizon. The cliffs of France then drew attention by changing from white to pink and to a brilliant luminous carmine. The sun image faded and the colouration of the clouds became a brilliant pink, and about 50° above the horizon, and in the same azimuth as the image which had gone, and there appeared a portion of an arc of rainbow colours." (R4) The spectral arc here could be a true rainbow formed by the sun below the horizon. The "image" of the sun might have been a countersun. (WRC)

X3. April 21, 1954. North Pacific Ocean. "The sun set at 0900, and at 0930 a beautiful rainbow effect was seen the the W, although the arc did not appear to have the sun as centre. It consisted of a greenish patch surrounded by three red quarter arcs enclosing two spectra. These arcs lasted for about 3 min." (R5)

X4. December 31, 1954. Indian Ocean. "Shortly after sunset a very noticeable set of violet rays were seen, which reached an altitude of 47° and extended over a considerable arc. Immediately above the horizon the sky was bright orange, obscured in parts by Sc and Cu. Above this was an arc of approximately 5° of greenish-white, topped by the violet rays, which became most noticeable 15 min after sunset; they retained maximum brilliance for a further 10 min and then commenced to fade. Within another 10 min the rays were invisible." (R6)

X5. No date given. Europe. "Returning from holiday in Tangier I was flying to London on an evening flight. The course was due north, and I experienced the pleasure of seeing the sunset from a high and distant vantage point. Whilst the plane itself

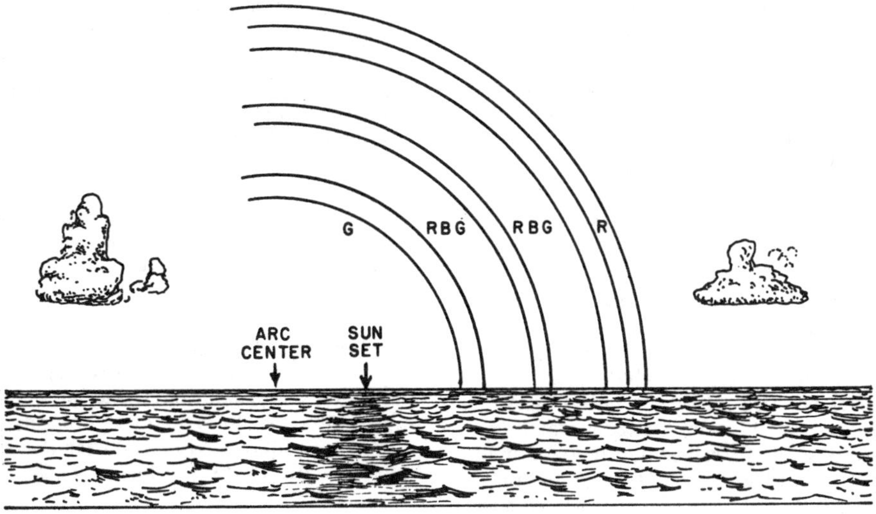

Anomalous bows were observed offset from sun's setting point. (X3)

was flying in darkness, and eastwards all was black and devoid of colour, the horizon to the west was immediately visible, being clearly defined by a spectrum. The red section met the silhouetted Earth, and the spectrum continued sky wards through orange, yellow and green to blue, the most prominent colour in the visible area. It deepened gradually to indigo then violet until, looking at about 20° to one's original plane of vision, the colour merged to black. The effect produced was that of a rainbow projected through darkness, and as the flight continued northwards the width of this coloured strip decreased until it was eventually no longer visible." (R7) One letter

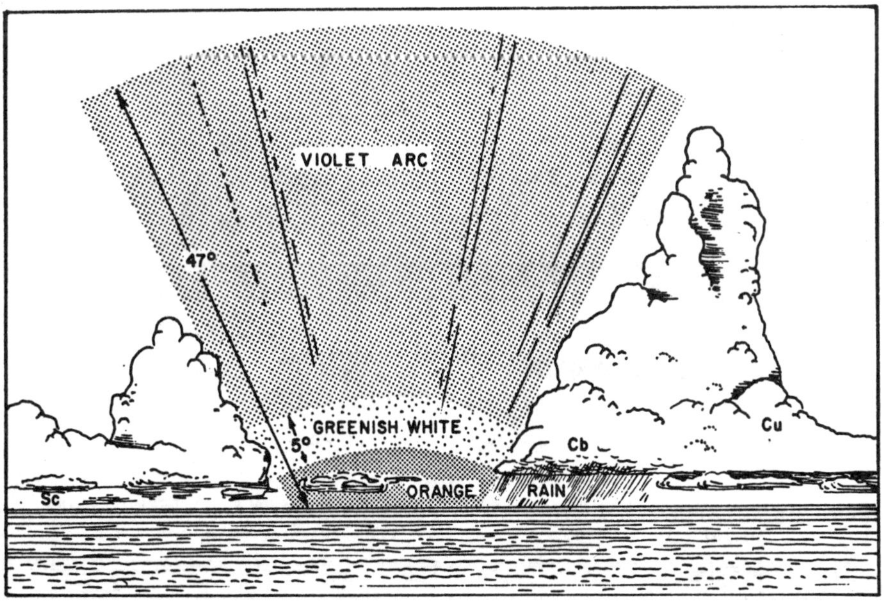

Peculiar arcs of color observed over the Indian Ocean (X4)

in response to this description thought it due to dispersion of the spectrum by atmospheric refraction. (R8) A second writer ridiculed this suggestion, insisting that the spectral arc was due to the multiple scattering of sunlight in the atmosphere. (R9)

X6. July 6, 1964. Mediterranean Sea. "At 1915 GMT approx., but definitely after sunset, the rings shown in the sketches appeared on the western horizon in the direction where the sun had set. They were distinct and the colour differentiation was well defined, but the shadow to the right was less so, though plainly seen. Within a short space of time the whole complex expanded, became lighter in colour and began to merge. With no apparent warning a yellow disc appeared superimposed upon the very light shadow. It is stressed that the disc appeared and did not rise from the horizon. It rapidly diffused, increasing in size and becoming pale in colour until it finally merged with the now rapidly dispersing blue green shadow. All traces disappeared by 1929. This disc was estimated to be about 32' in diameter and was seen at an altitude of 10°-15°. Sky cloudless." (R10)

X7. September 26, 1946. Gulf of Aden, Indian Ocean. "1454 G.M.T. (Sunset 1437 G.M.T.) A cobalt bow, 1° in width, appeared across the sky, emanating a little above the horizon on a bearing of 270°, and extending to an altitude of 68°, curved to the eastern horizon on a bearing of 90°. It appeared visible for about 30 minutes slowly losing intensity with the fading light, but gaining richly in colour against the 'earth shadow' to the eastward. The bow was transparent: α Lyrae was plainly visible and unimpaired in colour at an altitude of 65° in the centre of the bow.... Note.--- This is a very interesting observation and the cause of the bow is not evident. It may have been the shadow of something below the horizon, a cumulus cloud or a hill on the Arabian coast, but in this case the bow should have changed its position in the sky to some extent, with the gradual change of the sun's azimuth below the horizon." (R11) This observation should be compared with GES8-X2 and, obviously, with the so-called crepuscular rays (GEL2). (WRC)

References

R1. Morse, Samuel F.B.; "The Twilight Bow," American Journal of Science, 1:38:389, 1840. (X0)

R2. "Solar Phenomenon," Astronomical Register, 7:139, 1869. (X1)

R3. Backhouse, Thos. Wm.; "Pink Semi-circle Opposite the Sun," Symons's Monthly Meteorological Magazine, 18:187, 1884. (X0)

R4. Nichols, A.E.; "A Mock Sun and Rainbow Colours after Sunset," Meteorological Magazine, 55:149, 1920. (X2)

R5. Llewellyn, S.C.; "Iridescence," Ma-

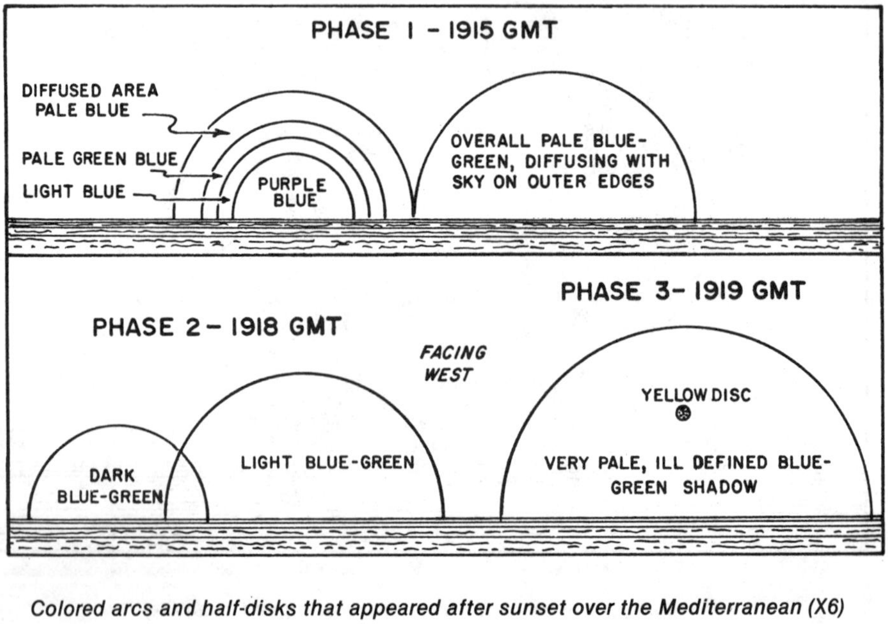

Colored arcs and half-disks that appeared after sunset over the Mediterranean (X6)

rine Observer, 25:103, 1955. (X3)
R6. McNeil, H.W.; "Sunset Rays," Marine Observer, 25:216, 1955. (X4)
R7. Hagan, David; "Astral Spectrum," New Scientist, 16:168, 1962. (X5)
R8. Bevan, Stanley; "Sunset Spectrum," New Scientist, 16:280, 1962. (X5)
R9. Dewhirst, D.W.; "Colours of the Twilight Sky," New Scientist, 16:402, 1962. (X5)
R10. Vaughan, J.; "Unidentified Phenomenon," Marine Observer, 25:123, 1965. (X6)
R11. Beck, R.B.; "Sunset Phenomenon," Marine Observer, 16:14, 1947. (X7)

GEL10 Low-Sun Shadow Bands

Description. Shadow bands observed on flat artificial surfaces and on featureless natural terrain for a few seconds just prior to sunset and after sunrise. The sun must set behind a mountain crest for the phenomenon to occur. The bands are parallel to the mountain crest, may move in either direction perpendicular to their length and may vary in speed. The phenomenon does not always occur, even under what seem to be ideal conditions, and the weather does have some undetermined effect, particularly on band speed.

Data Evaluation. The few reports uncovered are of high quality, representing careful scientific observations of the phenomenon. Rating: 1.

Anomaly Evaluation. The dimensions and spacings of the shadow bands are qualitatively consistent with the diffraction of sunlight by the mountain crest. Quantitatively, though, diffraction theory is wanting. Rating: 2.

Possible Explanations. Diffraction of sunlight by mountain crests; ripple-like structures in the atmosphere similar to those hypothesized to explain eclipse shadow bands (GES1).

Similar and Related Phenomena. Shadow bands observed during total solar eclipses (GES0).

Examples of Low-Sun Shadow Bands

X1. Circa 1906. France. A series of experiments. "Another series of interesting observations, which appear to throw light on the curious shadow band phenomena during total solar eclipses, have been recently described by Monsieur C. Rozet. In December, 1905, M. Amann saw distinct bars of light and shade on the surface of a partition, at the instance of the sun's appearance above a mountain. This led M. Rozet to make arrangements for obtaining a more definite determination, and he fitted up a white screen inside a room facing the point of sunrise. The bands were easily seen, and were generally straight, parallel, and not at all likely to be confounded with the irregular shadows produced by convection currents near the screen. As a result of 75 observations, he makes the following conclusions: (a) The orientation of the dark bands, on a screen of perpendicular rays, is constantly parallel to the part of the mountain edge over which the sun is rising or setting. (b) The direction of their displacement is always perpendicular to their orientation, but may be in one of two directions, direct or retrograde. (c) The velocity of the motion of the bands varies considerably from time to time, which may have some relation to the force of the wind, as the most rapid movements occur during high winds, and the slow ones during calm weather. (d) The bands cease to be visible two or three seconds before the sun sets, and may become visible several seconds after sunrise. If the disappearance of the sun takes place behind a vertical screen, the time of the apparition may be lengthened to 12-15 seconds. (e) At first the bands are wide and far apart, becoming sharper and straighter later. Their width usually was 3-4 cm., and distance apart 3-4 cm., but might vary from 1-20 cm. The width and distance apart appeared to vary with the velocity of translation. (f) The colour of the bands was generally of a uniform grey. The distances of the mountains over which the phenomena have been observed, have varied from 6-36 km., with elevations of

3°-22°. In spite of these variations, the features of the bands have been fairly constant." (R1)

X2. Various times. Western North America. "Observed Phenomena. At sunset when looking eastward over flat barren terrain, from a point having a higher angular elevation than the sun, faint, eastward-moving shadow bands, several miles wide, and extending an unknown, but relatively great, distance to the north and south, have been observed on several occasions. These bands were seen, in each instance, just as the sun dipped below the local horizon. Duration of the phenomena was less than 30 seconds. Estimated rate of eastward motion of the nearest shadow band was 40 miles per hour. Recession of the bands seemed to accelerate with distance. Conditions extant at time of observation are shown in Fig. 1. General similarity of these shadow bands to those occurring during a solar eclipse is notable. Field Conditions. Sunset shadow bands have been observed six times in fifteen years---once from the Pinacate Peaks, Sonora, Mexico; once from the summit of Navajo Peak (Boulder County), Colorado; twice from an airliner approaching El Paso, Texas, from the west; and twice from high points in the Dugway Mountains (Tooele County), Utah. Astronomical, topographic, and meteorological conditions were substantially identical in each instance. The angular elevation of the observation point was greater than that of the sun; there were high mountains to the west of the location of the shadow bands, which were seen on level, treeless land, with either sparse or no vegetative cover. The sky was absolutely clear (C. A. V. U.), the relative humidity very low, the temperature high, the wind speed near zero, and the lapse rate steep (nearly adiabatic). As shadow bands were seen rarely even under these conditions, it is believed that other critical conditions, still unknown, must exist. Explanatory Hypothesis. Because of their general similarity to eclipse shadow bands, sunset shadow bands are believed to be caused by diffraction of sunlight across mountain crests. Discrepancy between computed diffractive effects and observed phenomena is considerable, and led to a search for other possible causes. Rather extensive field meteorological investigations in the Salt Lake Desert failed to disclose any meteorologic cause for the shadow bands. Specifically investigated were air motion patterns, both normal wind and katabatic; regional variations in temperature and relative humidity; and the possibility of interference between thermally differing air strata. Absence of other plausible explanations, and qualitative, but not quantitative, agreement of diffraction theory with field observations, lead to the belief that sunset shadow bands are a diffraction phenomenon." (R2)

References

R1. Butler, Charles P.; "Shadow Bands at

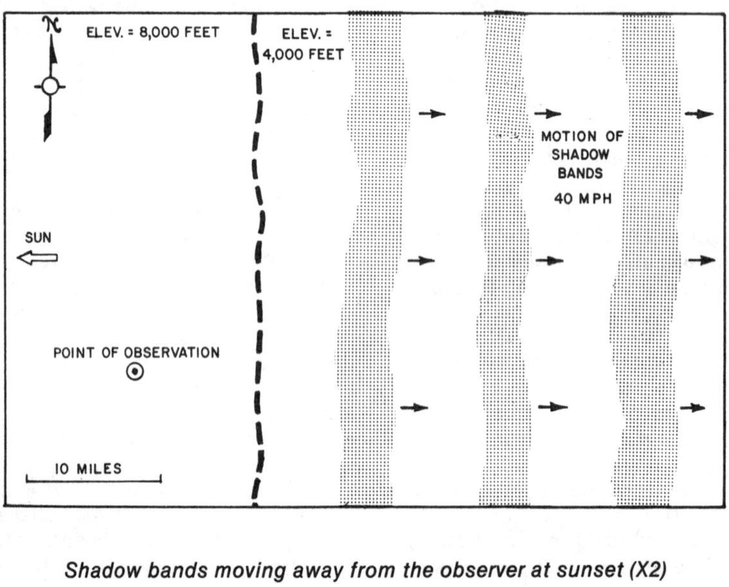

Shadow bands moving away from the observer at sunset (X2)

Sunrise and Sunset," Knowledge, 3:457, 1906. (X1)
R2. Ives, Ronald L.; "Sunset Shadow Bands," Optical Society of America, Journal, 35: 736, 1945. (X2)

GEL11 The Second Purple Light

Description. A semicircular region of purple light occasionally observed in the east 1 1/2 to 2 hours before sunrise, and in the west a similar period after sunset. The second purple light is enhanced by major volcanic eruptions.

Background. The first purple light is a well-recognized optical phenomenon just a few minutes before sunrise and after sunset. See X0 below.

Data Evaluation. Although few reports describing the second purple light have been found so far, the literature indicates that it is a rare but recognized phenomenon. It was frequently seen after the Krakatoa eruption in 1883. Rating: 1.

Anomaly Evaluation. The only scientific issue at hand here is which of several theories is correct. Rating: 3.

Possible Explanations. The first purple light is generally believed to be the reflection of sunlight from a layer of particles high in the atmosphere. The second purple light is usually considered to be a second reflection from the same layer; but modern data indicate that higher particulate layers exist, and that the second purple light may result from a single reflection from higher layers.

Similar and Related Phenomena. The Alpine Glow (GEL6), which seems to be synchronous with the appearance of the first purple light; false dawns (GLA25).

Examples of Anomalous Purple Light

X0. January 13, 1955. York, England. "The appearance of a very beautiful display of the Purple Light seen in the eastern sky from the outskirts of York may be worth recording. Shortly after 7.30 a.m. on 13 January the eastern sky was suffused by a considerable semicircular area of rosy-purple light. The display was very markedly brilliant, and a definite gap or period of semi-darkness occurred before the actual glow preceding the rising sun appeared." (R1)

X1. Photoelectric experiments. "By photoelectric observation of the twilight sky, Bigg has found evidence of scattering of the Sun's rays from an elevated atmospheric layer. Assuming primary scattering, he calculates its height to average some 81 km. and finds an association between the intensity of the effect and the incidence of meteor showers. He also observed scattering by a lower dust layer (10-25 km.) which is the cause of the purple light frequently observed above the western horizon after sunset and before sunrise. But after some great volcanic eruptions a striking feature of the brilliant sunsets has been the long-continued after-glows, with a second purple light appearing some time after the first had set and only fading away some 1 1/2-2 hr. after sunset." At first, the second purple light was thought to be due to a second scattering layer at a higher altitude. Later, though, the theory that only one scattering layer existed gained sway; and the second purple light was a consequence of a double reflection from this single layer. Bigg's data seem to support the original hypothesis of two separate scattering layers being the cause of the first and second purple light. (R2)

X2. May 19, 1910. Europe. A third purple light appeared at the time the earth was supposedly passing through the tail of Halley's comet. (R3)

References

R1. Ping, A. Wentworth; "Purple Light," Weather, 10:358, 1955. (X0)

R2. Deacon, E. L.; "The Second Purple Light," Nature, 178:688, 1956. (X1)

R3. Hogg, Helen Sawyer; "Halley's Comet and Unusual Atmospheric Phenomena," Royal Astronomical Society of Canada, Journal, 59:89, 1965. (X2)

GEL12 Moving Patches of Light on the Horizon

Description. Wedge-shaped patches of light that move along the horizon after sunset.

Data Evaluation. One sketchy observation only. Rating: 3.

Anomaly Evaluation. The data are so poor that assessment of this anomaly is difficult. Taking the observation at face value, no easy explanation (or even a hard one) come to mind. Rating: 2.

Possible Explanation. None.

Similar and Related Phenomena. Bright auroral patches on the horizon (GLA15).

Examples of Moving, Bright Patches on the Horizon

X1. October 8, 1954. Arabian Sea, Indian Ocean. "1800 S.M.T. (1445 G.M.T.) Just after the sun had disappeared below the horizon bright patches of light were seen to travel round the horizon in a clockwise direction as shown in the sketch. Each wedge was seen to sink into the horizon and start again." (R1)

References

R1. Porter, L.; "Abnormal Refraction," Marine Observer, 25:213, 1955. (X1)

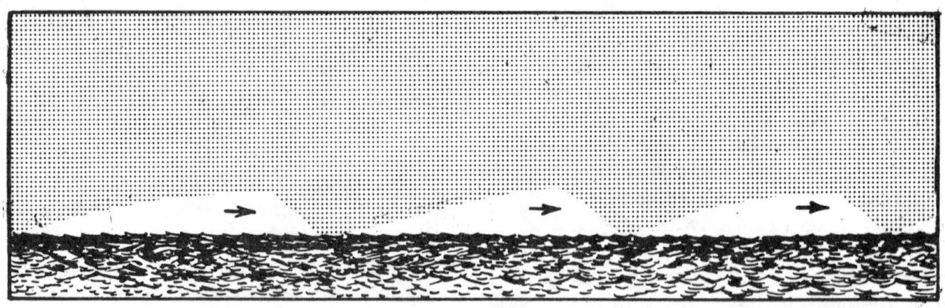

Patches of light moving around the horizon on the Arabian Sea (X1)

GEM THE MAGIC OF MIRAGES

Key to Phenomena

GEM0 Introduction

Many mirages are associated with the sea. Landfalls and other ships are seen long before they actually poke above the horizon. Inverted ships and shorelines, frequently distorted and even multiplied, are rather common. Good mirages also appear over the land. Flat plains, deserts, and ice fields are the best places to hunt for these atmospheric apparitions. The most common mirage of all is that of the sky seen in the air just above a hot road surface. But let us talk of less mundane things: great but nonexistent cities in the sky, lateral mirages, reflection mirages, and other atmospheric images that are hard to explain.

Mirages sometimes display highly magnified scenes. Islands and cities, hundreds of miles away and well below the horizon, appear larger than life in the sky. To magnify in this fashion, the atmosphere must behave like a lens---several lenses in the case of multiple mirages. Just how magnifying air lenses are formed, how they persist so long, and how they create such sharp images when the atmosphere is constantly changing; all are puzzles to the meteorologist. While some mirages are fuzzy and blurred, others show distant scenes with great clarity, just as one might see through rigid, precisely ground glass lenses. Just as in the cases of binocular and telescopes, the mirage lens often focusses its image in a very restricted area. Moving one's head up, down, or sideways a few feet may destroy the image. Lateral mirages also tax the science of meteorological optics because they require the formation of edge-up layers of air possessing different densities. Gravity should quickly destroy these unstable structures. Yet, lateral mirages are rather common, with as many as nine replications of a scene stacked up side-by-side.

One of the strangest of all mirages---one that seems almost impossible to meteorological optics----is the mirror or reflection mirage. Several observations exist where a scene is apparently reflected in the sky as if some giant mirror had been erected, with the observer located between the real object and the reflected image. These images are not like the shadowy Brocken Specters but well-defined images.

Making the problem more complex is the tendency of the human mind to embellish what the eye sees. These psychic aspects may help account for the fantastic visions of armies in the sky, unidentified cities fair, idyllic countrysides, and ghost ships plying foggy seas. Just how much does the human mind add to these scenes?

GEM1 Fata Morganas

Description. The magnification, multiplication, and gross distortion of distant rocks, ice, and other natural objects so that they appear superficially like buildings, armies, pastoral scenes, etc. The classical Fata Morgana is seen over the Strait of Messina, Italy. The notorious Silent City of Alaska, appearing regularly over Muir Glacier in Alaska, is also a Fata Morgana derived from ice formations.

Background. The term Fata Morgana derives from Morgan la Fey, the enchantress from Arthurian legends. Visions of her undersea city were believed to lure seamen to their deaths.

Data Evaluation. Fata Morganas appear in many parts of the world; but very few scientific studies have been found, especially in recent years. Like the will o' the wisp, no one doubts doubts their existence, but neither does anyone want to study them carefully. Rating: 1.

Anomaly Evaluation. A curious facet the the Fata Morgana phenomenon is the tendency of some observers to see cities, people, grazing flocks, great armies, and all sorts of fantastic images. Objective scientific accounts (X7) differ greatly from popular descriptions (X1, X2). In some respects the situation resembles that in UFO research in that bizarre things seem to be read into commonplace phenomena. If all Fata Morganas are truly only optical distortions of rocks, ice, etc., there is little that is anomalous. Rating: 3.

Possible Explanations. Much of what is seen in Fata Morganas can probably be blamed on abnormal refraction distorting natural, everyday objects. It is the human observer who creates people, animals, and other objects from the hints in the distorted images. This human trait is anomalous in itself and is dealt with in another Catalog volume (PH). Some Fata Morganas may be true representations of distant scenes, but an effort has been made to collect such possibilities in GEM2. To be complete here, it should be mentioned that some wild speculations are associated with Fata Morganas (holograms, etc.), but this is another subject (PH).

Similar and Related Phenomena. Telescopic mirages (GEM2); hallucinations (PH).

Examples of the Fata Morgana

X1. June 1833. Portbalintrea, Ireland. "Mr. M'Farland then mentioned that, in June 1833, he himself and a party of friends, when standing on a rock at Portbalintrea, perceived a small roundish island as if in the act of emerging from the deep, at a distance of a mile from the shore; at first it appeared but as a green field, afterwards it became fringed with red, yellow and blue; whilst the forms of trees, men and cattle rose upon it slowly and successively; and these continued for about a quarter of an hour, distinct in their outlines, shape and colour; the figures, too, seemed to walk across it, or wandered among the trees, the ocean bathed it around, the sun shone upon it from above; and all was fresh, fair, and beautiful, till the sward assumed a shadowy form, and its various objects, mingling into one confused whole, passed away as strangely as they came." (R1) M'Farland preceded this description with a short review of great historical mirages in the area. Many of these earlier apparitions were even more romantic and fantastic than the one he saw himself. (WRC)

X2. 1871. Firth of Forth, Scotland. "For some time past the atmospheric phenomena at the mouth of the Firth of Forth have been of a remarkably vivid and interesting character, and have attracted a great deal of attention. During the past week especially, scarcely a day has passed without exhibiting extraordinary optical illusions in connection with the surrounding scenery, both at sea and on shore. As an instance of the unusual nature of these phenomena, the whole of the Broxmouth policies, mansion-house, and plantations, were on one day apparently removed out to sea. One of the finest displays of mirage, however, occurred on Saturday afternoon. The early part of the day had been warm, and there was the usual dull, deceptive haze extending about half-way across the Firth, rendering the Fife coast invisible. The only object on the Fife coast, indeed, which was brought within the range of the refraction was Balconic Castle on the "East Neuk," which appeared half-way up the horizon, and in a line with the Isle of May. The most extraordinary illusions, however, were those presented by the May island, which, from a mere speck on the water, suddenly shot up in the form of a huge perpendicular wall, apparently 800 or 900 feet high, with a smooth and unbroken front to the sea. On the east side lay a long low range of rocks, apparently detached from the island at various points, and it was on these that the most fantastic exhibitions took place. Besides assuming the most diversified and fantastic shapes, the rocks were constantly changing their positions, now moving off, and again approaching each other. At one time, a beautiful columnar circle, the column seemingly from 20 to 30 feet high, appeared on the outermost rock. Presently the figure was changed to a clump of trees, whose green umbrageous foliage had a very vivid appearance. By and by the clump of trees increased to a large plantation, which gradually approached the main portion of the island, until within 300 or 400 feet, when the intervening space was spanned by a beautiful arch. Another and another arch was afterwards formed in the same way, the spans being nearly of the same width, while the whole length of the island, from east to west, seemed as flat and smooth as the top of a table. At a later period the phenomena, which were constantly changing, showed huge jagged rifts and ravines in the face of the high wall, through which the light came and went as they opened and shut, while trees and towers, columns and arches, sprang up and disappeared as if by magic. It is a singular fact, that during the four hours the mirage lasted, the lighthouse, usually the most prominent object from the south side of the Firth, was wholly invisible. The last appearance which the island assumed was that of a thin blue line half-way up the horizon, with the lighthouse as a small pivot in the centre; and the extraordinary phantasmagoria were brought to a close about seven o'clock by a drenching rain, which fell for two hours." (R2) The best examples of the Fata Morgana always seem to involve images of buildings, men, animals, trees, etc. (WRC)

X3. Occasional appearance. Strait of Messina. Italy. The "classic" Fata Morgana. "When the rising sun shines from that point whence its incident ray forms an angle of about 45 degrees on the sea of Reggio, and the bright surface of the water in the bay is not disturbed either by the wind or the current, the spectator being placed on an eminence of the city, with his back to the sun and his face to the sea---on a sudden he sees appear in the water, as in a catoptric theater, various multipled objects, such as numberless series of pilasters, arches, castles well delineated, regular columns, lofty towers, superb palaces with balconies and windows, extended alleys of trees, delightful plains with herds and flocks, armies of men on foot and horseback, and many other strange figures, all in their natural colors and proper action, and passing rapidly in succession along the surface of the sea, during the whole short period of time that the above-mentioned causes remain. But if, in addition to the circumstances before described, the atmosphere be highly impregnated with vapor and exhalations not dispersed by the wind nor rarefied by the sun, it then happens that in this vapor, as in a curtain extended along the channel to the height of about thirty palms and nearly down to the sea, the observer will behold the scene of the same objects not only reflected from the surface of the sea, but likewise in the air, though not in so distinct and defined a manner as in the sea. And again, if the air be slightly hazy and opaque, and at the same time dewy and adapted to form the iris, then the objects will appear only at the surface of the sea, but they will be all vividly colored or fringed with red, green, blue, and the other prismatic colors.'

So runs the classical description of the Fata Morgana, written in 1773 by the Dominican friar, Antonio Minasi, and since become the common property of encyclopaedists the world over.

Minasi was born not far from Reggio, and saw the Fata Morgana himself three times.

Development of the Fata Morgana over the Strait of Messina (X3)

His description is probably, in the main, accurate, though, as we shall presently see, he limits the time of occurrence of the phenomenon too narrowly in stating that the sun must be at an altitude of 45 degrees. He was the first writer to point out that the Fata Morgana occurs in two distinct varieties; viz., the marine Morgana, which appears to lie in or beneath the water, and the aerial Morgana, which extends upward to a considerably greater apparent altitude. Minasi's third variety, the irridescent Morgana, appears not to have been observed since his time. The atmospheric refractions to which the Fata Morgana, in common with other forms of mirage, is due are not usually attended with a sensible dispersion of light (i.e., separation of the prismatic colors), but that such dispersion may sometimes occur is by no means impossible. In fact, we shall have occasion later to mention a form of mirage, seen in another part of Italy, in which irridescent coloring is stated to be a common feature.

Minasi's attempts to explain the Fata Morgana are far less happy than his descriptions. In his day little was really known about abnormal refractions in the atmosphere, though he lived upon the eve of the important discoveries of Gruber, Beusch, Monge and Biot. The Straits of Messina are subject to strong tidal currents, which often run in opposite directions at a given time; that is to say, in midchannel the current sets to the north while along the shore it is running south, and vice versa. Minasi supposed that the surface of the water thus acquired, at times, marked differences of level, so that it behaved like a mirror lying, not horizontal, but tilted at a slight angle, or at several angles in different places, and reflecting objects along the Calabrian shore (i.e., the shore of the Italian mainland, on which Reggio lies). He therefore took the Morgana to be the reflected image of Reggio and the adjoining coast---the same coast from which the phenomenon is seen. In Minasi's picture which has been the basis of the greater number of the representations of the Fata Morgana shown in the textbooks and reference books, the city in the background is Reggio, while in the foreground are shown all three of the forms of the phenomenon that he has described. It was intended to be a generalized diagram of the Fata Morgana, rather than a faithful picture of its appearance at any one time.

Minasi's explanation is of course no longer accepted, and we now know that the terrestrial objects seen in the Morgana are the refracted images of such objects on the Sicilian coast, or in some cases on parts of the Calabrian coast remote from the place of observation. Moreover, if we except the altogether unconvincing narrative of the French traveler Jean Houel, there are no cases on record in which the Morgana has been seen from the Sicilian side of the straits." (R7, R8)

General observations. The Fata Morgana is sometimes accompanied by thunder. (R3) This item is in a collection of accounts of unusual sounds in nature, some of which border on the fantastic.(WRC)

X4. June. Yukon Goldfields, Canada. "A contributor to the Weekly Times and Echo, who has just returned from the Yukon Goldfields, in a very interesting article in the last number, incidentally gives an account of a remarkable mirage, which, it is said, always occurs in those regions towards the end of June or thereabouts. The puzzle, of course, is, what 'city' is it that is reflected? The correspondent quoted says: 'Early one morning as we were tumbling out of our cots, Joey (the Indian) came rushing in, exclaiming in his own picturesque fashion that the Manitou had shown him his celestial abode. We hastily scrambled out of the wigwam, and craned our necks to catch a glimpse of the lost Eden, or whatever it might be, and sure enough there unrolled itself to our astonished and enraptured eyes a most wonderful mirage. It looked like some immense city of the past. Domes, minarets, spires, belfries, houses, streets, everything was as complete as could be, but in vain we racked our brains to see whether we could locate the curious place. Not one of us could form the remotest idea in what part of the world this settlement could be in. Some guessed at Toronto, others Montreal, and one of us even suggested Pekin. But whether this city exists in some unknown part of the world on the other side of the North Pole or not remains to be proved. Nevertheless, it is a fact that this wonderful mirage occurs from time to time yearly, and we were not the only ones who witnessed the spectacle. Therefore, it is evident that it must be the reflection of some place built by the hands of men." (R5) One must always be suspicious of accounts written in the above style. Further, this mirage may be the same as that reported next entry. In any case it is in the same geographical region. (WRC)

X5. June and July. Muir glacier, Alaska. "We have received, through Mr. F. Napier Denison, of the Meteorological Office, Victoria, British Columbia, a copy of the Victoria Daily Times, January 26, 1901, which contains an illustration of a remarkable mirage known as 'The Silent City of Alaska.' It is said to appear every year on the gigantic glacier of Mount Fairweather. This phenomenon has engaged the attention of scientists, and up to the present time it has baffled all investigation. The scene has been known to the Alaska Indians of the locality for generations, and has been a common subject of speculation among them. The photograph was shown to some Alaska Indians by Capt. Foot of the Danube, who brought it down to Capt. Walbran, and they instantly recognised it as the famous city in the clouds. The phenomenon is seen between 7 and 9 o'clock between June 21 and July 10, and the scene never varies excepting for slight changes in the buildings and other prominent landmarks. It is believed that the mirage is a representation of the city of Bristol, England. That it is a seaport is shown by the mast of a vessel, while a tower, an exact duplicate of that of St. Mary Redcliff, appears in the background. The earthquake of last year broke up the Muir glacier over which the route to the mountain lies. It is a distance of about 15 miles from Muir glacier bay to the scene of the phenomenon. A scientific party from San Francisco will investigate it next May. The distance between Bristol and Mount Fairweather is about 2500 miles. The longest distance a mirage has been seen hitherto is 600 miles." (R6) In connection with this and the preceding item, which are both associated with mirages seen over snow and ice, it is instructive to examine three photographs in Greenler's book (R11) which show mirages appearing over the jumbled ice of the Arctic Ocean. There is superficial resemblance between the distorted ice images and city buildings. (WRC)

X6. February 14, 1936. Gulf of Mexico, North Atlantic Ocean. 1:05 to 2:10 P.M. "Observed two mirages, No. 1 bearing 018° and No. 2 bearing 294°. Ship's course 325°. The appearance of a fog bank lay on the horizon, the angular height of which was 2° 10'. The mirage to seaward, No. 1 (not reproduced), was directly opposite the sun, and slowly opened its bearing on the starboard bow, as the sun altered its bearing. This mirage was visible for 1 hour 5 minutes, it maintained the same appearance throughout, and always bore directly opposite the sun. It appeared as to headlands forming an entrance to a bay or harbour,

Fata Morgana seen in the Gulf of Mexico (X6)

and trees were clearly visible, also hills in the background. The fronts of these headlands were a sandy colour. When this mirage commenced to disappear it rose up from the horizon, and became fainter, until when it reached the top of the fog bank it disappeared instantly. The fog bank cleared away 9 minutes later.

The other mirage, No. 2, was over the land and appeared between two mountains, it was visible from 1.05 p.m. until 1.24 p.m., and appeared as a harbour, surrounded by large buildings, some of these buildings were inverted and appeared to resemble reflections of the other buildings. As this mirage disappeared, it changed its appearance, becoming narrower, until it was only two dark lines in the sky, separated by a white line. The white line was visible all the time, and separated the inverted buildings from the upright buildings." (R9) Note that the Yucatan Peninsula was some 200-300 miles east of the ship, raising the possibility that mirage No. 1 might be a telescopic mirage (GEM2). It cannot be a Fata Morgana because there is no shoreline from which to create distorted images. See also GEN9-X0. (WRC)

X7. February 1, 1966. West Hawk Lake, Manitoba, Canada. "The phenomenon was first noticed at 1:15 p.m., C.S.T., from the winter ice near the centre of West Hawk Lake, Manitoba. The writer was waiting to record the time of solar transit across the line joining the diamond-drill holes being drilled in the meteorite crater, in order to determine the azimuth of the line. While the mirage phenomena were quite evident to the unaided eye, these were followed in detail with 7 X 50 binoculars around a 50-degree arc of the northern shore of the lake, some 6000 to 8000 feet (1.8 to 2.4 km) from the observer. The mirage included common effects such as one or more bands of water apparently overlying the snow-covered surface of the lake near the distant shoreline. This was followed in time by the true Fata Morgana mirage in which strongly banded vertical striae were observed between well-defined upper and lower bounds, very similar to the central figure reproduced by Dr. Hogg (not reproduced) except that the horizon was a shoreline consisting of rocks and trees. The mirage was especially pronounced in a bay on the north-west shore which was shadowed by the high cliffs on the southern boundary of the bay. A large dark rock about 20 feet above the true shoreline, with a patch of snow near its centre, gave multiple rapidly varying images between the object and the depressed image of the shoreline. Individual trees could be observed at times in the Fata Morgana and once a pronounced horizontal wave was propagated outwards from tree to tree as if from a central blast. As the wave progressed successive tree trunks would suddenly appear to bow outwards and then remain bent for some time. The phenomena were observed for more than an hour." (R10) For once, we have a Fata Morgana without cities, armies, and phantom lands! (WRC)

X8. General observations. Based on considerable experience with Fata Morganas over European lakes. "Many times when viewing the fata morgana through a powerful glass I have observed that the objects are not really increased in size, but that the impression is produced by the superposition of several images of the same object, sometimes upright, sometimes upside down. I have counted as many as five of these images. As they are generally very close together and sometimes encroach one upon the other, it is very difficult to separate them by the naked eye, and therefore the illusion of an enlarged object is produced. Sometimes one portion alone produced many images.

Thus, I have sometimes seen boats with two hulls, while the sails looked perfectly natural; a few minutes later only one hull remained, but the sails appeared gigantic. It seems to result from these observations that the fata morgana is nothing more than a mirage of multiple images." (R4)

References

R1. M'Farland, Mr.; "On the Fata Morgana of Ireland," Report of the British Association, 1852, part 2, p. 29. (X1)
R2. "Extraordinary Mirage in the Firth of Forth," Symons's Monthly Meteorological Magazine, 6:98, 1871. (X2)
R3. Springer, Robert; "The Mysterious Sounds of Nature," Popular Science Monthly, 17:772, 1880. (X3)
R4. Delebecque, Andre; "Atmospheric Refractions at the Surface of Water," Monthly Weather Review, 24:371, 1896. (X8)
R5. "Curious Mirage in British Columbia---What City Can It Be?" English Mechanic, 66:81, 1897. (X4)
R6. "Extraordinary Mirage," Royal Meteorological Society, Quarterly Journal, 27:158, 1901. (X5)
R7. "The Fata Morgana of the Straits of Messina," Nature, 67:393, 1903. (X3)
R8. Talman, C. Fitzhugh; "The Real Fata Morgana," Scientific American, 106:335, 1912. (X3)
R9. Paterson, H.; "Mirage," Marine Observer, 14:11, 1937. (X6)
R10. Halliday, Ian; "An Observation of the 'Fata Morgana' in the Canadian Winter," Royal Astronomical Society of Canada, Journal, 61:74, 1967. (X7)
R11. Greenler, Robert; "The Fata Morgana," Rainbows, Halos, and Glories, New York, 1980, p. 165. (X5)

GEM2 Telescopic Mirages

Description. Mirages that either transmit images over very long distances (magnifying them in the process) or greatly magnify nearby objects. To be remarkable enough for inclusion in this Catalog, a mirage must operate over distances of 100 miles or more. An important characteristic of the telescopic mirage is its great sharpness and stability over many minutes.

Data Evaluation. Admittedly some of the older reports are fanciful, but several high-quality examples of telescopic mirages reside in the more recent scientific literature. Rating: 1.

Anomaly Evaluation. Stripped on embellishment, telescopic mirages transmit sharp, stable, magnified images at least several hundred miles, even for an observer on a vehicle (usually a ship). Since the physics of mirage action is well-established, there is little that is truly anomalous in telescopic mirages per se. Rather, the surprise is in the astounding stability of the atmosphere over several hundred miles of sea, ice, or prairie. Rating: 3.

Possible Explanations. As the textbooks are wont to tell us, mirages are a consequence of abnormal refraction of light in the atmosphere. It would be superfluous to repeat these elementary explanations here. The immense layers of stable air, where at least some turbulence would be expected, is a surprising situation in need of further study.

Similar and Related Phenomena. Distortion of astronomical objects near the horizon (GEL4); the green flash (GEL1); other mirages in this section (GEM).

Examples of Telescopic Mirages

X1. September 27, 1846. Liverpool, England. "The author states that during the exhibition of a panoramic model of Edinburgh, in the Zoological Gardens at Liverpool, on Sept. 27, 1846, about 3 P.M., an erect image of Edinburgh, depicted on the clouds over Liverpool, was seen by two residents in the Great Park at Birkenhead, for a period of forty minutes." (R1) This entry is questionable. The straight-line distance between Edinburgh and Liverpool is about 170 miles, so a mirage of Edinburgh at Liverpool is not impossible. It is the coincidence of the mirage and the model exhibition that raises suspicions. (WRC)

X2. January 22, 1854. Westphalia, Germany. The following description of a fantastic mirage resembles the embellished tales of some Fata Morganas. It is rather typical of the "armies in the skies" stories. Later more-scientific reports will confirm that mirages may be seen over distances of many hundreds of miles; so, in principle, the Westphalia apparition could have occurred. "If the east has its Fata Morgana, we, in Westphalia, have also quite peculiar natural phenomena, which, hitherto, it has been as impossible to explain satisfactorily, as to deny. A rare and striking appearance of this description forms now the subject of universal talk and comment in our province. On the 22nd of last month a surprising prodigy of nature was seen by many persons at Buderich, a village between Unna and Werl. Shortly after sunset, an army, of boundless extent, and consisting of infantry, cavalry, and an enormous number of waggons, was observed to proceed across the country in marching order. So distinctly seen were all these appearances, that even the flashing of the firelocks, and the colour of the cavalry uniform, which was white, could be distinguished. This whole array advanced in the direction of the wood of Schafhauser, and as the infantry entered the thicket, and the cavalry drew near, they were hid all at once, with the trees in a thick smoke. Two houses, also, in flames, were seen with the same distinctness. At sunset the whole phenomenon vanished. As respects the fact, government has taken the evidence of fifty eyewitnesses, who have deposed to a universal agreement respecting this most remarkable appearance." (R2)

X3. 1931. Marseilles, France. "Residents of Marseilles recently had the experience of actually looking 'around the corner' of the earth when they saw Mt. Canigou, 157 miles to the west, silhouetted against the setting sun. Actually, the straight line joining the peak with Marseilles passes almost 400 feet below the surface of the Mediterranean. However, the atmosphere of the earth refracts the rays of light around the curved surface. In February and October the sun sets behind the mountain, as seen from here, and the effect can be observed." (R5)

X4. May 9, 1937. South Pacific Ocean. "At 5.05 p.m. A.T.S. (0321 G.M.T. 10th) the islands of Tahiti, Moorea, Huahine Iti and Raiatea were seen. Only the peaks of Raiatea and Huahine Iti were visible, but Tahiti and Moorea were plainly recognizable. Very heavy rain had been experienced for four hours previously, ceasing at 4.30 p.m. A.T.S. The horizon cleared from east to north, and showed with unusual sharpness and clearness some time before the mirage came into being. Tahiti gave the appearance of being 35 miles distant, when in reality Tahiti Peak was 210 miles away. The sky was overcast except that it had cleared from E. to N. above the horizon. The mirage remained constant throughout, lasting for 15 minutes and disappearing at 5.20 p.m. A.T.S. (0336 G.M.T. 10th)." (R6) The magnification of Tahiti was roughly 6. (WRC)

X5. July 17, 1939. North Atlantic Ocean. "On July 17, the schooner was from its noon observation in sunshine found to be in latitude 63° 38' N and longitude 33° 42' W. The ship's three chronometers had been checked daily by the Naval Observatory signal, and the air was calm and the sea smooth. At 4 P.M. with sun in the southwest the remarkable mirage appeared in the direction of southwestern Iceland. The Snaefells Jokull (4,715 feet) and other landmarks well known to the captain and the mate were seen as though at a distance of twenty-five or thirty nautical miles, though the position of the schooner showed that these features were actually at a distance of 335 to 350 statute miles. A checking observation of the sun made at 6 P.M. gave the latitude at that time as 63° 42' N and longitude 33° 32' W. It was warm and rainy; the air had throughout been calm and the sea smooth. Captain Bartlett writes: 'If I hadn't been sure of my position and had been bound for Rejkjavik, I would have expected to arrive within a few hours. The contours of the land and the snow-covered summit of the Snaefells Jokull showed up almost unbelievably near.'" (R7)

X6. June 29, 1981. South Atlantic Ocean. "At 1700 GMT a vessel initially bearing 290°(T) at 6 n. mile. heading 025°(T) passed on our starboard side. When the bearing of the vessel changed to 320°(T) at 12.0 n. mile, the visual image had become approximately four times its normal size. When the bearing was observed to be 350° (T) at 16.0 n. mile the image had changed to approximately eight times normal size for that distance and gave the appearance of a large block of flats. On taking a final bearing of the vessel at 020°(T) at 20.0 n. mile the image had returned to what was considered normal for that distance." (R11)

X7. No date given. Charles City, Iowa. A mirage of the Mississippi bluffs, fully 100 miles away, was observed. (R4)

X8. No dates given. Near Yuma, Arizona.

Repeated mirages of San Diego. "In the case of the repeatedly seen mirages of the city of San Diego, the image is picked up at sea level, carried over a 4,000-foot-high mountain mass, and brought down again to near seal level some 150 miles away. Elevation and location of the first lens are unknown; those of the second lens may be guessed at, although not too accurately." Mirages of San Jose and San Francisco, both much farther away, have been reported from the same area, but these are of questionable reliability. (R9) The same study was reported in a digested form in Science Digest. (R10)

X9. No date given. Near Port Danger, on the South African coast. Passengers on a vessel saw a mirage of a well-known English man-of-war, which displayed great detail. Expecting to find the warship at Port Danger, they were surprised to learn that it was some 300 miles away at the time of the mirage. (R8)

X10. June. Mount Fairweather, Alaska. Some who have observed the mirage called "the Silent City of Alaska" claim that the purported city is really Bristol, England, located some 2500 miles away. (R3)

References

R1. Thomson, D.P.; "On an Extraordinary Mirage Witnessed at Birkenhead in Cheshire," Report of the British Association, 1847, part 2, p. 39. (X1)

R2. Macray, J.; "Fata Morgana," Notes and Queries, 1:9:267, 1854. (X2)

R3. "Extraordinary Mirage," Royal Meteorological Society, Quarterly Journal, 27:158, 1901. (X10)

R4. Brown, Howard C.; "The Prairie Mirage," American Midland Naturalist, 6:172, 1920. (X7)

R5. Science, 73:sup 14, May 1, 1931. (X3)

R6. MacDonald, D.; "Mirage," Marine Observer, 15:52, 1938. (X4)

R7. Hobbs, William H.; "A Remarkable Example of Polar Mirage," Science, 90:513, 1939. (X5)

R8. Liddel, Urner; "Phantasmagoria or Unusual Observations in the Atmosphere," Optical Society of America, Journal, 43:314, 1953. (X9)

R9. Gordon, James H.; "Operation Long Distance," Smithsonian Institution Annual Report, 1959, p. 327. (X8)

R10. "Some Mirages Nullify the Laws of Physics," Science Digest, 49:31, April 1961. (X8)

R11. White, A.H.; "Abnormal Refraction," Marine Observer, 52:80, 1982. (X6)

GEM3 Lateral Mirages

Description. The mirage image of a scene observed side-by-side with the actual scene instead of above or below, as is the usual situation. Multiple lateral mirages, with as many as nine images, have been observed. Lateral mirages are much rarer than vertical mirages.

Data Evaluation. Although rare, several good examples of lateral mirages are in the Sourcebook Project files. There seems no reason to doubt the reality of the phenomenon. Rating: 1.

Anomaly Evaluation. The production of lateral mirages requires an atmospheric lens standing on its edge, so to speak. This means that vertically oriented layers of different density must exist side-by-side. Such a configuration is unstable. The anomaly here, therefore, is the long-term stability of lateral mirages. Rating: 3.

Possible Explanations. Sheet-like plumes of different-density air rising from a heat transfer surface.

Similar and Related Phenomena. Doubled astronomical objects near the horizon (GEL4); mock suns resemble lateral mirages but are not due to mirage action (GEH13).

Examples of Lateral Mirages

X1. September 17, 1818. Geneva, Switzerland. "A curious case of lateral mirage was observed at Geneva on the 17th September 1818, at 10 o'clock, P.M. by MM. Jurine and Soret. A bark near Bellerive, at the distance of about 4000 toises from Geneva, was seen approaching Genevá by the left bank of the lake; and at the same time, an image of the sails was observed above the water, which, instead of following the direction of the bark, separated from it, and appeared to approach Geneva by the right bank of the lake,---the image moving from east to west, while the bark moved from north to south. When the image separated from the bark, it was of the same dimensions as the bark; but it diminished as it receded from it, so as to be reduced by one half, when the mirage ceased." (R1, R2)

X2. October 11, 1927. Wales. "At 17h. 5m. G. M. T. on October 11th, the sun had fallen behind the Snowdon range, and the range was backed by a large golden, fleecy cloud, on which was outlined a dark, well-defined shadow of Snowdon, persisting for some five or ten minutes. I find it difficult to understand how the shadow was cast in the direction of the sun, as the cloud was manifestly on the far side of the hills, the atmosphere was clear, and the eastern sky almost cloudless. At 17h. 20m. the cloud had disappeared, and the western horizon was clear except for some distant striations, when suddenly there appeared over Crib Coch a streaky golden outline of its summit, which faded out after perhaps 15 seconds or half a minute." This report by W. G. Emmett was followed by a comment by R. Corless to the effect that the phenomenon was probably an example of the rare lateral mirage, in which an object may be seen in two different directions. (R3)

X3. June 14, 1956. North Atlantic Ocean. 2145 G. M. T. "Shortly before sunset, with the sun at an altitude of 2° 35', bearing 298°, twin suns were observed side by side, quite clearly separated. The image was to the right of the true sun, there being no distortion of either. This phenomenon persisted for 2 min, after which the image gradually elongated in the direction of the true sun, eventually merging with it within 1/2 min..... Note. This is an extremely interesting observation of a lateral mirage and appears to be a unique one. For lateral mirage it is necessary for the air to be stratified into vertical masses of different densities. Ordinary abnormal refraction, which may produce an image vertically above the object, is produced by more or less horizontal distribution of layers of air of different densities. (R4) This example could just as easily have been classified with the double suns and moons of GEL4. There is also a resemblance to close, one-sided mock suns (GEH13), but the subsequent merging of the image with the true sun militates against this explanation. (WRC)

X4. No date or place given. A photograph of a nine-image lateral mirage is reported. The multiplicity of images seems to be associated with the nature of the heat transport from a vertical surface. (R5)

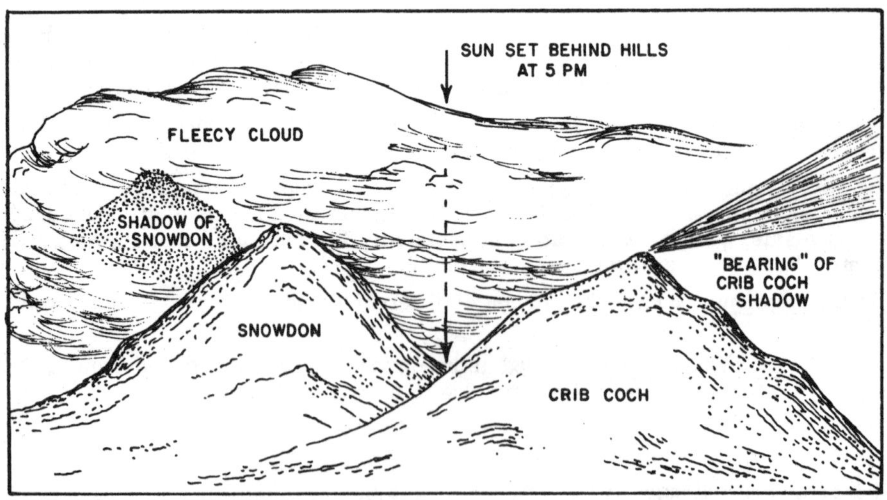

Possible lateral mirage of Snowdon mountain in Wales (X2)

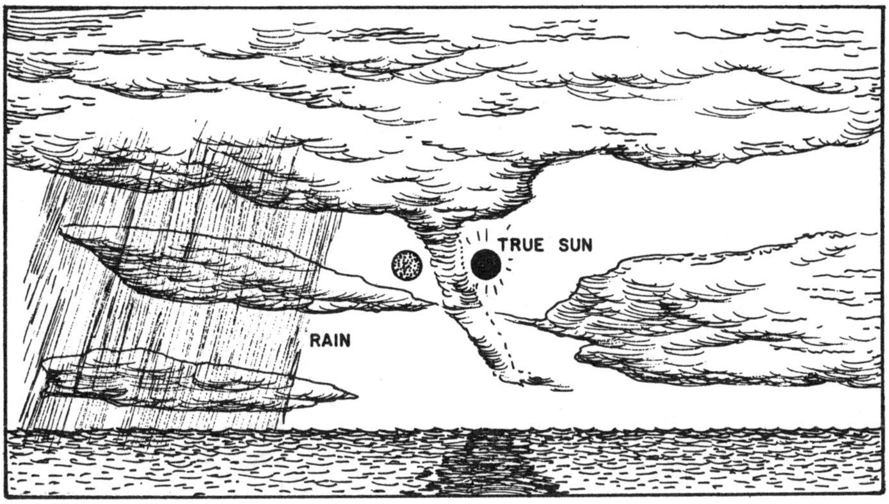

A double sun probably created by a lateral mirage (X3)

References

R1. "Case of Lateral Mirage," Edinburgh Philosophical Journal, 3:400, 1820. (X1)
R2. Brocklesby, John; "Magnified Images," Elements of Meteorology, New York, 1855, p. 185. (X1)
R3. Emmett, W.G., and Corless, R.; "A Sideways Mirage," Meteorological Magazine, 63:15, 1928. (X2)
R4. Evans, J.E.; "Abnormal Refraction," Marine Observer, 27:84, 1957. (X3)
R5. Fraser, Alistair B., et al; "Multiple-Image Lateral Mirage," Eos, 53:274, 1972. (X4)

GEM4 Multiple Mirages

Description. Two or more repetitions of the mirage image. The images may be erect or inverted, often occurring as pairs of erect and inverted images one above the other. As many as nine lateral mirage images have been photographed. The common case of an single erect-inverted pair of images of the original scene is not included here.

Background. Multiple mirages may be far more common than the scanty literature indicates. Most multiple mirages have been viewed with either telescopes or binoculars and would not have been appreciated with the naked eye. Many seemingly single mirages may really be slightly overlapping multiple mirages.

Data Evaluation. Few multiple mirages have been reported; but these are generally of good quality. Rating: 2.

Anomaly Evaluation. Multiple mirages imply the existence of several, relatively stable layers of air of different densities. That the detection of these mirages usually requires optical magnification suggests that the atmosphere possesses unappreciated fine structure. It is not clear how these many-layered sandwiches of air are created. Rating: 2.

Possible Explanations. The multiple images are the consequence of superimposed layers of air with different indexes of refraction.

Similar and Related Phenomena. Abnormal refraction near the horizon (GEL4).

Examples of Multiple Mirages

X1. April 13, 1869. England. "During the morning, and up to two o'clock, P.M., a dense fog had hung over the sea; but, apparently it was not very deep, as the sun's rays penetrated it pretty freely. At the hour mentioned, the fog opened toward the southeast, disclosing the cliffs on the French coast; and, in the course of a few minutes, the fog had disappeared, leaving the atmosphere in a state of unusual transparency. The French cliffs were apparently so lofty, and with every indentation, so clearly visible, that one might easily have imagined that they were but ten miles distant. On examining the objects in view through a small telescope, with a 25-power, it was at once apparent that this arose from something more than common looming. The French coast could be seen from near Calais, toward the east, far away, and many miles beyond Boulogne, toward the southwest; the land in the latter direction being ordinarily invisible, as it is situated below the horizon. Immediately under the erect image of the coast was an inverted one of almost double the height of the former. The light-house at Cape Gris-nez gave five images in a vertical line, the lowest erect, but somewhat magnified; above that, and separated from it a pair of images of the center and highest portion of the building only, one erect and the other inverted; and over these another pair, the inverted image being like the former one, but the erect image showing the whole building. Over Boulogne, in the air, were two images of the double funnels and the mast of a tug boat, the lower image being erect and the upper inverted, the two lines of smoke bending, the one upward and the other downward, and both toward the west, till they joined together." (R1)

X2. December 5, 1922. Cape Wrath, Scotland. "On the morning of December 5, 1922, about 10.30 A.M. G.M.T., Mr. John Anderson, lightkeeper at the Cape Wrath Lighthouse, Durness, observed a mirage of unusual character. Mr. Anderson focussed his telescope on a sheep which was grazing on top of a conical hill (height about 200 feet) and about a quarter of a mile away, and immediately noticed an unusual appearance in the atmosphere around. Upon swinging the telescope slightly upward, he observed that a belt of the atmosphere appeared to be land and sea, giving a perfect representation of the whole of the coast line from Cape Wrath to Dunnet Head. The appearance in the mirage was an exact replica of what would have been seen from a distance of about 10 miles out to sea. In a direction south of the lighthouse there were three repititions of the mirage one above the other, with the sea separating each pair. The entrance to Loch Eriboll and the other bays could be seen and easily recognised in the main mirage, though Cape Wrath itself was rather indistinct. The accompanying map shows the apparent position of the mirage and the outline of the coast (not reproduced) while the sketch gives a rough idea of how

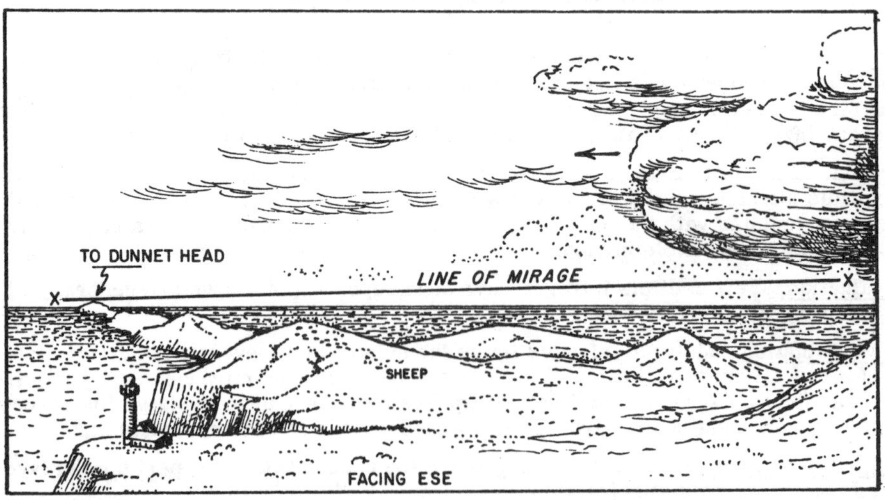

Setting of a multiple mirage at Cape Wrath (X2)

the country appeared to the observer. The mirage was hidden at one point by a hill. The mirage was practically invisible to the naked eye, and was visible only from a very restricted area. Mr. Anderson states that it was not visible at a distance of 20 yards either way from his original position, but was still visible 4 or 5 yards from that point. Mr Anderson estimates the apparent height of the image above the ground as about 1000 feet, in a southerly direction, while the distance from Cape Wrath of the triple image shown on the map is about 12 miles." (R3) It is interesting how much more complex mirages turn out to be when viewed through telescopes or binoculars. (WRC)

X3. December 7, 1958. South Pacific Ocean. "At 1000 G. M. T., when at an altitude of 35°, the whole of the moon's disc was very faintly visible, due to 'earth shine,' in addition to the crescent which was of normal brilliance. At 1007, the crescent appeared to become detached, and occupied a separate position, a small distance away from the moon's disc. At the same time, another crescent, slightly thinner than the original one, took its place; it was just as distinct and had the same colour (Fig. 2). At 1010 the separation between the two crescents increased further and at 1012 a third narrow crescent appeared between the other two. It was just as clearly defined and of the same colour as those which preceded it (Fig. 3). At 1014, low cloud covered the moon and no further observations were possible." (R4) This observation is especially interesting because the moon was so high (35°) that atmospheric turbulence near the surface can be ruled out. Several similar cases may be found in GEL4. (WRC)

X4. No date or place given. A nine-image lateral mirage has been photographed. (R5)

X5. General observations. Based on considerable experience with Fata Morganas over European lakes. Telescopic study of the mirages reveals that the common increase in image size is really due to the existence of many adjacent images of the original object rather than magnification per se. (R2)

References

R1. "Remarkable Mirage in the English Channel," Scientific American, 21:97, 1869. (X1)
R2. Delebecque, Andre; "Atmospheric Refractions at the Surface of Water," Monthly Weather Review, 24:371, 1896. (X5)
R3. Brunt, D.; "A Double Vertical-Reflection Mirage at Cape Wrath," Nature, 111:222, 1923. (X2)
R4. Townshend, C. R.; "Abnormal Refraction," Marine Observer, 29:178, 1959. (X3)
R5. Fraser, Alistair B., et al; "Multiple-Image Lateral Mirage," Eos, 53:274, 1972. (X4)

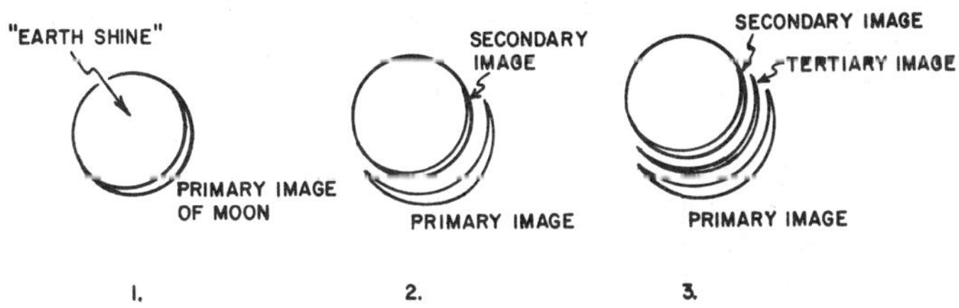

Multiple lunar crescents develop in the South Pacific (X3)

GEM5 Mirages Sensitive to the Observer's Position

Description. Mirages that disappear or change substantially when the observer moves a few yards (or even inches) vertically or horizontally.

Background. Other mirages detailed in this section maintain their sharpness and image content even though the observer moves many miles.

Data Evaluation. Only a very few reports mention position-sensitivity, but this neglect may only be because most observers do not think to look for this feature. There is no doubt, though, that this is a poorly observed phenomenon. Rating: 2.

Anomaly Evaluation. There are two fascinating implications of position-sensitivity: (1) The atmosphere contains considerable, unappreciated fine structure; and (2) The likelihood that innumerable mirages are observable only within small volumes and that only a few observers have been lucky enough to be in the right place at the right time. (Potential observers occupy only an infinitesimal fraction of the potential image volume!) Rating: 2.

Possible Explanations. Optical imaging is often very sensitive to observer position. It is all a function of the lenses and distances, as anyone who has experimented with an optical bench in a high school physics lab will testify. Atmospheric imaging is even more complicated.

Similar and Related Phenomena. Multiple mirages (GEM4).

Examples of Mirages Unusually Sensitive to the Observer's Position

X1. September 1919. Fair Isle, England. A surging mirage was visible only within a space of a few yards. (R3) Full text may be found in GEM8-X2.

X2. December 5, 1922. Cape Wrath, Scotland. A multiple mirage disappeared when the observer moved more than a few yards. (R4) Full text may be found in GEM4-X2.

X3. May 1, 1949. Hunstanton beach, England. "A rather unusual and conspicuous mirage was observed on May 1, 1949, from 14.00-16.30 B.S.T. over the Wash from Hunstanton beach. The weather was clear with a brilliant deep blue sky, light bluish-green just on the horizon. Visability was very good (40 miles or more) with a light N. to NE. wind. The sea was out, leaving behind a wet sandy beach which towards the southwest was shining with silverish blue reflected sunlight. To the northwest was shallow sea water, warmed by sun, of a dark dirty colour caused by sand, sea growths and shells below the surface.

The phenomenon occurred between N. and NE. at a distance of about 1300-1500 yds. from us. It was also observed by Mr. A.E. Wallis who first drew my attention to it. A smooth dark sheet of water changed sharply at this distance to a bright silver-blue, spreading towards the open sea and merging with the sky, no horizon line being discernable in this direction. The display was mainly over the sea since the beach narrowed in that direction. Most probably it was a distant wide sector of the beach with objects and people on it reflected. We could distinctly see little human figures walking to and fro far out on the sea but the 'permanent' objects changed position with time very little and this was probably due to the motion of the sun. A group of beach huts was visible on the right for about 20 minutes, one of them being of yellowish colour. With a decrease of the angle of view, i.e. by bending down to about 3 feet, the bright surface spread a little towards us, the objects and the illusive walkers faded and then completely vanished. The other people present on the beach over which we watched the mirage, did not disappear. With an increase of the angle of view the objects reappeared and the people again walked on the sea. By standing on small rocks (3-4 feet) it all became more clearly defined. The tide coming in developed small waves like cirro-cumulus cloud on the shiny surface of this phenomenon." (R5)

X4. General observations. "One very curious thing about mirage is that it depends very much upon the position of the eye; a few inches in the height of the eye may make all the difference. I remember myself, on the plains of India, observing a mirage which was only evident when I was at a particular height; there was only a vertical space of two or three inches in which the effect could be seen, so that these phenomena may easily escape notice. A singular effect may sometimes be observed at a particular spot on the south coast, and very likely at other places; when the waves come in on to a very hot beach, if you place the eye within about a foot from the ground and look parallel to the wavefronts, you can see an image of the wave two or three feet above the real wave. This may conceivably arise in this way: the wave may bring in some cold air, and if the wind were blowing a little off the heated beach there might be some heated air brought in as a layer above that cold air; that would give that rapidly diminishing density upwards which gives a ray with considerable curvature and with concavity presented downwards, and would certainly result in an image of the wave above the real wave." (R1, R2)

References

R1. MacMahon, P. A.; "Mirage," Nature, 59:259, 1899. (X4)

R2. "Mirage," English Mechanic, 68:584, 1899. (X4)

R3. "Surging Mirage," Meteorological Magazine, 58:12, 1923. (X1)

R4. Brunt, D.; "A Double Vertical-Reflection Mirage at Cape Wrath," Nature, 111: 222, 1923. (X2)

R5. Szczyrbae, S.; "May-day Mirage," Weather, 4:234, 1949. (X3)

GEM6 The Novaya Zemlya Effect

Description. In polar regions, the appearance of the sun's image above the horizon when the sun is actually several degrees below the geometrical horizon. These mirages are usually distorted and may be multiple.

Data Evaluation. The phenomenon is rare but several polar explorers have noted it. In recent years the mirage has been photographed. Rating: 1.

Anomaly Evaluation. While the sun may be seen above the visible horizon due to ordinary mirage action, a new theory involving the ducting of light has been proposed (see below). The Novaya Zemlya Effect thus does not seem particularly anomalous, although it is certainly curious. The major reason for including it in this Catalog is that its novel explanation may also be applicable to other optical phenomena occurring near the horizon, which are generally considered to be the consequence of simple refraction by atmospheric strata. Rating: 3.

Possible Explanations. As in conventional mirage theory, optical ducting carries light from a source below the horizon around the curvature of the earth to the observer. The important difference is that the light rays are not gradually bent by refraction but rather are reflected back and forth between the upper and lower boundaries of the atmospheric layer. Such a process is called ducting and is employed in optical fibers and wave guides. In ducting one does not expect sharp images of the source with single ducts.

Similar and Related Phenomena. Other mirages (GEM); low-sun phenomena (GEL), which may involve ducting, although no one seems to have explored this possibility.

Examples of the Novaya Zemlya Effect

X1. January 1597. Novaya Zemlya, Russia. Explorer Willem Barrents and his crew were astonished to see the sun appear two weeks early. It was actually about 5° below the geometrical horizon. (R1, R2)

X2. May 8, 1915. Antarctica. Shackleton sees the sun reappear seven days after it had set---supposedly for the entire winter. (R1)

X3. 1951. Antarctica. Several examples of the Novaya Zemlya Effect. (R1)

X4. May 16, 1979. Tuktoyaktuk, Canada. The sun did not set (apparently) but remained above the horizon as a rectangular strip of light. Other observations of the Effect included one with three pancake-like images one above the other. (R1, R2)

X5. General observations and theory. The Novaya Zemlya Effect is the consequence of the sun's rays being trapped within thermocline layers. Light propagation is like that in a duct; i.e., as in an optical fiber. (R3)

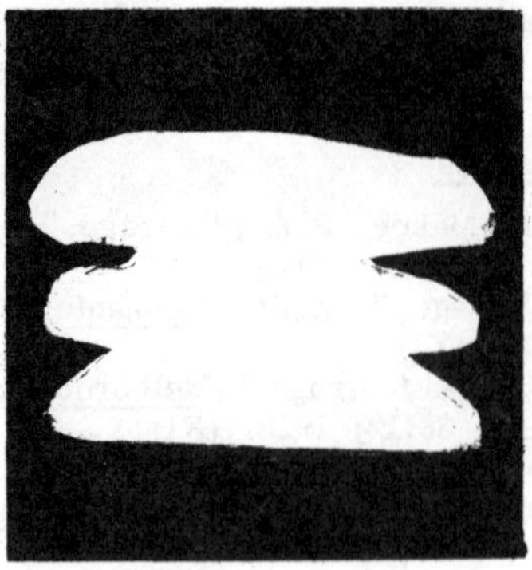

The Novaya Zemlya Effect (X4)

References

R1. "Polar Solar Mirages," Science News, 118:248, 1980. (X1, X4)
R2. "New Light on Novaya Zemlya Polar Mirage," Physics Today, 34:21, January 1981. (X1-X4)
R3. Lehn, W.H.; "Novaya Zemlya Effect: An Arctic Mirage," Optical Society of America, Journal, 69:776, 1979. (X1-X5)

GEM7 Illusions of a Flat Earth

Description. The apparent flatness of the earth's surface under appropriate atmospheric conditions.

Background. This entry is not entirely frivolous. Situations like the one described here arise occasionally in science; and it is well to consider the factors involved. Two aspects of the Bedford Canal Experiments, and others like them, are instructive: (1) Under appropriate conditions well-understood natural phenomena can distort human perception and experimental measurements so much and so subtly that some of the holiest tenets of science may seem to be challenged; and (2) Scientists have strong inclinations to ridicule and/or ignore data that challenge said tenets. Such issues become quite clear in the context of such "ridiculous" assertions that the earth is really flat!

Data Evaluation. A single experiment peformed by a nonscientist. (There are doubtless others of this type, but they have not been found in the literature surveyed so far.) Rating: 4.

Anomaly Evaluation. We would have a major scientific anomaly if the earth's curvature could not be demonstrated through optical measurements over distances of several miles. Rating: 1.

Possible Explanations. A small amount of refraction along the Canal's surface could, in effect, make the earth's surface seem flat. This type of mirage is similar to that in operation when one sees an object below the horizon, such as a ship, only on a much smaller scale.

Similar and Related Phenomena. Those ordinary mirages seen on roadways and at sea when one may see ships still below the horizon; surging mirages (GEM8) in which dynamic bulges appear on the earth's horizon.

Examples of False Indications of a Flat Earth

X0. Historical aside. In 1870 a carefully monitored experiment was performed along the Old Bedford Canal, in England. A flat-earther named John Hampden had claimed that optical instruments sighted along the several-mile length of one Canal section would fail to show any curvature of the earth's surface. A.R. Wallace (of evolution fame?) bet Hampden that the curvature would be detected. 500 pounds were at stake. From the observations made along the Canal, the judges decided in Wallace's favor. The same John Hampden had earlier been stymied by similar experiments using the famous Eddystone Light. (R1)

X1. Circa 1900. Bedford Canal, England. A repeat of the 1870 experiment. "'A train of empty turf-boats had just entered the canal from the river Ouse, and was about proceeding to Ramsey. I arranged with the captain to place the shallowest boat last in the train, and to take me on to Welney Bridge, a distance of six miles. A good telescope was then fixed on the lowest part of the stern of the last boat. The sluice-gate of the Old Bedford Bridge was 5ft. 8in. high; the turf-boat moored there was 2ft. 6in. high, and the notice board was 6ft. 6in. from the water. The sun was shining strongly upon them in the direction of the south-south-west; the air was exceedingly still and clear; and the surface of the water smooth as a molten mirror, so that everything was favourable for observation. At 1.15 p.m. the train started for Welney. As the boats gradually receded, the sluice-gate, the turf boat and the notice board continued to be visible to the naked eye for about four miles. When the sluice-gate and the turf boat---being of a dark colour---became somewhat indistinct, the notice board---which was white---was still plainly visible, and remained so to the end of six miles. But on looking through the telescope all the objects were distinctly visible throughout the whole distance. On reaching Welney Bridge I made very careful and repeated observations, and finding several men upon the banks of the canal, I called them to look through the telescope. They all saw distinctly the white notice board, the sluice gate, and the black turf boat moored near them. Now, as the telescope was 18in. above the water, the line of sight would touch the horizon at one mile and a half away---if the surface were convex. The curvature of the remaining four miles and a half would be 13ft. 6in. Hence the turf boat should have been 11ft., the top of the sluice gate 7ft. 10in., and the bottom of the notice board 7ft. below the horizon.'

My recent experiment affords undeniable proof of the earth's unglobularity, because it rests not on transitory vision; but my proof remains printed on the negative of the photograph which Mr. Clifton took for me, and in my presence, on behalf of J. H. Dallmeyer, Ltd. A photograph cannot 'imagine' nor lie!" (R2)

Several readers of the English Mechanic quickly responded to the effect that the lack of curvature must have been due to the refraction of light along the canal. (R3) No one, however, repeated the experiment. (WRC)

References

R1. "Is the World Round?" Eclectic Magazine, 12:497, 1870. (X0)

R2. Blount, E. A. M.; "The Flat Earth: Another Bedford Canal Experiment," English Mechanic, 80:137, 1904. (X1)

R3. Watson, Bernard H., et al; "The Flat Earth: Another Bedford Canal Experiment," English Mechanic, 80:160, 1904.

GEM8 Dynamic Mirages

Description. Moving distortions of the horizon or distant scenery. These surges or heavings may be correlated with passing weather fronts. Not included here are those small-scale perturbations due to surface heat convection.

Data Evaluation. If looked for, this phenomenon might turn out to be widespread; but few people have bothered to mention it in the scientific literature. Rating: 2.

Anomaly Evaluation. Dynamic mirages should not surprise anyone familiar with the vicissitudes of surface weather. Some wavering of the horizon and scenery distortion due to passing atmospheric disturbances is to be expected. Although these effects are sometimes surprisingly large---and certainly curious---they are hardly anomalies. They may, however, be related to other horizon phenomena (see below) and thus worth recording. Rating: 4.

Possible Explanations. The passage of weather front, gravity waves, convection cells, etc., introduce density variations and thus changes in the refractive index. Image distortion naturally follows.

Similar and Related Phenomena. Bands crossing halo displays (GEH11); visible sound waves (GES2); moving patches of light on the horizon (GEL12); nocturnal lights (GLN1) some of which may be refracted artificial lights.

Examples of Dynamic Mirages

X1. December 14, 1917. Lavina, Montana. A range of mountains about 20 miles northwest was strongly distorted, with the distorted area moving west to east at a rapid rate. Within a few minutes the local temperature began to rise rapidly, indicating the arrival of a Chinook. (R1)

X2. April 1919. Fair Isle, England. "Coming from the east toward the station about 8 a.m., I observed peculiar shades flittering at about an altitude of 22 to 30 degrees in the western sky. The sun was very bright and the sea carried a deep roll, so to speak, with a glassy surface. Both of which, I consider, played an important part in this phenomenon. By the eye alone I could form no conception, so resorted to the telescope. By its aid, to my great surprise, I could see spread out a panorama wonderful to behold. The islands of North Ronaldsbay, Sanday and Westray, were stretched out, mapped and clothed in their natural beauty, against the curve of the western sky. But this most astonishing phenomenon increased in interest when I observed a reversed view, equally as clear, of North Ronaldsbay and part of Sanday only, immediately under, as if growing from one base; and, as sure as the timely rise and fall of the fountain ball, so did this strange sight rise and fall as if governed by the movement of some sighing bosom. Each movement downwards seemed to separate each object at its base, and growing more distorted on the upward movement things gradually assumed their natural aspect, when I could make out many of the houses on North Ronaldsbay, Star Point Lighthouse and farms on Sanday, also Noup Head, Westray. Noup Head Lighthouse is difficult to pick up with a good telescope in the clearest of weather from an altitude of 400 feet, but on this occasion it was clearly seen. This phenomenon kept clear for about an hour and, as it disappeared slowly, it seemed to sink down behind the horizon, leaving the high objects a little longer to do their farewell bow." (R2)

X3. March 29, 1948. Cape Town, South Africa. "The photographs (not reproduced) were taken from Camps Bay looking along the coastline towards the Cape of Good Hope. The phenomenon first appeared at 1400 G.M.T. and lasted for 7 minutes, during which period the whole horizon in this direction appeared to be traversed by an enormous swell. The 'crest', with slight variations in shape, moved steadily along the horizon from south to west and finally merged into the usual horizon line. The remainder of the horizon was clear and completely undistorted throughout." (R3)

References

R1. Pound, Robert T.; "Visible Weather," Scientific American, 118:147, 1918. (X1)

R2. "Surging Mirage," Meteorological Magazine, 58:12, 1923. (X2)

R3. Pegler, G.G.; "Mirage," Marine Observer, 19:10, 1949. (X3)

GEM9 Mirror or Reflection Mirages

Description. Apparent reflections of the observer and his surroundings in the sky, or the mirroring of objects on one side of the observer on the other side. In other words, natural mirror action.

Data Evaluation. Several fascinating cases of mirror mirages have been uncovered; a few are rather old and perhaps embellished, others suffer from questionable association of objects and images. As with Fata Morganas, the human imagination may play a role here. The credibility of the examples often depends entirely upon positive identification of the image. Rating: 3.

Anomaly Evaluation. It is difficult to see how sharp images of a scene could be reflected by dust, water droplets, water vapor, or any other constituents of the atmosphere---shadows, yes, but mirror-sharp images, no. Rating: 2.

Possible Explanations. Misidentification of the image is most likely. Sheets of water vapor between air masses have been suggested, but no one has explained how clear, stable images might be created by them. Highly displaced lateral mirages might account for some reflection mirages, but these would require extreme density differences.

Similar and Related Phenomena. Brocken Specters (GEI1); lateral mirages (GEM3); Fata Morganas (GEM1); cloud arches (GWC2); anomalous crepuscular rays (GEL2).

Examples of Mirror Mirages

X1. December 17, 1826. Migne, France. "It was Sunday, and 3000 persons were engaged in the exercises of the Jubilee. As a part of the ceremony, a large red cross, twenty-five feet high, was planted beside the church, in the open air. Towards the close of the day, while one of the clergy was addressing the multitude, and reminding them of the miraculous cross beheld in the sky by Constantine and his army, a cross was seen at that moment in the heavens, directly before the porch of the church, and at the height of two hundred feet above the ground. Its length was nearly one hundred and forty feet, its breadth from three to four, and it shone with a bright silvery hue, tinged with red. The assembled multitude were struck with awe, many regarding it as a miracle, and such was the extraordinary sensation produced throughout the country, that a committee was appointed to investigate this phenomenon. From the circumstances detailed in their report it is evident, that the cross in the sky was the magnified image of the cross before the church, and reflected from the concave surface of some atmospheric mirror. The image possessed exactly the shape and proportions of the wooden cross, it was tinged with the same color, and the state of the air at the time was favorable to the formation of such images." (R2) Most of the crosses-in-the-sky reported by the superstitious are probably halo phenomena. (WRC)

X2. No date given. American Salt Desert, California. "During the subsidence of this tempest, there appeared upon the plain, one of the most extraordinary phenomena, I dare to assert, ever witnessed. As I have before stated, I had dismounted from my mule, and turning it in with the cabsilada, was walking several rods in front of the party, in order to lead in a direct course to the point of our destination. Diagonally in front, to the right, our course being west, there appeared the figures of a number of men and horses, some fifteen or twenty. Some of these figures were mounted and others dismounted and appeared to be marching on foot. Their faces and the heads of the horses were turned towards us, and at first they appeared as if they were rushing down upon us. Their apparent distance, judging from the horizon, was from three to five miles. But their size was not correspondent for they seemed nearly as large as our own bodies and consequently were of gigantic stature. At first view I supposed them to be a small party of Indians (probably the Utahs) marching from the opposite side of the plain. But this seemed to me scarcely probable, as no hunting or war party would be likely to take this route. I called to some of our nearest men to hasten forward, as there were men in front coming towards us. Very soon the fifteen or twenty figures were multiplied into three or four hundred, and appeared to be marching forward with the greatest action and speed. I then conjectured that they might be Captain Fremont and his party with others, from California, returning to the United States by this route, although they seemed to be too numerous even for this. I spoke to Brown who was nearest to me, and asked him if he noticed the figures of men and horses in front. He answered that he did and that he had observed the same appearances several times previously, but they had disappeared and he believed them to be optical illusion similar to the mirage. It was then, for the first time, so perfect was the deception, that I conjectured the probable fact that these figures were the reflection of our own images by the atmosphere, filled as it was with fine particles of crystallized matter, or by the distant horizon covered by the same substance. This induced a more minute observation of the phenomenon, in order to detect the deception, if such it were. I noticed a single figure, apparently in front in advance of all the others, and was struck with its likeness to myself. Its motions, too, I thought, were the same as mine. To test the hypothesis above suggested, I wheeled suddenly around, at the same time stretching my arms out to their full length, and turning my face sidewise to notice the movements of this figure. It went through precisely the same motions. I then marched deliberately and with long strides several paces; the figure did the same. To test it more thoroughly, I repeated the experiment, and with the same result. The fact then was clear. But it was more fully verified still, for the whole array of this numerous shadowy host melted entirely away, and was no more seen." (R1) Note the close resemblence to the Brocken Specter (GEI1).

X3. 1887. Near Paris, France. A reflection of the buildings and other structures in the clouds. (R3)

X4. October 24, 1914. Mawnan, England. "A most curious phenomenon was to be seen from the Cornish coast at Mawnan, a headland at the mouth of Helford River, five miles from Falmouth, from 3.45 to 4.15 in the afternoon of Saturday, October 24, 1914.

In a shallow cloudbank stretching across the horizon from south-east to south gradually appeared what seemed to be a line of coast, with woods, trees, fields, hedges, and houses in their natural colours.

At first I thought it was merely a peculiar arrangement of clouds, but soon I recognized that it was a complete reflected panorama of this coast, and I gradually identified St. Keverne with its church spire, the Helford River, Mawnan Church, Rosemullion, Falmouth Harbour, and St. Anthony.

Shortly before the whole scene faded away, a reflection of Pendennis Head with the castle and the military huts, emerging like some huge ship from a fog bank, appeared most distinctly and was visible for several minutes. Every detail of the whole scene was reversed as in a looking-glass and slightly magnified, for although the panorama seemed to be some twenty miles away, everything appeared as it would to an observer from the distance of only a mile or so. The colours were a little less bright than those of the real coast at the time.

The scene as a whole was visible all the time, but the details of a small portion only were quite distinct at any one time, commencing in the south and gradually moving towards the east, the effect being that of a powerful searchlight being very slowly turned from right to left on the opposite shore of a large lake. Nothing of it could be seen through a telescope; at least I found it impossible to get it into focus.

I presume that the reason that the reflectect objects were seen reversed and not inverted, as is usual in mirages, was the fact that my point of observation was about 300 feet above sea-level, and I was therefore, as it were, looking down upon the mirage, the reflecting medium of which was between the horizon and myself, whereas in most instances mirages are the reflection in the sky of objects beyond the horizon.

A curious point I noticed was that the reflection of some of the objects was not reversed precisely to correspond with my view of the real objects. For instance, I could see St. Keverne Church itself, but its reflection seemed to be from an entirely different point of view. This would, I think, mean either that the cloudbank or medium of reflection was irregular in its reflecting surface and reflected different points at different angles; or that, perhaps more probably, the cloudbank acted like a huge mirror from far beyond the object, and showed me the reversed view of the coast as it would appear if observed from the point of the reflection; and, since most of the scenery, viz. Helford River, Rosemullion, Pendennis Castle, and the fields, etc., inland were quite out of my range of vision, and yet were all reflected, I am inclined to think that the latter theory is the correct one.

Another curious point was that although the sun was behind me and therefore the coast was in the shade, the reflection was not a silhouette, as one would have expected, but the details showed up distinctly as though a brilliant light were shining on them." (R4, R5)

X5. February 14, 1936. Gulf of Mexico, North Atlantic Ocean. Two mirages observed, almost exactly 180° apart, one over a land background, the other over the sea. The mirage over the sea bore superficial resemblence to the landform in the opposite direction. See GEM1-X6 for the full account. (R6)

X6. May 31, 1952. Indian Ocean. Two cumulus clouds, similar in appearance and 180° apart, were joined together by a cloud arch passing overhead. See full account in GWC2-X3 plus the illustration. (R7)

X7. December 2, 1961. Red Sea, Indian Ocean. "Between 1440 and 1445 GMT the unusual refraction effects shown in the sketch were observed (sketch not reproduced). A Blue Funnel ship was seen on the horizon towards the south at 1440, with the funnel much enlarged due to refraction; simultaneously a 'watery looking' image of the same vessel was seen on the horizon in the WNW. At 1443 the sun was obscured by hills some 63 miles distant and 3,000-7,000 ft. high. The vessel and its image now looked larger than they did 3 min. previously. The image disappeared at 1445, which was, by calculation, the time of sunset.... Note. We cannot explain this observation, except by suggesting that the 'watery looking' image was that of another ship well below the horizon (i.e., images of two similar ships were observed). (R8) The foregoing report was followed by a letter stating that conventional mirage theory could not account for the observation. (R9)

X8. November 26, 1963. Off Cape Vincent, North Atlantic Ocean. "At 0315 GMT when the vessel was 4 miles due south of the Cape St. Vincent light a false loom was seen on the reciprocal bearing of the lighthouse, which was so real in appearance that one would have expected the light itself to have been sighted in that direction at any moment. The loom was flashing ever 5 sec. as does the St. Vincent lighthouse. As the nearest land was some 240 miles away in the direction of the loom, it was evident that the St. Vincent light itself was undergoing some effect of abnormal refraction. This was borne out by the fact that as St. Vincent drew astern of our port quarter, the loom drew ahead on our starboard bow. ....Note. We have now had three occasions on which images have been observed in a reciprocal direction from the object. On two previous occasions it has been suggested that a sharp horizontal increase in moisture content with a sharp fall in temperature along the line of vision, extending over a large vertical plane or one inclined towards the observer in the atmosphere, can act like a mirror causing an observer to see an image of a scene or object behind him." (R10)

X9. April 17, 1974. North Pacific Ocean. "At about 1930 SMT (0930 GMT), shortly after sunset, the sky was observed to be divided into two distinct halves by what appeared to be crepuscular rays. These rays spread from the western horizon across to the eastern horizon in a band approx. 5° wide. The sky was deep blue and the rays were arranged in the order of pale green, pale orange, cream, pale orange and pale green. At the time of sunset there were, on the western horizon, medium and towering Cu clouds, the eastern horizon being clear. By following these rays across to the east a well-defined reflection of the Cu cloud on the western horizon could be seen at an altitude of approx. 15° above the eastern horizon where the rays converged." (R11) This example and its illustration are very similar to those of X6. (WRC)

References

R1. "Mirage in the Great Salt Desert," Scientific American, 4:131, 1848. (X2)
R2. Brocklesby, John; Elements of Meteorology, New York, 1855. (Cr. L. Gearhart) (X1)
R3. "Reflexion d'Objets Terrestres sur les Nuages," Ciel et Terre, 14:187, 1887. (X3)
R4. Horton, Percy H.; "Mirage on the Cornish Coast," Royal Meteorological Society, Quarterly Journal, 41:71, 1915. (X4)
R5. "Mirage on the Cornish Coast," Knowledge, 12:152, 1915. (X4)
R6. Paterson, H.; "Mirage," Marine Observer, 14:11, 1937. (X5)
R7. Townshend, C.R.; "Unusual Cloud Formation," Marine Observer, 23:76, 1953. (X6)
R8. Johnston, R.; Marine Observer, 32: 183, 1962. (X7)
R9. Botley, Cicely M.; Marine Observer, 33:221, 1963. (X7)
R10. Etches, J.A.; Marine Observer, 34: 180, 1964. (X8)
R11. Baker, J.; "Crepuscular Rays," Marine Observer, 45:68, 1975. (X9)

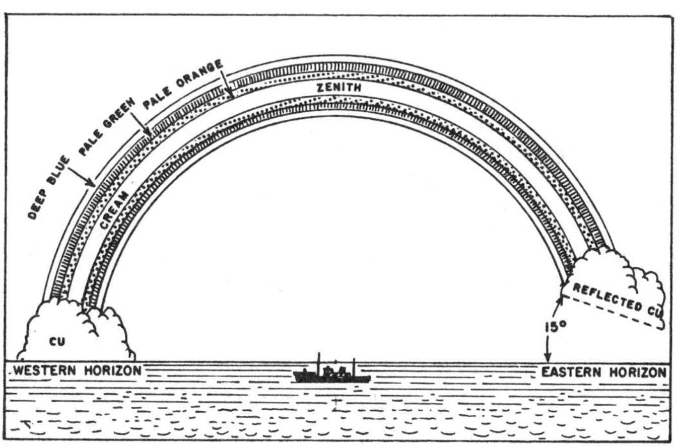

Possible reflection mirage of cloud in the North Pacific (X9)

GER RADIO AND RADAR ANOMALIES

Key to Phenomena

GER0 Introduction

Our eyes see only a very narrow portion of the electromagnetic spectrum. It is therefore reasonable to assume that the major share of atmospheric anomalies go undetected by the observer not possessing special instruments. When radio and radar came along earlier in this century, the operators of this primitive equipment discovered a whole new universe of atmospheric phenomena. These idiosyncracies are not as widely known as mirages, halos, rainbows, etc., because they can be detected only by a select few. Of course, no one can call them beautiful or awe-inspiring; they are simply instrument readings. One reads of them in the technical journals. As in the case of their visual counterparts, almost all radio and radar anomalies are probably explainable in terms of reflection, refraction, and frequency dispersion of electromagnetic waves.

The long-delayed radio echos first detected in the 1920s have received the most publicity, some of it sensational in character. Even in today's overcrowded electromagnetic environment, long-delayed echos are frequently heard by amateurs and professionals alike. The delays are so long (several seconds) that the reflector (if that is what causes them) must be located outside the moon's orbit. Even alien space probes carrying radio-repeaters have been proposed to explain these echos. Natural clouds of ionized matter are more likely, but no one knows for sure. The signals may just get "stored" in ionospheric ducts for a few seconds and then be released later to wondering radio operators.

The apparent effects of the positions of the moon and planets upon radio transmission are also quite controversial. In the case of the planets in particular, there seems to be no natural force strong enough to affect terrestrial radio propagation. More likely the culprit is solar activity, as measured by sunspot number, which can also be loosely correlated with planetary positions (according to some researchers, at least); but this just transfers the mystery to some unrecognized planetary force that affects solar activity. Here is a genuine scientific puzzle---the possibility of a hitherto undetected physical force! No wonder these planetary correlations are viewed with suspicion.

In the case of radar, human blindness at long wavelengths is particularly frustrating because radars are always detecting "angels" and other spurious targets that cannot be matched to known ships, aircraft, or topographic and meteorological features. Birds and insects, singly or en masse, reflect enough radar energy to create some angels. In addition, the bubble-like and wave-like microstructure of the atmosphere---all quite invisible to man---can cause additional false radar targets. Nevertheless, some solid-appearing radar targets cannot be accounted for.

GER1 Long-Delayed Radio Echos (LDEs)

Description. Echos of radio signals received more than a second after the original transmissions. Time delays may be as long as 5 minutes, but most are shorter than 30 seconds. LDEs have been heard in the frequency range 810 kHz to 144 MHz. On rare occasions multiple echos of the original signal are noticed.

Background. LDEs were originally discovered by European scientists experimenting with short-wave radio in the late 1920s. After a hiatus of nearly 40 years, radio amateurs began compiling observations of this phenomenon. LDEs turned out to be rather common and still rather mysterious. About the same time, scientists again began experimenting to determine just what caused the echos.

Data Evaluation. Both radio amateurs and professional scientists have recorded LDEs in large numbers for over 50 years. The literature is extensive, and only a fraction of it is recorded below. Rating: 1.

Anomaly Evaluation. Originally it was believed that the signals bounced back from sheets or tongues of solar plasma far out beyond the moon's orbit, but current thought is that the terrestriallly emitted signals are somehow trapped in ionospheric ducts for some seconds and then released, perhaps after hundreds or thousands of trips along the duct. Attractive though this picture may be, no one knows just why some signals are detained and then released over the original transmitting area. Many details remain to be worked out. Rating: 2.

Possible Explanations. As mentioned above, some radio signals may be captured in ionospheric ducts, where they travel back and forth, reflecting off ionized layers with very little loss of energy, until they escape. Some long-delayed signals may be simple reflections from the moon and from accumulations of matter and ionized gases in the vicinities of the Lagrangian points in the earth-moon system. (The Lagrangian points are small regions in space where gravity and centrifugal force combine to create dynamically stable situations, where dust and gas can collect.)

Similar and Related Phenomena. Visual observations of matter collected at the earth-moon Lagrangian points (AL).

Examples of Long-Delayed Radio Echos

X1. Summer 1927. Norway. In a letter to Nature from the well-known geophysicist Carl Stormer. "On Feb. 29 of this year I received a letter from Engineer Jørgen Hals, Bygdo, Oslo, in which he says: 'I herewith have the honour to advise you that at the end of the summer 1927 I repeatedly heard signals from the Dutch short-wave transmitter station PCJJ (Eindhoven). At the same time as I heard the telegraph signals I also heard echoes. I heard the usual echo, which goes round the earth with an interval of about 1/7 second, as well as a weaker echo about 3 seconds after the principal signal had gone. When the principal signal was especially strong, I suppose that the amplitude for the last echo 3 seconds after lay between 1/10 and 1/20 of the principal signal in strength. From where this echo comes I cannot say for the present. I will only herewith confirm that I really heard this echo.'" (R2)

X2. Spring-summer 1928 . Oslo, Norway, and Eindhoven, Holland. "In the issue of Nature for Nov. 3 a short note by Prof. Carl Stormer appeared under this title. Prof. Stormer there described the observations made by him and Engineer Jørgen Hals, Oslo, of some remarkable echoes heard several seconds after the original signals---which were emitted from the short wave transmitter PCJJ (λ= 31.4 metre) Hilversum specially for the experiment---reached the receiver at Oslo. These special signals were first sent in March 1928. Since then the experiment has been repeated over and over again, sometimes twice and often four times a week. A continuous watch for these echoes was also kept at Eindhoven, Holland, in two different places, either by myself or an assistant, or by both of us. We did not hear any of these long period echoes for several months.

Then suddenly, on Oct. 11, I got a telegram from Prof. Stormer stating that very fine echoes had been heard that afternoon. Thereupon I immediately arranged the same night a series of test signals to be sent consisting of three short dots in rapid succession given every 30 seconds between 20 and 21 o'clock local time. I listened with my assistant to the 120 signals. Thirteen echoes were observed by both of us, the times between the signals and the echoes being: 8, 11, 15, 8, 13, 3, 8, 8, 8, 12, 15, 13, 8, 8 sec. The (radio) frequency of an echo was always exactly equal to the frequency of the signal, which fact could be easily verified as the signals were 'unmodulated', and therefore the receiver was kept oscillating. The combination tone thus formed had exactly the same pitch, whether the original signal was received or the echo. The frequency of the local oscillator was slightly changed a few times after a signal was received, and then the echo came in causing a slightly varied pitch of the combination tone. When thereupon the receiver was left unaltered, the next real signal caused exactly the same pitch in the receiver as the last echo.

The echoes I heard were rather weak, and though their oscillation frequency could be easily identified to be the same as the frequency of the direct signals, the three dots of the original signal could not be recognised in the echo, the latter being of a blurred nature, except in the one case where the echo came in 3 seconds after the signal, when the three dots of the original signal were very plainly audible in the echo as well.

Thereupon I suggested to Prof. Stormer to count the signals in the further experiments, so that echoes heard in Oslo and in Eindhoven could perhaps be identified. Up to Oct. 24 neither in Oslo nor in Eindhoven were echoes heard. However, on that date, between 16 and 17 G. M. T., echoes were again observed both in Oslo and at two different places (3 km. apart) at Eindhoven. The frequencies of the two oscillating receivers at Eindhoven were adjusted at different sides of the carrier frequency of the signal in order to eliminate as far as possible the risk of regarding stray signals as echoes. Prof. Stormer kindly sent me the observations made that day at Oslo where 48 echoes were noted. Receiver No. 1 at Eindhoven (with two observers) noted 4 very weak echoes, and receiver No. 2 at Eindhoven registered 5 echoes. A part of the simultaneous observations are plotted in the accompanying graph (Fig. 1). The timing of the ⦿ observations was done with a stop watch, while for the observations ◎ the second hand of an ordinary watch was used. As the echoes often lasted more than 1.5 sec., there is no doubt that some echoes were heard practically simultaneously in the three places referred to above. Therefore, though they are often difficult to observe, there is not doubt that the echoes really exist, as they have been heard by several observers at different places and a few times even simultaneously. (R2, R3)

X3. General observations. From the article's summary: "The long delayed echoes must therefore be due to signal waves having travelled very long distances (about $3 \times 10^5 t$ km, t being the retardation in seconds) either along bounding surfaces between ionized and empty parts of space outside the earth's atmosphere but 'within the

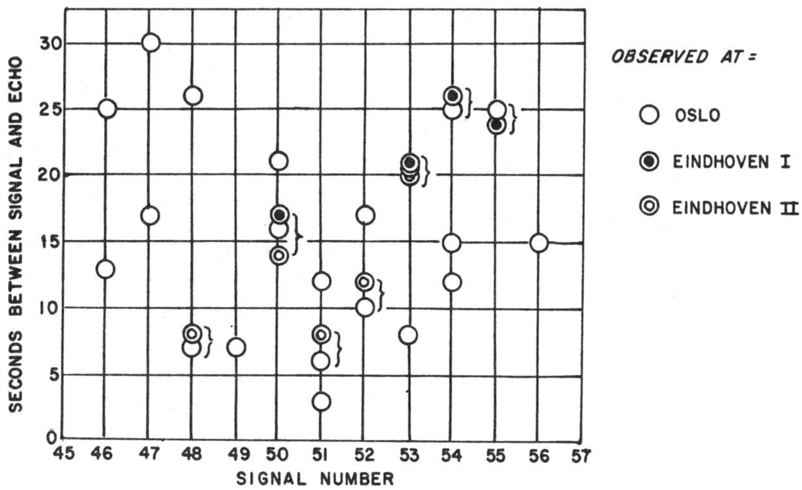

Long-delayed radio echos observed by Stormer in 1928 (X2)

earth's magnetic field,' or in free space, having been subjected to one or more reflections at such boundary surfaces. The geometry of these ionized layers which causes delays up to 30 (possibly 60) sec. will mainly be controlled by the earth's magnetic field as shown by Stormer. Beside these echoes originating within the earth's magnetic field the writer has predicted that echoes might occur with delays of several minutes by reflections from ionization-bands outside the influence of the earth's magnetic field. Such long-delayed echoes---3 min. 15 sec. and 4 min. 20 sec.---have recently been reported by Jørgen Hals." (R4, R5)

X4. 1930. Indochina. "... two observers in Indo-China observed two thousand echoes from a relatively small emitter station. The echoes came about 30 sec. after the signal and their amplitude was sometimes as great as one-third the signal. Some of the experiments recorded prove conclusively that they were echoes. It seems as if the space outside the earth's orbit was traversed intermittently by very unstable streams of electrons. This may explain the great variety of echo times observed." (R6)

X5. General summary of status of long-delayed radio echos. Stormer suggests that the emitted signals penetrate the ionosphere and are reflected back from a belt of electrified corpuscles ejected by the sun and formed into a vast toroid by the earth's magnetic field. (R7)

X6. General summary of experiment planned to investigate long-delayed radio echos on a worldwide basis. (R8) No results from this experiment have been located as yet. (WRC)

X7. Summary of observations by many individuals, especially amateur radio operators. "Following is a summary of the conclusions which can be derived from the reports taken as a group:
1. Multiple-second 'coherent' signal echoes, either phone of c.w., appear to be real, and are observable for short periods of time at highly infrequent intervals.
2. They are aubible both on a station's own signals, and on signals of other stations.
3. They have been observed at 7, 14, 21, and 28 MHz., but apparently not at higher frequencies.
4. They either occur most frequently (or perhaps are most easily heard) when a given band is just 'opening up'---i.e., when sky-wave propagation to some point on earth is just becoming possible.
5. They seem to be audible when long-distance propagation is good, and when geomagnetic activity is low. (The presence of long-path as well as short-path propagation, or signals from stations at antipodal locations, is apparently a good omen.)
6. Stations reporting LDEs typically have been ones having antennas well up in the air, at locations reasonably good for DX, but other than that no exceptional facilities seem to be required.
7. An active ham who DXes one or two hours

a day, may expect to hear an LDE once a year, on the average.
8. The LDEs appear to be one single echo, rather than several successive ones.
9. No Doppler shift is perceptible.
10. The sound of the echo resembles that of a DX signal (i. e, it apparently involves long-distance multipath propagation.)
11. The strength is usually weak, although some reports have put it at 83 or more.
12. Echo strength always decays with time, rather than the other way around.
13. The total time interval during which the echo effect can be heard is remarkably short ---usually no more than a few minutes.
14. There is some indication that LDEs may be heard more frequently on signals which have travelled through the northern and southern auroral zones." (R9)

X8. 1967-1970. Stanford, California. "Summary. Observations have been made under controlled conditions of what are believed to be very long delayed radio echoes returning from the ionosphere about 10 sec after transmission. Ionosonde records have suggested a new mechanism for the phenomenon. It is proposed that the signals travel through the ionosphere at very low group velocity under conditions in which collisional attenuation is offset by weak beam-plasma instability." (R10)

X9. September 22, 1969. No place given. "On September 22, 1969 (GMT), WB6VKV was in QSO with WA3KQA on 20-meter c. w. At 0100 plus or minus one minute, he stood by. The frequency was QRM-less, and WA3KQA was unaccountably slow in coming back. When the receiver came alive, a signal could be heard in the background noise on WB6VKV's frequency, about S2 but clearly copiable. Although band conditions were normal for six o'clock on a California afternoon, and not much by way of DX could be heard, the signal sounded as if it had come a long way. It sent: 'WA3KQA de WB6VKV K', and was an exact repeat of WB6KVK's transmission.

Sounds spooky? You bet. This was WB6VKV's initiation into a very exclusive club" those who have heard long-delayed echoes, or LDEs.....

Following are the author's present views:
(a) It appears that LDEs tend to be reported at times when magnetic activity is low.
(b) According to psychologists, there is a chance that one- or two-second 'echoes' may be a trick of the imagination. But the longer ones don't seem to be readily explainable as 'internally' generated.
(c) There appear to be two classes of unusual echoes. One is observable at 3.5 MHz over short distances, and involves one- or two-second delays. The other usually is characterized by longer delays and seems to be associated with long-distance propagation. It is observable at higher frequencies, especially at times of band opening and closing.
(d) On balance, the evidence suggests that both effects are real and associated with the ionosphere......

The diversity of reported effects is impressive. In longitude, the locations at which LDEs have been heard range from Africa to the Marshall Islands; in latitude, from Peru to Canada. The reported delays range from fractions of a second to over five minutes. The radio frequencies at which unusual echoes appear extend all the way from 810 kHz up through 144 MHz." (R11) This article includes a large table of LDE observations from all over the world. This table was expanded in a later article. (R12)

X10. A new theory formulated. "The three experimenters suggest that because of inhomogeneities in the F layer, part of the energy in the transmitted wave is converted into 'plasma waves' which are propagated much more slowly---at perhaps only one km/s. Travelling through a 10-km thick region of the F layer near its density peak these could introduce a 10-second time delay, being eventually 'coupled out', again as radio waves by another inhomogeneity." (R13)

X11. More theories. LDEs of about 2 1/2 seconds delay could be reflections from the moon. Echoes delayed 8 seconds could be reflected from a cloud of ionized gas located at the L2 Lagrangian point. (R14) The L2 point is located at a stable region far beyond the moon, along the sun-earth-moon line. (WRC)

X12. Still more theories. Possible reflection from gas and dust at the Lagrangian points is analyzed, as is the speculation that the LDEs originate from an artificial radio repeater placed at one of the Lagrangian points. (R15, R16) Another paper insists that the echos are natural and very likely reflections from the Lagrangian points. (R17)

X13. July 7, 1974. Asum Odense, Denmark. While using the Moon Bounce technique at 1,296 MHz, signals reflected from the moon with the normal 2.6-second delay were followed by second echos 2 seconds later. The following day a severe radio blackout occurred coincidentally with the appearance of a large sunspot. A large streamer of ionized gas from the sun may have caused the observed second echos. (R18)

X14. A lengthy review of LDEs, plus the suggestion that they may arise from signals being trapped in ducts in the ionosphere. (R19)

X15. Statistical analysis. A search for periodicities in LDEs revealed that most were integral multiples of 138 milliseconds (the time required for a radio signal to circuit the earth). This observation strengthens the theory that LDEs result from signals that are trapped in ducts in the ionosphere. (R20)

X16. July 1980. College, Alaska. "Abstract. "A high-latitude, earth-based high-frequency (9.9 MHz) radiowave backscatter technique for the real-time detection of Long Delayed Echoes was tested at College, Alaska. Over 2000 apparent LDE were detected and recorded from 15 to 22 July, 1980. A positive correlation was obtained between the number of reflections per unit time and the probable entry of enhanced solar plasma into the magnetosphere and/or local disturbed magnetic conditions. The modes and paths by which the signals propagate are yet to be established. However, regardless of their origin, the echoes seem to be genuine and these observations are therefore unique in the study of LDE radiowave propagation phenomenology." (R21) Two important features of the study are: (1) The correlation with solar plasma entry into the magnetosphere, and (2) The confirmation that multiple echos from the same signal occur, suggesting that many ionospheric ducts may be activated. (WRC)

X17. Historically interesting observations. While studying round-the-world radio signals (erroneously called 'echos' by many investigators of the time), with delays of 1/7 second, researchers at the Naval Research Laboratory noted many echos with much shorter delays. These were explained as reflections from ionized regions in the vicinities of the magnetic poles. (R1)

References

R1. Taylor, A. Hoyt, and Young, L. C.; "Studies of High-Frequency Radio Wave Propagation," Institute of Radio Engineers, Proceedings, 16:561, 1928. (X17)

R2. Stormer, Carl; "Short Wave Echoes and the Aurora Borealis," Nature, 122:681, 1928. (X1, X2)

R3. van der Pol, Balth, and Appleton, E. V.; "Short Wave Echoes and the Aurora Borealis," Nature, 122:878, 1928. (X2)

R4. Pedersen, P. O.; "Wireless Echoes of Long Delay," Institute of Radio Engineers, Proceedings, 17:1750, 1929. (X3)

R5. "Long Delayed Radio Echoes," Nature, 124:164, 1929. (X3)

R6. "Wireless Echoes," Nature, 126:381, 1930. (X4)

R7. "Short Wave Radio Echoes," Nature, 133:560, 1934. (X5)

R8. "To Make Survey of Delayed Echoes of Radio Signals, Science News Letter, 26:23, 1934. (X6)

R9. Villard, O. G., et al; "Long-Delayed Echoes...Radio's 'Flying Saucer' Effect," QST, 53:38, May 1969. (X7)

R10. Crawford, F. W., et al; "Possible Observations and Mechanism of Very Long Delayed Radio Echoes," Journal of Geophysical Research, 75:7326, 1970. (X8)

R11. Villard, O. G., et al; "A Long-Delayed Echo AR," QST, 59:30, February 1970. (X9)

R12. Villard, O. G., Jr., et al; "LDEs, Hoaxes, and the Cosmic Repeater Hypothesis," QST, :54, May 1971. (X9)

R13. "Radio Echoes---But Wait for Them," New Scientist, 49:411, 1971. (C10)

R14. Clark, C. R.; "Two Possible Explanations for LDEs," QST, :40, November 1971. (X11)

R15. Ridpath, Ian; "Long-Delayed Signals May Echo from Moon's Ghost," New Scientist, 64:9, 1974. (X12)

R16. Lunan, D. A.; "Long-Delayed Echoes," New Scientist, 64:292, 1974. (X12)

R17. Lawton, A. T., and Newton, S. J.; "Long Delayed Echoes: The Trojan Ionosphere," British Interplanetary Society, Journal, 27:907, 1974. (X12)

R18. Rasmussen, Hans Lohmann; "Ghost Echoes on the Earth-Moon Path," Nature, 257:36, 1975. (X13)

R19. Muldrew, D. B.; "Generation of Long-Delay Radio Echoes," Journal of Geophysical Research, 84A:5199, 1979. (X14)

R20. Goodacre, A. K.; "Observations of Long-Delayed Echoes on 28 MHz," QST, 64:14 March 1980. (X15)

R21. Freyman, Robert W.; "Measurements of Long Delayed Radio Echoes in the Auroral Zone," Geophysical Research Letters, 8:385, 1981. (X16)

GER2 The Moon's Effect on Radio Propagation

Description. The changes in signal strengths, usable frequencies, fading conditions, and background noise with either the position of the moon in the sky at the receiving station or the phase of the moon.

Background. The long-distance transmission of lower-frequency radio signals depends upon their reflection from the ionosphere. At a certain critical frequency, however, reflection from the ionosphere ceases and the radio waves pass out into space. Signals above this critical frequency are limited to line-of-sight communication unless retransmitted by satellites. Several factors control the critical frequency, one of which seems to be lunar phase.

Data Evaluation. Several analyses have been found showing that the moon indeed does affect radio propagation, both through its position in the sky and phase. The bibliography below, however, must be only a small fraction of what is probably available in the unsurveyed literature of radio engineering. Rating: 1.

Anomaly Evaluation. The correlation of the moon's phase with the critical frequency may only be a confusion with the similar solar synodic period (see GEZ2), which is known to affect terrestrial ionization levels. Possibly no anomaly exists as far as this facet of the phenomenon is concerned. As for the correlation of radio propagation with lunar position at the receiver's location, lunar tidal effects on the ionosphere may be adequate to explain all data. Rating: 3.

Possible Explanations. In the early history of this question, electrostatic charge on the moon and lunar ionizing radiation were considered. Today, the moon's tidal effect on the earth's ionosphere is considered the primary lunar influence on radio propagation. Much smaller effects occur when the moon passes through the earth's magnetospheric tail disturbing it, and during eclipses when it partially shields the earth from the sun's ionizing radiation.

Similar and Related Phenomena. The moon's effect on the geomagnetic field (GEM2); the moon and the weather (GWS1); the correlation of meteor frequency with lunar phase (AY).

Examples of the Moon's Effect upon Radio Propagation

X1. Statistical analysis. 1926-1930, United States. A long series of systematic measurements of the signal strength of a Chicago radio station, operating at 770 kHz, at locations in eastern Massachusetts, showed a strong dependence upon the moon's position in the sky. The effect, as shown in the illustration, peaked at lunar altitudes of -30°, or when the moon was below the horizon. The author speculated that the moon might

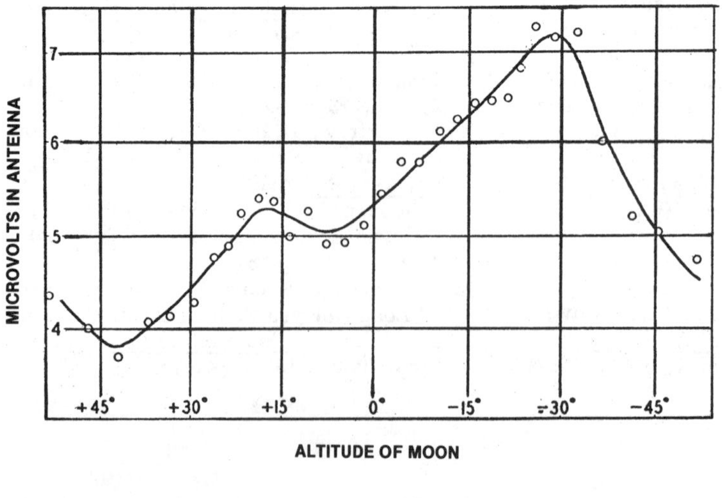

Effect of moon's altitude on radio reception (X1)

be electrically charged, since the force of gravity did not seem sufficient to perturb the ionosphere enough. (R1)

X2. Statistical analysis. 1925-1934, Hopedale, Labrador, Canada. A log of radio reception quality indicated that the height of the ocean tide correlated well with times of good and bad reception, especially fading. The tides, of course, follow the moon. (R2)

X3. 1935. United States. Radio transmission between Chicago and Massachusetts improved noticeably when the moon entered the earth's shadow during a lunar eclipse. (R3)

X4. Statistical analyses. United States and New Zealand. "From time to time, some observational evidence has been forthcoming suggesting that the strength of the signals received from radio transmitting stations over fixed paths is under certain conditions dependent upon the position of the moon at the time of observation. In particular, H. T. Stetson has investigated this matter, and in 1944 (R3) described the results of the analysis of some eight years of measurements taken at Boston on the strength of signals received from the Chicago broadcasting station. Contrary to his expectations, the evidence strongly indicated that the strength of the signals was dependent upon the age of the moon, although the effect was of a complex nature. It will also be recalled that a paper published in 1939 (Proc. Roy. Soc. A, 171, 171), E. V. Appleton and K. Weekes showed the existence of a lunar tide effect in the height of Region E of the ionosphere. The tide was found to be semi-diurnal, with an amplitude variation of about ± 1 km. If these results are interpreted according to the simple theory of the formation of the ionized regions, they indicate a relative air-pressure oscillation several thousand times the measured relative pressure oscillations at ground level.

In a recent communication to the Editors, Mr. P.A. de G. Howell, 77 Glandovey Road, Fendalton, Christchurch, N.W., New Zealand, claims to have observed during 1938-39 and 1944-45, a correlation between the variation in long-distance transmission conditions at short wave-lengths and the phases of the moon. It was observed that there was a minimum of background noise and high signal strength with little tendency to fade for about two or three days on either side of full phase, those conditions changing to a maximum of noise with poor signals and fading around the time of the new moon. The cycle of occurrences is easily confused with those associated with the solar period of similar duration, with which, however, it moves in and out of phase as the months progress. While both Dr. Stetson and Mr. Howell suggest that the effects are the result of changes in the ionization in the earth's upper atmosphere, it is clear that the whole subject requires further investigation before definite conclusions can be reached as to the cause of the effect." (R4, R5)

X5. Statistical analyses. Peru and Australia. "Not just the sun, but also the moon affects reception of shortwave radio broadcasts. Three or four days after the moon is new or full, the band of frequencies that can be used to get shortwave broadcasts through to distant stations is slightly narrower in some parts of the world than at other times. At Huancayo, Peru, where the effect is quite pronounced, the usable frequencies at noon from November to February averaged 1.7 megacycles lower three or four days after new and full moons than after the quarters of the moon. This was reported by Dr. A.G. McNish of the National Bureau of Standards at the joint meeting in Washington of the International Scientific Radio Union, American Section, and Institute of Radio Engineers.

Just as the moon raises tides in the ocean, so it produces tides in the atmosphere. That portion of the earth's atmosphere facing the moon is actually pulled toward the moon, and the earth in turn is pulled away from the atmosphere on the opposite side. Thus the atmosphere bulges out at these two regions. This football effect also works on the ionosphere, the series of invisible layers 50 to 250 miles above sea level that bounces radio waves back to earth and enables us to hear distant broadcasts. But this direct tidal effect alone does not adequately explain the low critical frequencies, Dr. McNish pointed out. Such an explanation has been advanced by D. F. Martyn of Australia who last year independently reported that the critical frequencies vary with the phases of the moon." (R6) The critical frequency referred to above is that frequency at which radio waves begin to penetrate through the ionosphere. See GHS for a discussion of the true tide-raising forces, as opposed to the simplistic explanation above. (WRC)

References

R1. Stetson, Harlan T.; "On the Correlation of Radio Reception with the Moon's Position in the Observer's Sky," Terrestrial Magnetism and Atmospheric Electricity, 36:1, 1931. (X1)

R2. Science, 80:sup 7, August 24, 1934. (X2)
R3. Science, 82:sup 10, August 2, 1935. (X3)
R4. Stetson, Harlan T.; "On the Observational Evidence for Changes in Ionization of the Upper Atmosphere Attributable to the Moon," Terrestrial Magnetism and Atmospheric Electricity, 49:9, 1944. (X4)
R5. "Effect of the Moon on Radio Wave Propagation," Nature, 159:396, 1947. (X4)
R6. "Moon Affects Reception," Science News Letter, 53:309, 1948. (X5)

GER3 Thunderstorm Modification of Radio Propagation

Description. The changing of radio propagation conditions by thunderstorms, as determined by enhanced communication with distant stations, or the direct measurement of ionosphere heights by radio reflection.

Background. This phenomenon should be related to observations of luminous phenomena high above thunderstorms (GLA9) and increases in the ionization level in the sporadic E layer during thunderstorm passage (GWC8).

Data Evaluation. Only a very few reports are in the files, but it is anticipated that surveys of radio engineering journals will yield more. Rating: 2.

Anomaly Evaluation. Generally, thunderstorms are considered phenomena of the lower atmosphere; but this phenomenon suggests that thunderstorms influence layers of the ionosphere over 50 miles up---an unexpected deviation from the conventional picture. Rating: 2.

Possible Explanations. Somehow, thunderstorms may "short-circuit" the earth's surface and the ionosphere, changing the height of the ionosphere in the process.

Similar and Related Phenomena. See Background above; the correlation of thunderstorms with cosmic-ray intensity (GLL16).

Examples of Thunderstorm Modification of Radio Propagation

X1. Frequent occurrence. Eastern United States. "Morgantown, West Virginia, U.S.A., is situated on the western slope of the Appalachian Mountains, which run in a south-westerly direction. These mountains cause a great deal of variation in the signal strength of the broadcasting stations along the Atlantic coast as received in Morgantown, and also affect the short-wave band.

One of my students, Mr. A.W. Friend, has been operating a short-wave station here for many years. He informs me that on account of the high hills near his home, his station cannot be heard in the south-eastern sector of the United States except after a thunderstorm. He can hear the amateur short-wave stations in the southern states, but they can never hear him if the weather is fine; but after a thunderstorm he can remain in contact with them for several hours.... Mr. Friend's observations strongly support Prof. C.T.R. Wilson's theory that some of the ionization in the ionosphere is due to thunderstorms. Not all abnormal ionization arises from local thunderstorms, for I have often observed increased ionization in the E-layer after sunset during the winter months when there were no thunderstorms within a thousand miles." (R1)

X2. General observations. "Thunderstorms shatter the radio 'mirrors' of electrical particles hundreds of miles above the earth and bring 'bad luck' to radio listeners in the form of poor reception. This, in substance, is the report of J.A. Ratcliffe, of the Cavendish Laboratory of the University of Cambridge, before the British Association.

The radio 'mirrors,' which reflect radio waves and make possible transmission over long distances, are in the ionosphere. Each mirror is a layer of air atoms split apart and in this way electrified. Radio waves bounce off the under side of these layers and are reflected back to earth.

Mr. Ratcliffe used radio signals to study the height of such reflecting layers during thunderstorms. The method is essentially an 'echo' one and similar to the system of determining the depths of the ocean by sending sound waves and waiting for the echo.

The one difference is that radio instead of sound waves were employed.

It was found that one electrical layer, designated by the symbol E, was 78 miles above the earth before a thunderstorm arrived. During the storm the electrical discharge pushed the layer down to 65 miles. With the cessation of the storm the height of the layer rose again to 78 miles, but within fifteen minutes it mounted to 93 miles. The electrical reflecting surface, therefore, undergoes great oscillatory motion during the storm.

Because of the up and down motion the fidelity of radio reception varies during the thunderstorm just as a beam of light reflected from a still pool is greatly different from that reflected by the water in the pool when waves are stirred up in it." (R2, R3)

References

R1. Colwell, R. C.; "Effect of Thunder storms upon the Ionosphere," Nature, 133:948, 1934. (X1)

R2. "Thunderstorms Shatter Radio 'Mirrors,'" Science, 80:sup 6, August 14, 1934. (X2)

R3. "Thunderstorms Shatter Radio 'Mirrors,'" Science News Letter, 26:169, 1934. (X2)

GER4 Stable Patterns of Electromagnetic Radiation

Description. Relatively stable patterns of electromagnetic field strength measured over the ocean. The signal source is usually a land-based broadcasting station. Here, "stability" means at least several hours, perhaps longer.

Data Evaluation. Only one set of experiments has been described. Rating: 3.

Anomaly Evaluation. Although the source of this curious pattern has not been pinpointed, it seems likely that the signal strength is modulated by semipermanent patterns of electrical currents in the ocean or sea floor. In other words, no geophysical theories are seriously challenged. Rating: 3.

Possible Explanations. Oceanic currents, including those that create the stable slicks and calms, carry electrical charge and in turn modify electromagnetic patterns above the water. The same thoughts may apply to offshore crustal currents, which are known to be strong in some regions.

Similar and Related Phenomena. Stable earth-current patterns (EC); slicks and calms (GHC2).

Examples of Stable Patterns of Electromagnetic Radiation

X1. Permanent feature. Off Los Angeles, North Pacific Ocean. "During a flight test program performed by LTV for another purpose, it was recognized that there exists a unique low and medium frequency electromagnetic radiation pattern over the surface of the ocean. This radiation pattern appears to be quite stable and is found as variations in the signal from radio stations. Statistical analysis showed correlation between patterns taken on consecutive runs over the same area to be as high as 0.9 over a one-hour period. Figures 1 and 2 (only Fig. 2 reproduced) clearly show the repeatability of the pattern during successive runs over the same areas. Figure 1 shows two patterns (runs 2 and 5) from two broadcast stations, recorded approximately twenty minutes apart, while Figure 2 shows a set of successive runs, arranged one above the other, from one station. The day-to-day repeatability is not known.

The amplitude of the variations shown in the figures represents only the top few percent of the signal strength in the area; zero has been greatly suppressed. The equipment and experimental techniques will be explained before the 'wiggles' are discussed further.

These patterns were obtained by accurately flying back and forth over the same track over the open ocean. A pair of Collins URR-23A multiband AM receivers, modified to output their delayed AGC voltage to a seven-channel Ampex FM tape recorder (with response to DC), and a Texas

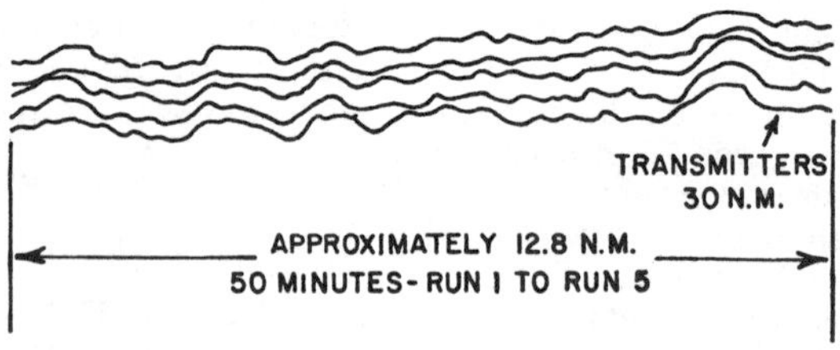

Stable patterns of radio signal amplitude off California (X1)

Instruments dual pen recorder, were used as a sensitive field strength recording system. A dead reckoning computer, radar, careful drift corrections, and smokelights were used as navigational aids to obtain repeat accuracies of a few hundred yards, at altitudes of around 1000 feet. The Navy supplied the aircraft and crew.

The data were played back from the magnetic tape through a 0.1 cps low pass filter, onto rectilinear Sanborn charts. Large photo positives were then made from these, in which the downwind (fast) runs were enlarged slightly to match the upwind (slow) runs, thus providing a series of signal strength patterns scaled to a constant ground speed. The transparencies permit these traces to be overlaid or compared in any fashion (Figures 1 and 2) or digitized for computer analyses.

What produces this pattern? It cannot be considered to be random noise, due to the close correlation of the patterns on different runs. This pattern is considered to be a natural phenomenon which has never been recognized before to our knowledge, but which can be duplicated by anyone using adequate equipment and techniques.

The exact cause of this radiation pattern is presently unknown. It was noticed during the program that the nature of the pattern varies with the transmitted frequencies which are used; see Figures 1 and 3 (not reproduced). The effects are not particularly noticeable above 5,000 feet, but the pattern is generally consistent (on the same flight path) in the range from 500 to 2000 feet as shown in Figure 3. The pattern has been recorded primarily from two, clear channel, 50 KW Los Angeles broadcast stations, but is equally noticeable from other stations nearer out usual flight area, southwest of San Diego. It has also been observed, but not mathematically correlated, on the Atlantic and Gulf coasts." (R1)

References

R1. Curtis, George D.; "An Electromagnetic Radiation Pattern over the Ocean," Undersea Technology, 5:29, August 1964. (X1)

GER5 Unidentified Radio Signals and Disturbances

Description. Radio signals or disturbances of unknown origins.

Background. Both Tesla (1899) and Marconi (1920) claimed they detected curious radio signals that might have come from Mars. The public was excited by these suggestions, but scientists treated them with disdain. Today, however, the search for intelligent radio signals from outer space is a recognized scientific activity.

Data Evaluation. Details of the Tesla and Marconi observations are lacking, but most of the later observations have been carefully described in the scientific literature. Rating: 1.

Anomaly Evaluation. Since we can only conjecture as to the sources of these radio emissions, the evaluation of their scientific significance is impossible---they could be many things (see below). Tentative rating: 2.

Possible Explanations. Unidentified human terrestrial transmissions (claimed years later to be the source of Marconi's sginals); regular natural disturbances in the ionosphere or along the transmission paths (ephemeral earth satellites would be included here); interference of signals with themselves at antipodal points; and, of course, alien signals are not impossible!

Similar and Related Phenomena. Geomagnetic disturbances (GEM1).

Examples of Unidentified Radio Signals

X1. 1899. Colorado Springs, Colorado. "In 1899, Nikola Tesla established an experimental station at Colorado Springs, where he would be free to pursue certain electrical investigations, unrestrained by the limits imposed by his New York City laboratory. In an article by him, published after his return to New York in the following year, he presented a rather thrilling account of signals received while alone one night in the experimental station: 'The changes I noted were taking place periodically and with such a clear suggestion of number and order that they were not traceable to any cause then known to me. I was familiar, of course, with such electrical disturbances as are produced by the sun, Aurora Borealis and earth currents, and I was as sure as could be of any fact that these variations were due to none of these causes... It was some time afterward when the thought flashed upon my mind that the disturbances I had observed might be due to intelligent control.... The feeling is constantly growing on me that I had been the first to hear the greeting of one planet to another.... ' I have acquired the original of Tesla's statement to the Red Cross Society, dated Christmas (January 7), 1900, which I believe is the first expression of the possibility of radio communication with other planetary communities. The announcement was harshly criticized. This was particularly unfortunate since, because of the attacks, Tesla never produced detailed information concerning the results of his intended investigations and the chance reception of signals." (R3)

Following is a sample of the reaction to Tesla's claim. "It would be interesting and possibly vastly entertaining to be supplied with the process of ratiocination by which Mr. Tesla deduces from the existence of certain puzzling electrical disturbances his 'conviction, amounting almost to knowledge, that these disturbances had been launched at our particular planet from some other planet (preferably Mars), that was desirous of intercourse. Signor Marconi has suggested that these disturbances (which seem to have worked with such brilliant results upon Mr. Tesla's imagination) were due to atmospheric electricity, which is especially active at such a high altitude as that of Mr. Tesla's laboratory; and Sir Norman Lockyer pertinently asks why, if electrical energy had been transmitted from Mars, it should have made its presence manifest in Colorado only, since all magnetic observatories in the world would have been simultaneously aware of it?" (R1)

X2. Circa 1920. No place given. "Mars is again in the public eye. For several weeks past there has been almost no end to the opinions of the leading scientists and radio experts of this and other countries regarding the mysterious wireless signals received by Marconi, the inventor of practical wireless telegraphy, at several of his stations. Marconi has stated publically that the signals, because of their exceptionally long and fixed wave length, as well as their marked similarity to the Continental-Morse code, lead one to believe that they are perhaps generated from some point outside the earth, and if so, most likely Mars.

At this late date when so many dreams of yesterday have become stern realities, it would be foolhardy indeed to discredit Marconi's statement on the face of it; for, to be sure, there is no definite proof either way to show that he is right or wrong in his hypothesis. Examining the facts in the case, we learn that the signals are of an exceptionally long wave length---a wave length far in excess of those generally employed for long-distance work. Next, we learn that the wave length is practically constant, which would naturally lead one to believe that the signals are artificially generated. Then we are told that the signals are received with equal strength at stations spaced several thousand miles apart, which is a truly remarkable fact, since, so far, any signals generated with our most powerful transmitters rapidly lose strength the farther they have to travel. Finally, the signals are said to simulate the Continental Morse code, which is that almost universally used in radio work."

Editorial comment follows to the effect that Marconi's signals probably have a natural origin. Also, why did other stations around the world not hear them? (R2)

X3. May 31, 1962, June 11, 1962, June 12, 1962, and June 25, 1962. Salisbury, Australia. "A peculiar phenomenon, to our knowledge not reported before, has been observed on the signals of Station GBR on 16 kc/s. The signals of this Station, monitored for frequency standardization purposes, are mixed with a suitable frequency derived from a 100 kc/s standard oscillator and a record, as the one shown in Fig. 1 (not reproduced), is produced. The number of beat cycles over 24-h intervals is counted and the drift of the local source is determined. At about 12.30 h local time (9 1/2 h ahead of U.T.) on May 31, 1962, the number of beat cycles suddenly increased and this faster beat persisted for about half an hour, as shown in the figure. Similar occurrences were recorded on June 11 at approximately 12.30 h local time, on June 12 at 09.45 h, and on June 25 at 09.40 h. These lasted approximately 15, 30 and 8 min respectively. All possibility of instability of local sources must be ruled out as the same derived signal is mixed also with the signals of the United States transmitter NBA on 18 kc/s, and no trace of a similar effect has been noticed on the record of this station. The beat frequency, while the described phenomenon lasted, was of a high stability not likely to be possible to derive from a local source switched on for this period." (R4)

May 31, 1962, and June 25, 1962. Lower Hutt, New Zealand. Similar results to those obtained at Salisbury (3,200 km away) were noted at 16 kc/s. (R5)

April 12-13, 1963, April 14, 1963, and April 16, 1963. Lower Hutt, New Zealand. Signal interference at 18 kc/s recorded on the above dates, similar to the earlier disturbances at 16 kc/s. The U.S. station at Boulder, Colorado, noticed nothing at 18 kc/s at these times. (R6) However, a station at Deal, New Jersey, recorded perturbations of the 18 kc/s signal at the same times as the New Zealand station, especially on April 16. (R8)

In 1964, G. A. Isted (R7) suggested that the New Zealand anomalies might be explained in terms of antipodal signal interference and changes in propagation conditions over one or both great circle routes. But the simultaneous interference in New Jersey weakens that explanation. (R8)

November and December, 1965. Many stations around the world. Simultaneous episodes of interference at 16 kc/s, lasting in the typical case a few minutes. The interpretation of these data is that there was without doubt an interfering transmitter somewhere in the world. (R8)

Alternate theory. The various records of interference at both 16 kc/s and 18 kc/s may be explained in terms of reflections from ephemeral earth satellites. (R9) See AGO, in the Catalog of Anomalies, for optical evidence and details. (WRC)

X4. Frequent occurrence. Caspian Sea and British Columbia, Canada. "British scientists are mystified by signals emanating from regions around British Columbia and the Caspian Sea that are turning off high voltage supplies in the UK research satellite, Ariel 6. After each satellite orbit over the areas, the high voltage of two experiments is switched off while a third supply is completely unaffected. The commands controlling the switching consist of a 148.25 MHz carrier with a 5 kHz subcarrier carrying the command signal so the interface has to be highly specific in order to affect two supplies but not the third. Dr Len Culhane, Chief Project Scientist in Mullard Space Science Laboratory's X-ray astronomy group, has no explanation. 'We think the signals are being generated spuriously but we can't figure out how.'.....

Two other features of the interference puzzle the experimenters. First the interference occurs only when the Sun is shining on British Columbia or the Caspian. And second the experimenters have discovered that the satellite can be 'immunised' against being switched off by beaming pure carrier frequency at it just before it enters the mysterious zones.....Alan Wells of the Mullard group thinks that the likely cause is a jolt effect---a ground based radio signal combines with electrical noise in the space craft to produce the coding signals that activate the equipment switches. The possibility that the interference has been caused by vhf blasting signals coming from a dam construction site has been ruled out as has the possibility that interference is random." (R10)

References

R1. "That Message from Mars," Scientific American, 84:34, 1901. (X1)
R2. "Those Martian Radio Signals," Scientific American, 122:156, 1920. (X2)
R3. Anderson, Leland I.; "Extra-Terrestrial Radio Transmissions," Nature, 190: 374, 1961. (X1)
R4. Rohan, P., et al; "Sudden Changes in Amplitude and Phase of the Very-Low-

Frequency Signal from Station GBR at Salisbury, South Australia," Nature, 197:783, 1963. (X3)
R5. Allan, A.H.; "Sudden Changes in Amplitude and Phase of 16 kc/s Very Low-Frequency Signal Received in New Zealand," Nature, 199:582, 1963. (X3)
R6. Allan, A.H.; "Interfering Radio Signals on 18 kc/s Received in New Zealand," Nature, 201:1016, 1964. (X3)
R7. Isted, G.A.; "Anomalous Reception of Very-Low-Frequency Signals in Australia and New Zealand," Nature, 202:994, 1964. (X3)
R8. Reder, F., et al; "Interfering VLF Radio Signals Observed on GBR-16.0 Kc/s Transmissions during November and December 1965," Nature, 213:584, 1967. (X3)
R9. Bagby, John P.; "Radio Anomalies Associated with an Ephemeral Satellite Still in Orbit," Nature, 215:1050, 1967. (X3)
R10. Schwartz, Joe; "Mystery Beams Affect UK Satellite," Nature, 280:95, 1979. (X4)

GER6 The Humming Earth

Description. A 7.8 Hz radio hum emitted by the earth-as-a-whole.

Data Evaluation. A few sentences in a scientific magazine. Rating: 3.

Anomaly Evaluation. Since an appealing physical explanation has been offerred (see below), this phenomenon is listed for its curiosity value. Rating: 3.

Possible Explanations. As suggested in X1, the earth's geomagnetic cavity may be excited by lightning and radiate like a magnetron at its natural frequency, which is close to 7.8 Hz.

Similar and Related Phenomena. Cavity oscillators (such as magnetrons); natural lasers in planetary atmospheres (AV, AM).

Examples of Terrestrial Radio Hums

X1. General observations. "The Earth gives off a radio hum at the subsonic frequency of 7.8 cycles per second according to scientists at the Massachusetts Institute of Technology. This hum they believe to be produced by lightning exciting the electrical resonance of the gigantic cavity bounded on one side by the Earth's surface, and on the other by the lower layers of the ionosphere." (R1)

References

R1. "The Humming Earth," New Scientist, 9:763, 1961. (X1)

GER7 Polar Radio Blackouts

Description. The impossibility of transmitting shortwave signals at night in the polar regions for periods of several days at a time.

Data Evaluation. A popular account in a science magazine is the only reference found to date, but once again the radio engineering journals are probably the major sources of such phenomena, and they have not been surveyed as yet. Rating: 3.

Anomaly Evaluation. Normally radio blackouts occur when high fluxes of solar radiation increase ionization levels in the ionosphere. In this phenomenon, however, the blackouts occur at night, so the presumed source of increased ionization has not been identified. Rating: 2.

Possible Explanations. One possible explanation involves ions that are normally negatively charged while in the dark and neutral when in sunlight. When the sun disappears, these ions reform and make the ionosphere more absorbent. The episodic nature of the polar blackouts does not seem to be explained by this theory.

Similar and Related Phenomena. None.

Examples of Polar Blackouts

X1. General observations. "For periods sometimes lasting several days at a time, short wave radio communication in the polar regions becomes impossible at night---an event known as polar blackout. During these periods the lower region of the ionosphere, the D region, becomes highly absorbing to radio waves, a phenomenon which can be accounted for by an increase in the ion concentration of the region during the hours of darkness. Normally, the ionizing radiation of the Sun increases ion concentration in daylight, so the reverse process is something of a puzzle." (R1)

References

R1. "Complex Ion May Cause Polar Blackout," New Scientist, 40:206, 1968. (X1)

GER8 Around-the-World Transmission of High-Frequency Signals

Description. The around-the-world propagation of high-frequency radio waves.

Data Evaluation. A single series of observations has been reported. Rating: 3.

Anomaly Evaluation. High frequency radio waves (37 MHz in the reported study) normally are not reflected from the ionosphere, limiting them to line-of-sight communication. The forward scattering (not reflection) of high-frequency signals does add to their utility in long-distance communication, but around-the-world transmissions would require many high-loss scatterings. No adequate explanation seems at hand for this phenomenon. Rating: 3.

Possible Explanations. Repeated forward scattering.

Similar and Related Phenomena. None.

Examples of Around-the-World Transmission of High-Frequency Signals

X1. No date given. Great Britain. "An unusual record of around-the-world radio reception at very high frequencies is reported by G. W. Luscombe of the Radio Research Station, Ditton Park, Slough, Bucks, England. Further studies of the mysterious method by which the signals are propagated are underway, since there is at present no satisfactory explanation for why the ionosphere would reflect radio waves of 37 megacycles by the roundabout path.

The reception in Great Britain of the delayed signals from Gibraltar was strongest between 9:00 a.m. and 10:30 a.m., Universal Time, Mr. Luscombe reports in Nature (June 20). He believes the reason for this is that the great circle path on which the radio waves would have traveled is close to the twilight zone during these hours. The great circle path is longer than the direct one, accounting for the delay. The delayed signals are heard about 140-thousandths of a second after the direct transmission. They were discovered in studies of very high frequency forward scatter propagation, which is a recently developed method of long-distance communications." (R1)

References

R1. "Around-the-World Radio Reception Reported," Science News Letter, 72:73, 1957. (X1)

GER9 Periodic Fading of Satellite Radio Transmissions

Description. Episodic rhythmic fading of radio signals sent back to earth from artificial satellites. Solar eclipses may initiate or enhance the phenomenon.

Data Evaluation. A study of satellite transmissions recorded in Australia is the only source so far of descriptions of this phenomenon. Rating: 2.

Anomaly Evaluation. Episodes of periodic fading may be caused by short wave trains of ionospheric disturbances. The existence of such wave trains is not considered anomalous, but their origin(s) have not been pinned down completely. Rating: 3.

Possible Explanations. Short wave trains, apparently similar to sound waves, in the ionosphere cause periodic reductions in radio signal strength sent through the ionosphere to earth from artificial satellites. Potential sources of the waves are solar eclipses (a thermal effect); earthquakes; large meteors; and powerful volcanic eruptions. Man-made sources might be nuclear weapon detonations; rockets; and even experimental, high-altitude aircraft.

Similar and Related Phenomena. Eclipse shadow bands (GES1); visible sound waves (GES2); lines crossing halo displays (GEH11); earthquake-generated disturbances of the ionosphere (GER12); air waves (GSW).

Examples of Periodic Fading of Satellite Transmissions

X1. Occasional occurrence. Australia. "We report here our observations of large-magnitude periodic fadings during the eclipse of 23 October 1976, which indicate that the eclipse is associated with generation of wave disturbance affecting radio-satellite transmissions in the upper VHF. The waves had a 40-km wavelength, and propagated from north to south with a velocity of 556 m s^{-1}." (Experiment details omitted.) "The distinct and prolonged periodic fadings are an unusual feature of the radio-satellite experiment conducted over several years at Brisbane. Amplitude records obtained from 150 satellite passes in October, 1976, indicated the presence of periodic fadings in 3% of passes. The periodic fadings recorded at other times than the eclipse period had a less well-defined quasi-sinusoidal structure and their magnitude was ≮ 3 dB peak-peak. Also their duration was limited to one satellite pass only, unlike the events (during the eclipse) which were recorded during three consecutive satellite passes." (R1)

References

R1. Hajkowicz, L. A.; "Periodic Fadings in VHF Radio-Satellite Transmissions during the Solar Eclipse on 23 October 1976," Nature, 266:146, 1977. (X1)

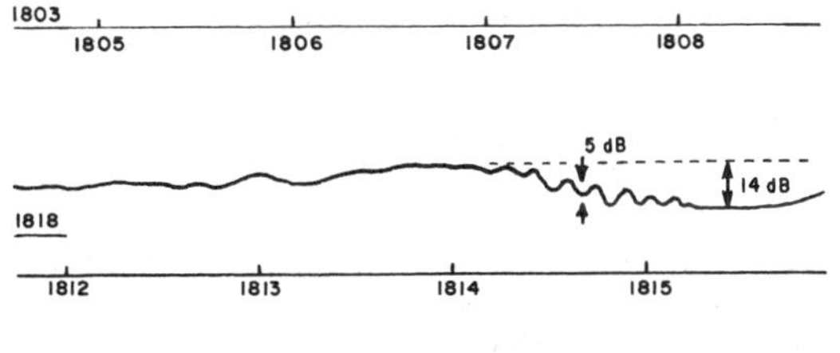

Ripples observed in amplitudes of VHF satellite transmissions during solar eclipse (X1)

GER10 Sudden Disappearance of High-Frequency Radio Transmissions

Description. Sudden fadeouts of high-frequency radio signals for periods of roughly ten minutes to twenty minutes. The only reports available occurred 54 days, or two solar rotations, apart.

Data Evaluation. Several incidents reported with scant details. Rating: 3.

Anomaly Evaluation. The (supposedly) natural phenomenon responsible for these fadeouts has not been identified. The periodicity of 54 days is especially puzzling, for it is hard to imagine a long-lived solar feature that would affect the ionosphere only every other solar revolution. No further mention of this phenomenon has been found in the literature, so we may be dealing with a transitory phenomenon. Rating: 2.

Possible Explanations. Thunderstorm activity could upset ionosopheric conditions on a nonperiodic basis (GER3); so could earthquakes (GER12).

Similar and Related Phenomena. Thunderstorm shattering of radio mirrors (GER3); earthquake disturbances of the ionosphere (GER12).

Examples of the Sudden Disappearance of High-Frequency Radio Transmissions

X1. March 20, 1935; May 12, 1935; July 6, 1935; August 30, 1935. Worldwide phenomenon. "Readers of Science may be interested in the possible occurrence of a cosmic phenomenon between October 21 and 25. The phenomenon has evinced itself as a remarkable vagary of radio transmission occurring recently at regular intervals, separated in time by twice the sun's rotation period. It is a worldwide phenomenon, or more accurately semi-worldwide, as it involves all high-frequency radio transmission over the illuminated half of the globe and not the dark half. Depending apparently on some solar emanation lasting only a few minutes, its thorough elucidation appears to call for the study of such cosmic data as solar activity, terrestrial magnetism, atmospheric ionization, aurora, earth currents, etc., as well as high-frequency radio reception.

The radio evidence of the phenomenon is a sudden disappearance of high-frequency long-distance radio signals for several minutes, the complete process of fading out and reappearing occupying about 15 minutes. It occurred on March 20, May 12, July 6 and August 30. The time intervals between these dates are close to 54 days, twice the period of rotation of the sun; the next such occurrence should therefore be around October 21 to 25." (R1 R2) Nothing further on this subject has been discovered in the literature surveyed so far. (WRC)

References

R1. Dellinger, J.H.; "A New Cosmic Phenomenon," Science, 82:351, 1935. (X1)

R2. "New Cosmic Phenomenon Reported by Radio Authority," Science News Letter, 28:300, 1935. (X1)

GER11 The Correlation of Radio Propagation Quality with Planetary Positions

Description. The apparent variation of radio propagation quality with various configurations of the planets. The correlation is complicated, depending upon various angles between the inner and outer planets.

Background. The suggestion that the planets may influence terrestrial affairs always generates considerable emotion. This is, of course, a consequence of science's long struggle to avoid all astrological trappings. Throughout this Catalog of Anomalies the subject of possible planetary influence will be found cropping up, often in unexpected places. It would be unwise to ignore this situation, onerous though it may seem, or to consider the matter as "settled", even though no causal relationships are recognized. One only has to recall that solar and lunar influences on geophysical phenomena were once derided, too, only to be reinstated when valid physical links between the earth and these luminaries were found.

Data Evaluation. Excellent radio forecasting results are claimed for techniques employing planetary configurations. However, these forecasts also involve the use of other conventional (i.e., accepted) factors. It is difficult to evaluate the reality of this phenomenon without long series of forecasts both with and without the use of planetary configurations. The situation is clouded further by suggestions of statistical flaws in the forecasting methods. Rating: 3.

Anomaly Evaluation. A real cause-and-effect connection between planetary configurations and radio propagation quality would be a scientific event of great importance, for all possibility of such a relationship is now denied. Rating: 1.

Possible Explanations. This phenomenon could be a secondary effect in the sense that the planets are also claimed to affect solar activity, which in turn influences the ionosphere and radio propagation characteristics. This, however, is not an "explanation" because there is no accepted causal relationship between solar activity and the positions of the planets ---the problem is merely transferred to category AS in the Astronomy portion of the Catalog of Anomalies.

Similar and Related Phenomena. Solar activity correlated with planetary configurations (AS); the planets and the weather (GWS8); the planets and earthquakes (GQS7).

Examples of the Correlation of Radio Propagation Quality with Planetary Positions

X1. Statistical analyses, "Many years of investigation of the vagaries of radio propagation phenomena led the author in the 1940's to a study of the correlation between planetary interrelationships and high-frequency radio signal behavior. That early study was suggested by the results of certain investigators of planetary phenomena, whose work indicated a connection between the interrelationship of the planets and the incidence and nature of sunspots. Although validity of the planet-sunspot relationship has been debated pro and con, the question is still unsettled; however, the evidence of those who supported this connection indicated that study of the correlation of planetary phenomena and radio disturbances might reveal information of value to high-frequency forecasters.

.....Technique. The technique used for the determination of the heliocentric angular separation of the planets as they circle the sun was described in detail by the author in the previous papers (R1, R2); however, the fundamentals will be repeated here briefly for for the guidance of new readers.

The heliocentric (sun-center) position of each of the sun's nine planets can be found in the American Ephemeris and Nautical Almanac published by the United States Goverment Printing Office, Washington 25, D. C. With the information given in this book, it is a simple matter to calculate the angular separation of each of the planets from any other planet.

The early papers placed more emphasis on some angles than others and stated that separations of 0°, 90°, and 180° were associated most frequently with radio disturbances, and that 30°, 60°, and 120° separations were associated with good radio signals.

Research soon indicated that some 0°, 90°, and 180° contacts did not coincide with a radio disturbance, some coincided with a moderate disturbance, and some with severe disturbances. Continued research showed that the failures were usually with single contacts of 0°, 90°, and 180° between two planets only, such as Mercury being 90° ahead of Jupiter.

It was further determined that the more severe disturbances occurred when two fast planets came into a 0°, 90°, and 180° arrangement with one or more slower planets ---thereby creating what can be called a multiple configuration. An example would be Mercury 90° ahead of Jupiter, while Venus is 0° or 180° from Jupiter. The fast planets are Mercury, Venus, Earth, and Mars: all planets further from the sun are considered to be the slow planets.

It was also found that single contacts of 0°, 90°, and 180° would be effective if there were other planets at certain harmonic angles simultaneously related to either of the two main components of a configuration.

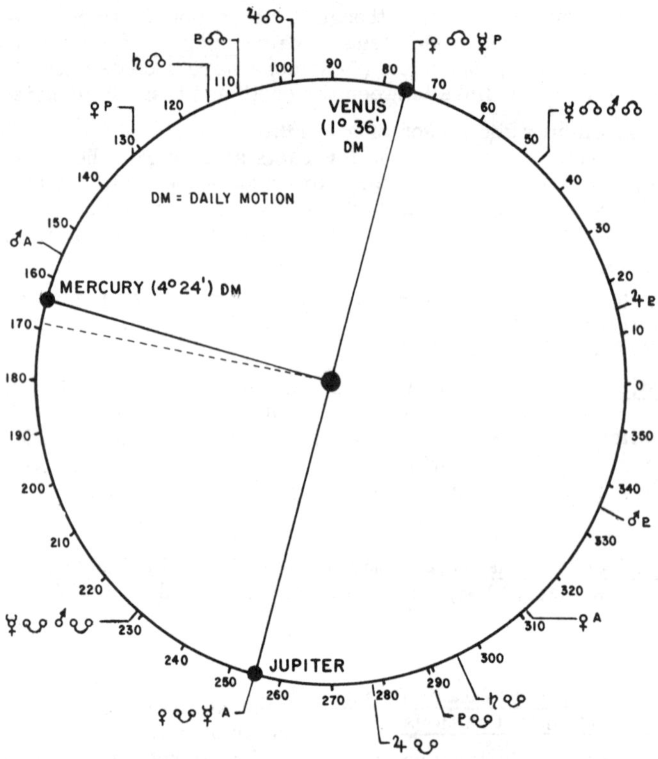

Heliocentric arrangement of Venus, Mercury, and Jupiter on February 23, 1948, which correlated with a severe degradation of radio propagation (X1)

The harmonics of the entire solar system can be found by dividing 360°, 180°, and 90° by 2, 3, 4, 5, and 6, which produces harmonics down to 18° and 15°. Under some conditions, half of 18° or 15° can also assume a role of importance. Usually, the harmonics themselves are of no importance unless there is a series of them occurring almost simultaneously---and then only when two of the planets are fast. These disturbing harmonic relationships have been narrowed down to contacts between Venus, Earth, and Mars, along with one or more of the others.

Continued research of the 30°, 60°, and 120° contacts associated with quiet conditions has produced the information that these contacts will work best if the angles are between slow planets. Frequently, there will be a disturbance when these angles occur between Venus, Earth, and Mars---particularly if they are at the same time harmonic to other planets.

The subject is extremely complicated and difficult to produce in a statistical form, since no nine-planet combination will be reproduced by the solar system in several hundred thousands of years. A combination can also occur in which Mercury is 90° ahead of Jupiter, while Jupiter is 60° from Venus and Venus is 18° from Mars. The next time Mercury is 90° from Jupiter, harmonics will not necessarily exist between Jupiter, Mars, and Venus. This applies to all multiple configurations and makes it impossible to treat the subject by standard statistics."

Nelson has employed such technique in forecasting radio propagation conditions for RCA. He claims that "These forecasts have maintained an accuracy of close to 90 percent for several years." (R3)

Some mathematicians and statisticians have faulted Nelson's work, calling the correlations "illusory." "If you have a set S1 of events happening quite randomly in time (for instance, planetary configurations) and another set S2 of events (for instance, radio disturbance days), and if you then associate each day of the second set with the nearest day of the first, you will obtain a distribution curve with a central peak just as impressive as Nelson's curves, whether or not the events are causally connected. Your technique of counting any date of set S2 from the nearest date of set S1 is not justified; it automatically results in a peaked curve!

More recently, Ianna and Margolin (R6) obtained a similar peaked curve by using simulated data. Their randomicity tests show without any doubt that the apparent relation between radio disturbances and planetary positions is an artifact of Nelson's counting methods." (R7)

Nelson claims very high success in his predictions based upon what seems to be a statistically faulty method. If he is truly successful, and if his predictions can be generated by a random process of some sort, we have a very remarkable situation! (WRC)

X2. Statistical analysis. Peak electron density of the ionosphere. "...it was, originally, our intention to demonstrate clearly the lunar influence upon the peak electron density by combining the observations of (summer) days with almost equal soli-lunar conditions, i.e. same lunar phase in a given month. This was tried with a long series of regular observations. When doing so we found the lunar effect always was appearing. However, the similarity between such days was not the same for all combinations. It was particularly good when the most important planets, Jupiter and Venus, were 'far away', roughly speaking on the other side of the Sun. If this condition is not fulfilled the similarity, though always visible, is not so strong. The different cases have been discussed above. It was found empirically that the lunar effect is, apparently, slightly disturbed by a kind of planetary influence which behaves like a gravitational one." (R5) This item bears on the subject because radio propagation depends upon electron density in the atmosphere. The similarity to Nelson's work is obvious. (WRC)

References

R1. Nelson, J.H.; "Shortwave Radio Propagation Correlation with Planetary Positions," RCA Review, 12:26, March 1951. (X1)
R2. Nelson, J.H.; "Planetary Position Effect on Short-Wave Signal Quality," Electrical Engineering, 71:421, 1952. (X1)
R3. Nelson, John H.; "Improved Disturbance Forecasting for the High-Frequency Band," Cycles, 13:155, 1962. (X1)
R4. Dewey, Edward R.; "Forecasting Radio Weather: An Interview with John H. Nelson," Cycles, 19:278, 1968. (X1)
R5. Harnischmacher, E., and Rawer, K.; "Lunar and Planetary Influences upon the Peak Electron Density of the Ionosphere," Journal of Atmospheric and Terrestrial Physics, 43:643, 1981. (X2)
R6. Ianna, Philip A., and Margolin, Chaim J.; "Planetary Positions, Radio Propagation and the Work of J.H. Nelson," Skeptical Inquirer, 6:32, Fall 1981. (X1)
R7. Meeus, Jean; "On the 'Correlation' between Radio Disturbances and Planetary Positions," Skeptical Inquirer, 6:30, Summer 1982. (X1)

GER12 Earthquake-Induced Ionospheric Disturbances

Description. The disturbance of various types of radio propagation utilizing the ionosphere after severe earthquakes.

Data Evaluation. Propagation disturbances have been reported for only two earthquakes so far. However, the data were recorded by scientific equipment employed in the study of the ionosphere and seem to be of high quality. Rating: 2.

Anomaly Evaluation. Earthquakes occur at ground level and below---some 50-plus kilometers below the ionosphere. The anomaly here involves the causal connection between earthquakes and ionospheric disturbances (as measured by propagation disturbances). It seems that earthquakes can create strong air waves (acoustic waves) (see GSW2), which can propagate long distances. The ionosphere is within easy reach of such air waves, so that these disturbances do not seem particularly anomalous. Rating: 3.

Possible Explanations. By diaphragm action, earthquakes launch strong air waves in all directions. Some of these penetrate the ionosphere, changing its height and structure, and thus upsetting radio propagation.

Similar and Related Phenomena. Air waves propduced by earthquakes and other natural phenomena (GSW); lines crossing halo displays (GEH11).

Examples of Earthquake-Induced Disturbances of the Ionosphere

X1. March 28, 1964. Essentially a worldwide phenomenon. "The Alaska earthquake, which occurred at 5:36 P.M. local time on Friday March 27, 1964 (0336, March 28, 1964 UT), initiated a large traveling ionospheric disturbance. This disturbance undoubtedly affected many different types of ionospheric sensors; however, the data reported here are limited to ionosondes operating at College, Alaska; Adak, Alaska; Palo Alto, California; and Maui, Hawaii.

The ionosonde data show indications of the perturbation in two different forms. The form depends on the distance between the observing station and the epicenter of the earthquake.

The ionograms from the two Alaska sites show the propagating disturbance as a perturbation of the F-layer trace. The less affected of these two sites was the closer station at College, Alaska, where the disturbance appeared as an abnormal spread in the trace near f_0F_2 and f_xF_2 that lasted for at least 10 minutes (Stanley and Hummel, private communication). The traveling disturbance produced a remarkably perturbed ionosphere over the sounder located at Adak, Alaska.....the ionograms starting at 0355 UT and continuing for approximately 45 minutes were quite disturbed, and the identification of the ordinary and extraordinary modes is not always clear. An abnormal increase in critical frequency is apparent on the sounding made at 0355 UT.

The disturbance is indicated at the more remote stations (Palo Alto, California, and Maui, Hawaii) by an abnormal increase, possibly preceded by a slight decrease, in the critical frequency." (R1)

Boulder, Colorado. "The purpose of this note is to draw attention to certain ionospheric disturbances observed near Boulder, Colorado, around the time of the Alaskan earthquake, which occurred about 0335 UT, March 28, 1964. These ionospheric observations consisted of the spectral analysis of ionospherically propagated radio signals by the method of Watts and Davies. Observations were made at Boulder on frequencies of 4 and 5 Mc/s, with vertical propagation, and on 10 Mc/s from WWVH in Hawaii, about 5000 km from Boulder....

The earthquake occurred at about 0336 UT near Anchorage, Alaska. The seismic disturbance first reached the Boulder area at 0343 UT (as recorded at Golden, Colorado, by the Colorado School of Mines). From the records shown in Figure 1 (figures not reproduced), there is no indication of any ionospheric disturbance associated with the arrival of the seismic wave at Boulder. However, the beginning of a disturbance was observed on the Boulder 4-Mc/s and WWVY 10-Mc/s signals at about 0352 UT.... Shortly before 0400 UT, a major disturbance occurred on all three frequencies. Following the subsidence of this large disturbance, the Boulder 4-Mc/s and WWVH 10-Mc/s signals show recurring small oscillations, while the Boulder 5-Mc/s record is dominated by the ground wave." (R2)

General observations. The ionospheric disturbances described above were probably the consequence of acoustic-gravity waves launched into the F-region by the mechanical motion of the quake. (R3)

X2. June 21, 1967. College, Alaska. An experiment was underway using the partial reflection method to study horizontal motions in the ionosphere at a frequency of 3 MHz. In the wake of an earthquake felt at College, rapid, periodic fading of the signal was noticed, with periods of 0.5 to 2 seconds. The disturbance was thought to be caused by acoustic waves created by the earthquake. (R4)

References

R1. Leonard, R.S. and Barnes, R.A., Jr.; "Observation of Ionospheric Disturbances following the Alaska Earthquake," Journal of Geophysical Research, 70:1250, 1965. (X1)

R2. Davies, Kenneth, and Baker, Donald M.; "Ionospheric Effects Observed around the Time of the Alaskan Earthquake of March 28, 1964," Journal of Geophysical Research, 70:2251, 1965. (X1)

R3. Row, Ronald V.; "Evidence of Long-Period Acoustic-Gravity Waves Launched into the F Region by the Alaskan Earthquake of March 28, 1964," Journal of Geophysical Research, 71:343, 1966. (X1)

R4. Rai, D.B., and Kisabeth, J.L.; "Ionospheric Irregularities Caused by Acoustic Waves," Nature, 216:568, 1967. (X2)

GER13 Easier Radio Transmission in One Direction Than Its Reverse

Description. Observations that VLF (Very Low Frequency) signals often seem to travel more easily in one direction than along the same path in reverse.

Data Evaluation. A short assertion of this fact in a science magazine is the only evidence for this phenomenon. Rating: 3.

Anomaly Evaluation. The ionosphere is electromagnetically complex, depending for some of its properties on the geomagnetic field. Since the components of the geomagnetic field would make different angles with the directions of propagation in the two cases, asymmetry could be introduced with little conceptual difficulty. Rating: 3.

Possible Explanations. The reflectivity of the ionosphere may vary with direction of propagation, possibly as a consequence of the geomagnetic field.

Similar and Related Phenomena. None.

Examples of Easier Radio Transmission in One Direction than Its Reverse

X1. General observations. "A curious fact has been observed with the transmission of very low frequency radio waves, is that they are often received better when they travel in one direction than in the reverse direction." (R1)

References

R1. "Why VLF Waves Travel Better in One Direction," New Scientist, 22:563, 1964. (X1)

GER14 Radar Dot Angels

Description. Radar echos typical of small, discrete targets from regions where nothing can be seen visually. Dot angels may remain stationary or move with or against the wind, but usually at low velocities.

Background. Radar is subject to various "spurious" echos from local terrain, vessel superstructures, thermal inversions, second-trace ghosts, etc. It is easy to be misled by these false echos and postulate fantastic objects (UFOs) when insects or atmospheric bubbles are much more likely.

Data Evaluation. To date only a few papers from the mainstream scientific literature have been collected. One must recognize that radar engineering has a literature of its own which has not yet been assimilated. Even so, it is apparent that dot angels are common in radar work and well-substantiated in print. Rating: 1.

Anomaly Evaluation. Whenever a dot angel appears on a radar screen, it is an easy matter to list many possible explanations from invisible natural phenomena to equipment malfunctions. Some of the natural phenomena revealed, particularly the "upward discharges" (X3) and atmospheric wave structures (X7), are remarkable and constitute the basis for adding dot angels to the Catalog. This atmospheric microstructure is probably linked to some of the puzzling visible anomalies discussed in other Catalog sections (see below). Unappreciated though these atmospheric structures may be, they do not seriously try any of the laws of atmospheric science. Rating: 3.

Possible Explanations. Atmospheric microstructures (upward discharges, bubbles, waves); insects; birds; current rips and upwellings; marine phosphorescence (GLW14); meteors; auroras; ground clutter; vessel superstructures; ghost echos (GER16); equipment malfunctions.

Similar and Related Phenomena. Radar-stimulated marine phosphorescence (GLW10); radar detection of marine phosphorescence (GLW14); low-level auroras (GLA4); rocket lightning (GLL11); eclipse shadow bands (GES1); lines crossing halo displays (GEH11); long-delayed radio echos (GER1).

Examples of Unusual Dot Angels

X1. May 8, 1949. Indian Ocean. "At approximately 0200 hours the radar was switched on. With the 'Heading Marker' checked echoes were recorded of all ships in the vicinity, but additional echoes were also observed. The false echoes were clear, firm and constant. One such echo was picked up 10° on the starboard bow at a distance of 10 miles. The echo remained on the same bearing as the distance decreased, though the size of the echo became smaller until at a distance of 3 miles it finally disappeared altogether. Although a strict visual watch was maintained no object could be seen on this bearing. Other false echoes were also apparent at the time and all proper objects, such as ships and eventually the lighthouse, were seen in their correct relative positions. As implied in the paragraph below, the visibility was good and even a darkened native sailing craft would have easily been seen.

Later, on nearing the beam bearing of Daedalus Reef Lighthouse, an echo was again picked up 10 miles distance and 10° on the starboard bow. The bearing of this echo also remained constant but its size gradually diminished until at approximately 3 miles distant it too disappeared. The accompanying sketch gives an indication of the relative positions at this moment. All echoes, false and true, changed their bearing and distance on the screen as might be anticipated. Bearings and distances of Daedalus Reef Lighthouse agreed with those obtained by normal navigational methods." The radar was working normally before and after the false echos. Radar experts, who commented on the situation, ruled out echos from the ship's superstructure and "second-trace" echos. (R1)

X2. May 11, 1953. North Atlantic Ocean. "Vessel steaming in dense fog. At approximately 2100 S. M. T., radar gave echoes on apparent objects that we were unable to identify. Echoes were similar to those given by sea clutter but were in two distinct groups. Firstly, one group to port, distant 1 1/2 miles, and about half an hour

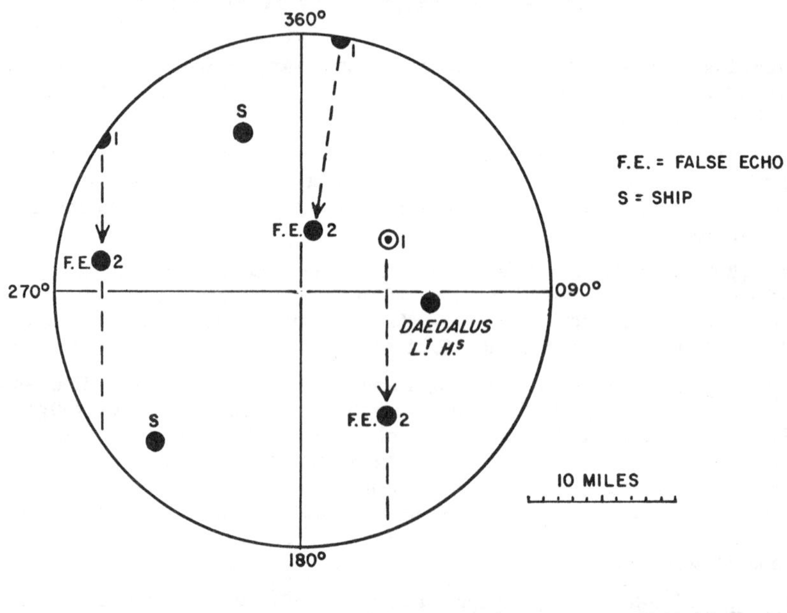

False radar echos (angels) detected on the Indian Ocean (X1)

later another group, this time to starboard, distant 1/2 mile. Speed was reduced each time accordingly. Nothing was sighted. After the objects (apparent) had been astern for about half an hour, the sea temperature started to rise until reaching 65° at 2300." The ocean temperature changes suggested the possibility of turbulence or current rips, which might have produced echos. (R2)

X3. Summer and fall 1955. Ithaca, New York. "Abstract. VHF radar observations during the summer and fall of 1955 at Ithaca, New York, gave useful information not only about lightning, meteors, and aurora, but also about the possible existence and characteristics of 'upward discharges' from the top of the troposphere to the bottom of the ionosphere.

A radar operated at 27.85 Mc/sec was used. A feature was the photographing from magnetic tape records that were played back to reproduce the original oscilloscope presentation. This technique proved to be useful in permitting more detailed analysis than had previously been possible. Photographs are presented that show examples of echoes (in an amplitude-vs-time display) that we have attempted to associate with 'upward discharges.' (Photographs of a simple meteor echo and of a meteor echo embedded in an auroral echo are also included.)

Detailed analysis led us to dissociate many of our echoes from meteors. Their rise and decay speeds, their duration, their flatness, their ranges of appearance, and the observed repetitions suggested that 'upward discharges' from troposphere to ionosphere could have been responsible for the reflection of radio waves. Nevertheless, meteors cannot be ruled out of a complete picture of the phenomenon. Meteors may play a role in the generation of 'upward discharges' as triggering agent."

.....

"The general characteristics of the echoes we attribute to 'upward discharges' are the following: (a) They are discrete and generally last less than 0.5 second. (b) They show no preferred range. The maximum range observed in the absence of aurora was around 900 km. (c) They sometimes rise from noise level to maximum value in two repetition periods of the radar, or 40 milliseconds. Calculations of the velocity required to produce the first Fresnel zone of a column of ionization (deduced from the rise time of the echo and its range) give values in some cases greater than 100 km/sec. (d) They usually decay very rapidly from maximum amplitude (at least five times noise level) to less than noise level in 100 milliseconds. The decay is not exponential. (e) They sometimes show a tendency to repeat themselves at the same range. (f) They generally occurred for two hours around midnight, but did not occur every night. During the months of September, October, and November, 1955, they occurred on more than half the nights. Very few were observed during December. As many as 400 echoes in one hour were seen. No correlation with any known phenomenon was established." (R3) These so-called "upward discharges" may be related to low-level auroras (GLA4) and rocket lightning (GLL1).

X4. May 25, 1960. North Atlantic Ocean. "A radar echo was plotted at 0340 GMT and it was found to be that of a stationary object; no lights were seen, nor were flashing signals answered. The echo was picked up at 9 miles and showed fairly well; it was intermittent at 5.2 miles, the nearest approach, and when abaft the beam it was not detectable beyond 7 miles. The weather was cloudless and there was an auroral glow, but no moon." (R4) An iceberg was suggested as the target.

X5. General observations. Abstract. "Recent independent observations of 'dot' angels are drawn together to provide a coherent picture of their physical structure. The angels act as point targets with durations proportional to beam width (for vertically pointing radars). The echoes are strongly coherent, indicating a smooth specular-like surface. Dual frequency measurements in the band 1 to 3 cm indicate that their radar cross sections are proportional to between the first and second power of wavelength. A range square dependence is also indicated. When tracked, the cross sections show strong enhancement at the zenith, decreasing roughly symmetrically on either side. Doppler measurements indicate that they almost always ascend with speeds of about 1 m sec^{-1}.

The observations are shown to be consistent with specular reflections from the hemispherical, concave-downward cap of a rising thermal convective bubble. The specular point of the cap appears only briefly in a vertically pointing beam and so the echo appears to be from a point source. The concave-downward surface provides at least partial focusing, thus accounting for both the magnitude of the cross sections and their range square dependence as the bubble expands with altitude. Furthermore, only rising bubbles have such favorably disposed upper surfaces. The dual frequency data indicate that the transition zone across the bubble cap is extremely sharp, of the order of 0.5 cm. The sharp transition, the smoothness of the cap, and the slow ascent rates suggest that the flow is laminar." (R5)

X6. General observations. Noting that dot angels have been ascribed to birds, insects, and atmospheric refractivity perturbations, the authors describe their experiments in tracking insects with radar. "It has been found, for example, that the 'dot angel' echoes appear to be highly localized point targets; the targets, at a given time and point in space, have nearly equal cross sections; the targets can be tracked for as long as 35 minutes; they frequently have a significant velocity relative to the mean wind velocity, and they are not observed in winter. Moreover, no single coherent atmospheric surface could be found to explain the combination of cross-section observations.....The observations presented here confirm the view that insects are highly localized point targets which can be tracked for long periods and which may or may not have a significant velocity at a large angle to the wind direction.....The data thus strongly suggest that 'dot angel' echoes having characteristics similar to those described." (R6) Obviously, this study is at variance with X5. (WRC)

X7. Experiments with high-resolution radar. "So-called radar 'angels' have correctly been attributed to both birds and insect colonies but there appeared to be a third type of unexplained radar echo---that caused by atmospheric perturbations in refractive index on too small a scale for conventional radar systems to analyse. (David) Atlas, (Juergen) Richter, and E. E. Gossard have now amply established this additional source of 'angels'.

Their experiments reveal the presence of unsuspected layers, no more than a few metres thick, associated with temperature inversions, in which turbulent eddies of the size of a few cm occur. These turbulent layers reflect 10 cm radar waves with a wavelength dependence quite unlike that of rain storms or ordinary clouds, and with reflectivities 10 to 100 times greater than for normal radar echoes.

Examination of the layers, whose origin remains mysterious, shows that they are ruffled both by long-period gravity-controlled waves, and by much faster waves which roll up and break like sea-waves.

According to Atlas, the thin layers form, initially flat, at the junction of underlying moist air and overlying dry air. Although the inversion possesses inherent hydrostatic stability, very fine-scale turbulence seems to arise, transporting air of low refractive index downwards and vice versa. Radar waves of 10-cm length 'see' only eddy textures with a granulation of five cm. These microscale eddies, or 'blobs' as Atlas termed them, enable detailed mapping of the changing shape of the newly discovered layers.

This development is illustrated by some typical observations the team made last July on a layer 265 m high. After 20 minutes of undisturbed flatness, the layer began to corrugate into waves that rolled over and eventually broke and dissipated. The largest wave observed reached a peak-to-peak amplitude of 150 m. As these remarkable waves broke across the shingleless beaches of the sky---either backwards or forwards, according to the circumstances---it was clear that their lower echoing strata were being stirred up by eddies some 40 to 50 m across. .....Dot-like echoes on the team's records are made by insects---probably houseflies at heights of around 350 m---which appear to rise and fall in concert with the laminar waves." (R7) One sees immediately a connection between these waves and the enigmatic eclipse shadow bands (GES1) and lines moving across halo displays (GEH11). (WRC)

References

R1. "Unusual Report on Radar Performance from the T. S. S. Clan Davidson on a Voyage from Suez to Aden," Marine Observer, 20:212, 1950. (X1)

R2. Goodman, F.; "Unidentified Radar Echoes," Marine Observer, 24:81, 1954. (X2)

R3. Rumi, G. C.; "VHF Radar Echoes Associated with Atmospheric Phenomena," Journal of Geophysical Research, 62:547, 1957. (X3)

R4. Marrow, J. K.; "Unidentified Radar Echo," Marine Observer, 31:61, 1961. (X4)

R5. Atlas, D.; "Angels in Focus," Journal of Research NBS, 69D:871, 1965. (X5)

R6. Glover, Kenneth M., et al; "Radar Observations of Insects in Free Flight," Science, 154:967, 1966. (X6)

R7. "Break, Break, Break, in the High Clear Air...," New Scientist, 46:319, 1970. (X7)

GER15 Radar Ring and Line Angels

Description. Radar echos in the shapes of long lines, arcs, and circles in regions which no obvious physical targets exist. Circular or ring angels often appear as multiple ripples concentric around the radar transmitter. Some ring and line angels move along with the radar, which is frequently on a ship; others are stationary.

Data Evaluation. Several good, well-detailed accounts of ring and line echos are available. As mentioned in GER14, the literature of radar engineering will doubtless yield many more. Rating: 1.

Anomaly Evaluation. As with radar dot angels (GER14), a multitude of natural and electromechanical causes exist for ring and line angels (see below). Although radar experts perhaps seem a bit too quick to write off strange echos as "atmospheric discontinuities", it does appear that prosaic explanations can be found for just about all ring and line angels. Actually, as in GER14, it is the atmospheric discontinuities themselves that require further explanation. Rating: 3.

Possible Explanations. Flights of birds; locust swarms; interference between echos from sea clutter; atmospheric ducting; atmospheric discontinuities; second-trace echos from distant linear targets; equipment malfunctions.

Similar and Related Phenomena. Radar stimulation of marine phosphorescence (GLW10); radar detection of marine phosphorescence (GLW14); radar dot angels (GER14).

Examples of Ring and Line Angels

X1. Frequent occurrence. England. Using a high-power, L-band radar. "The ring angel commences as a strong point echo; it then extends into an annular ring, to be followed by other rings at approximately equal intervals of time. The whole phenomenon has precisely the same appearance as the ripples on a pond expanding from a point of disturbance. In the case of the rings shown on Fig. 1 (not reproduced), eighteen rings were emitted from the centre with an average periodicity of 2 1/2 min. and a maximum diameter of 20 miles was achieved.

In the course of the present work we have observed sixteen such incidents..... The ring angels tend to occur around certain active regions..... The ring located at Ipswich has yielded five incidents during the past four months. Detailed examination of the map and of the areas involved has not revealed any topographical feature which might be a contributory factor to ring formation; nor are they regions of unusual bird roosting.

It will be noted that no ring angels have been observed over the sea, and that the ranges of the rings from the radar station, when taken in association with the radiation pattern of the radar, suggest that the reflecting or scattering elements of the ring occur at low altitude only. It has been possible to make some height measurements with the aid of a 10-cm nodding height-finder radar. These results, if interpreted as heights in the ordinary manner, indicate that the altitude of the rings does not exceed 2,000 ft..... The rings we have observed have all occurred at the sunrise period, and in most cases the presence of the ring centres has been noted for a few minutes before sunrise. This correlation with sunrise is quite vividly seen on the radar display, since the sun shows itself as a radial noise point upon the plan position indicator and ring angels and solar noise then exist simultaneously upon the tube. We have noted as many as three incidents occurring simultaneously about the various active centres." (R1)

Further study of ring angels, backed up by trips to the various sites, demonstrated conclusively that the English ring angels were due to flocks of starlings dispersing episodically from roosting sites. Similar ring angels observed in Texas were attributed to redwinged blackbirds. (R2)

X2. June 18, 1962. Red Sea, Indian Ocean. "At 1720 GMT shortly after moonrise an echo was observed on the PPI (Radiolocator Mk. IV), 12 miles off the starboard quarter of the ship, similar to that returned by a rain squall. It increased in size and closed the ship, so that by 1750 it stretched from either beam in a semi-circle round our stern at a range of about 10 miles. At 1820 the echo suddenly split astern of the ship, the strongly defined left arm sweeping away in a clockwise direction, while the nearer right hand arm continued to close our starboard quarter, coming to within 9 miles of the ship. By 1850 the two arms had joined together again but the strength of the echo as a whole had diminished, and it gradually receded astern, being completely lost on the PPI by 1915. In view of the dusty condition of the atmosphere at the time as indicated by the rising moon, the echo was assumed at first to be due to sand, but this idea was discounted when the m.v. Border Castle, known to be some 15 miles away on our starboard quarter, was contacted by W/T. This vessel reported that it had encountered no sandstorms and that she had experienced a similar echo astern of her, at the same time as our contact. The possibility of the image being a second-trace echo from land seems unlikely, for the sides of the Red Sea are between 64 and 80 miles away from the ship at this position." A radar expert ventured that the echos were probably due to "atmospheric discontinuities". (R3)

X3. June 29, 1963. Red Sea, Indian Ocean. "Between 1430 and 1630 GMT while on passage through the Red Sea, the radar interference shown in the diagram was experienced. It took the form of concentric bands about the ship, generally at a distance of about 6-8 miles away. For the most part there were eight bands but in some places 12 could be counted. The furthest away was the broadest, 3-4 cables, while the nearest was the narrowest, being about 1/4 cable wide. The spaces in between the bands also increased with increasing distance away. The exact shape of the bands was never constant but was at all times changing slowly. Sometimes a part moved toward the ship, sometimes it moved away; occasionally a part would fade, or bands would virtually joint, at times becoming quite solid. This form of interference persisted for about two hours. Both before and after this time, the interference showed no definite form but quite often it resembled second trace echoes from the mountains on either side though we doubted that this was really the case as the shape was continually changing. At one stage, the passing of a ship in the opposite direction caused the bands to join and become solid, as shown in the diagram. Ships going in the same direction as ours made no difference to the shape of the bands." An explanation was offered that involved backscattering from the sea taking two paths; one along the surface, the other reflecting from a low layer of the atmosphere; and combining to produce the interference pattern. (R4)

X4. June 22, 1964. Red Sea, Indian Ocean. "An unusual radar echo was observed, in the form of an irregular ring around the vessel. At first this was thought to be a second trace return, but subsequent observations of the echo seemed to disprove this

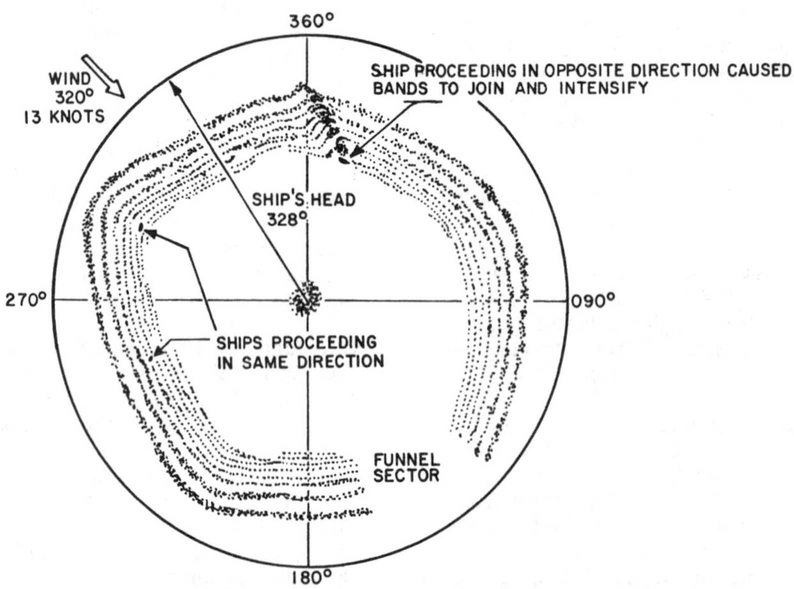

Ring angels appearing on the radar scope of a vessel in the Red Sea (X3)

and our opinion is that it marked the limits of the dust cloud under which we were proceeding.....An interesting feature is the distinct double echo 3 miles outside the original one, astern and on both quarters. The ring subsequently extended outwards---principally in a s'ly direction---becoming fainter as the distance increased. At the same time the sky overhead cleared quite considerably, revealing 3/8 C_H2. The echo finally disappeared at a range of 25-30 miles." A radar expert once again blamed atmospheric discontinuities. (R5) The reader should remark that examples X2-X4 all occurred in June in the Red Sea. (WRC)

X5. July 20, 1970. North Pacific Ocean. "At 1705 GMT a long line was seen on the radar running 005-185°, extending for over 40 miles in length at a distance of 10 miles from the ship on the port side. At 1714 it was 8 miles away. Visibility at the time was very good, no special cloud formation or change in the sea was observed and there was no precipitation. At 1719 the line was 6.4 miles from the ship and at 1723 it was 5.5 miles. On getting a little closer the width of the line was approx. 1 cable, fixed by radar range. At 1740 the vessel entered the line; there was no change in wind direction or speed, in sea or air temperature, or in pressure. The only visible change was a distinct break in the cloud which had been a continuous layer of St at 300 ft with ragged shreds. Through the break could be seen Cc cloud. The radar continued to show this line until 1810 when it slowly faded out. The several points of interest are the two breaks in the line, nothing unusual being actually seen except for the break in the cloud, no change in wind or temperatures on passing through the line and, of course, its very great length. Afterwards a true plot was done on the line's movements and it can be seen that the speed was a some stages was nil and at others very slow. The conclusion is that it was not atmospheric but how can one explain the break in the clouds?" This curious line echo was ultimately dismissed as an atmospheric discontinuity. (R6) The break in the clouds would certainly underscore this conclusion. (WRC)

X6. January 28, 1973. North Atlantic Ocean. "At 1825 GMT, after coming on watch, a curious effect was observed on the radar screen when the vessel was about 70 miles north of Cap Vert. A ring of echoes, similar to land and about 4 miles deep, was forming in a concentric circle about the centre spot of the screen. For the first hour or two the echoes varied in diameter from 13 to 15 miles. Ship echoes were still

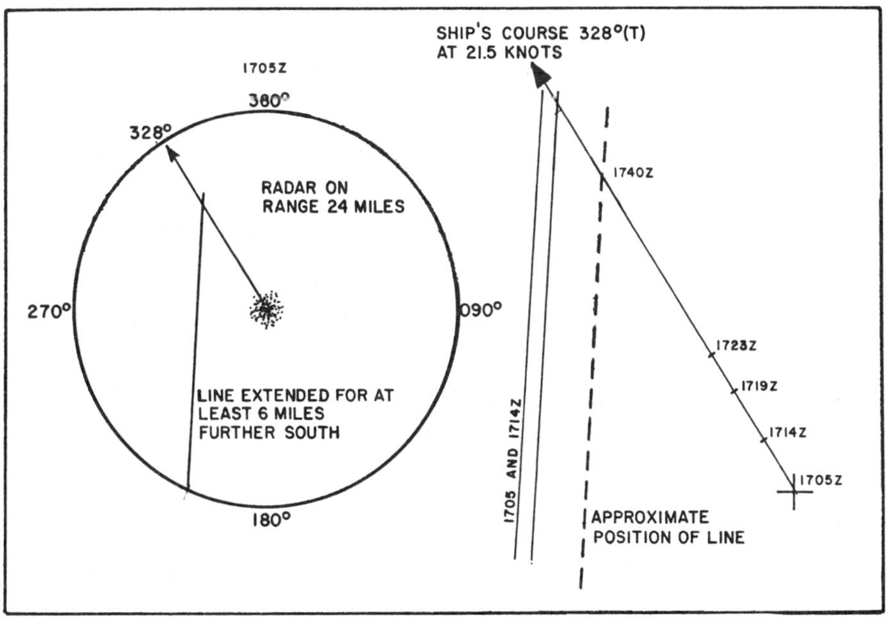

A linear radar angel over 40 miles long observed in the North Pacific (X5)

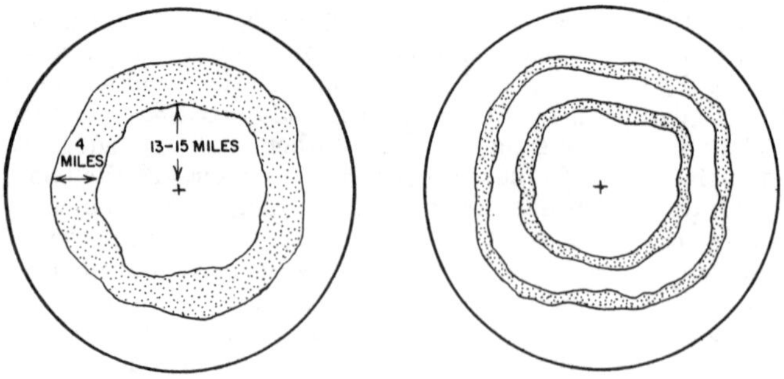

Radar ring angels near Cape Verde that may have been due to dust (X6)

displayed, apparently unaffected. After several hours the echoes became less intense and a second concentric ring formed; shortly afterwards the top half of the circles began to fade. Just before going off watch all that was left of the effect was the odd patch of echoes still concentric about the centre spot. (This effect had been observed by the 4-8 watch the night before though not recorded.)" Since the atmosphere around Cape Verde was heavy with dust at the time, it was thought that the echoes were due to particulate matter in the atmosphere. (R7) Cape Verde is, of course, noted for its dust "fogs" (GWD4). Some of the Red Sea ring angels reported above also involved the presence of dust. (WRC)

References

R1. Eastwood, E., et al; "'Ring Angels' over South-East England," Nature, 183: 1759, 1959. (X1)

R2. Eastwood, E., et al; "Radar 'Ring Angels and the Roosting Movements of Starlings," Nature, 186:112, 1960. (X1)

R3. North, J.; "Unidentified Radar Contact," Marine Observer, 33:66, 1963. (X2)

R4. Clarke, H.S.; "Radar Interference," Marine Observer, 34:65, 1964. (X3)

R5. Ramsay, J.R.; "Radar Echo," Marine Observer, 35:115, 1965. (X4)

R6. Cowen, E.; "Radar Phenomenon," Marine Observer, 41:94, 1971. (X5)

R7. Weston, J.H.B.; "Radar Echoes," Marine Observer, 44:13, 1974. (X6)

GER16 Radar Ghosts

Description. Radar echos of distant targets that appear close, or smaller repetitions of normal echos appearing alongside the normal echos. In this phenomenon, real targets do exist but the radar shows them displaced in range or azimuth.

Data Evaluation. Ghosts with the incorrect ranges are rather common, being called second-trace echos; but examples of azimuthal ghosts are hard to find. Rating: 2.

Anomaly Evaluation. Ghost echos with erroneous ranges (appearing close when they are actually far away) can almost always be shown to be second-trace echos. No natural or equipment anomaly is involved here; the real targets are so far away that they appear on the traces of subsequent radar cycles. Azimuthal ghosts are more mysterious, with no easy explanation, save for "equipment malfunctions". Rating: 3.

Possible Explanations. Second-trace echos; equipment malfunctions.

Similar and Related Phenomena. Radar dot angels (GER14); ring and line echos (GER15).

Examples of Ghost Echos

X1. November 1, 1952. Off Cape Cod, North Atlantic Ocean. "2345 G.M.T. When in position 41° 42'N, 69° 27'W, ships were observed at a distance of 50 miles (maximum radar range). Nantucket Island, Cape Cod and other land in the vicinity showed a clear outline having their correct bearing and distance from the vessel. This condition was observed for about 14 hours. Odd reflections appeared, having the character of normal vessels on the 15 mile range. About four or five images of vessels would appear at a distance of 8 to 10 miles in a straight line bearing NW, each image having smaller images on each side. These images would approach the PPI center until they were lost in the sea return and then similar images would appear at a distance of 10 miles.

These ghost images continued for about 8 hours, a distance of 125 miles from Nantucket Shoals Buoy No. 10, on a course of 343° towards Portland. Radar interference was also present in the vicinity of these images, extending from the outer edge of the screen to the center, having an arc of 10° to 15° width. The interference appeared as small white dots arranged in even rows all the same size." (R1)

X2. September 18, 1964. Mediterranean Sea. "0215 GMT. On sighting the looms of both Cap Corse and Revellata Point lights on the NW corner of Corsica, the radar was switched on to check the ship's position. The coast of Corsica appeared on the screen at approx. 35 miles which checked with the DR position. However the coast of Italy to the NW'ward appeared equally as strong on the screen at a distance of 17 miles when the ship was actually 55 miles from that coast. The lights already mentioned were at distances of 44 and 32 miles respectively and both are 23-mile lights. The Italian coast remained on the screen until about 0235 and did not appear again." The echos from the Italian coast at 17 miles, rather than the actual 55 miles, probably represent a second-trace echo from mountains about 43 miles behind the coastline. (R2) A second trace echo is an echo from a very distant target originating from a radar pulse one cycle previous. (WRC)

References

R1. Mullock, E. A.; "False Images on Radar," Marine Observer, 23:207, 1953. (X1)

R2. Jones, C.; "Radar Performance," Marine Observer, 35:117, 1965. (X2)

GES SHADOW PHENOMENA

Key to Phenomena

GES0 Introduction

Disturbances propagating through the atmosphere may become visible in unexpected ways. Shock waves from explosions and supersonic jets, usually invisible to the eye, are sometimes traced across the sky with geometric precision if atmospheric conditions are just right. Nothing mysterious here; aeronautical engineers photograph shock waves every day in wind tunnels using optical instruments to detect their different air properties. Yet, naturally visible shock waves are rare and, because most are observed under the stress of battle, quite disconcerting to the viewer.

Definitely more mysterious are the solar eclipse shadow bands that drift eerily past observers along the path of totality. They are usually explained as either shadows of ripples in the atmosphere (analogous to those on the bottom of a wind-ruffled swimming pool) or diffraction patterns caused by light bending around the moon's edge. Several peculiarities of eclipse shadow bands imply that neither explanation is correct. So what are they? There is no universally accepted theory. Probably they are related in some way to an invisible and unexplored fine structure woven into the atmosphere. The long-lived "aerial blobs" described below also hint at this unseen atmospheric fabric, often quite regular in structure, which is cast like an invisible net across the clear sky.

GES1 Eclipse Shadow Bands

Description. Long dusky strips and aligned splotches that appear on the ground, on buildings, and on other light-colored backgrounds during solar eclipses. Eclipse shadow bands are usually solid, though not intense. Some appear discontinuous; some display fine structures of tiny lines; others shimmer. Widths range from fractions of an inch to several inches---as do the interband spacings. Although most shadow bands move across the landscape at rates up to about 40 miles per hour; some are stationary, and still others change their directions of motion repeatedly. Several cases of two or more sets of superimposed bands with different velocities, spacings, etc., have been reported. Eclipse shadow bands generally occur a few minutes before and after totality, but some are seen during totality. Appropriate adjectives for the bands are "variable" and "inconsistent".

Background. Almost everyone today thinks that eclipse shadow bands are simply shadows cast by inhomogeneities in the atmosphere. The suggestion that they might be diffraction patterns created by the moon is rarely discussed any more, because the fastest bands move at only about 1% of the speed of the moon's shadow. It must be emphasized that observations of the shadow bands contain many contradictions; and one should not write the phenomenon off as explained until all facets are accounted for.

Data Evaluation. Shadow bands are observed during almost all solar eclipses that are total, with the annular and partial eclipses contributing but a few reports. The literature is nevertheless not massive, for most scientists concentrate their attentions elsewhere during eclipses. Shadow bands seem to be written off as one of nature's miscellaneous productions and not worthy of serious attention. The reports cataloged below represent only those dealing with unusual features of the bands. Many other accounts exist. Rating: 1.

Anomaly Evaluation. Although most scientists consider eclipse shadow bands to be completely understood, the examples below suggest otherwise. The phenomenon seems almost infinitely variable and full of inconsistencies---like the weather in a way. Despite these elusive, often contradictory aspects, it still seems likely that they are merely manifestations of atmospheric inhomogeneities and nothing more profound. Rating: 2.

Possible Explanations. Shadows cast by patterns of inhomogeneities in the atmosphere. A few still like the diffraction pattern explanation.

Similar and Related Phenomena. Patterns seen in the areas illuminated by searchlights (GES6); crescents and other images formed by the foliage during eclipses (GES5); marine phosphorescent bands and wheels (GLW), which structurally at least bear an uncanny resemblence to eclipse shadow bands (see X6, the Newport spoke structure).

Examples of Eclipse Shadow Bands

X1. September 7, 1820. No place given. "The appearance of moving fringes of light seen on a whitened wall during the total eclipse of Dec. 31, 1861, is remembered to have been seen by M. Goldschmidt in the annular eclipse of Sept. 7, 1820. He perceived them fully two and a half minutes before the annulus was formed. He was walking at the time in a direction from east to west, when he saw the moving shadows coming toward him slowly. The movement was not rapid, and the aspect like the shadows of smoke in sunshine; the forms being rhomboids of four or six inches in diameter, mixed up with ribbon-shaped shadows. The inner spaces were filled with round dots mixing gradually with the other in gray transparence. M. Goldschmidt saw this strange apparition whilst he walked for about one hundred steps, when the annulus was suddenly formed, the light of the sun running round the moon like a fluid." (R1) The peculiar shapes of the shadows may be connected with the annular nature of the eclipse. (WRC)

X2. January 1, 1889. California. "In the eclipse of August 19, 1887, observed in Russia, it was cloudy and no shadow-bands were seen, but in the eclipse of January 1, 1889, observed in California, with a clear sky, the bands were well-defined, though an attempt to photograph them failed. They were more prominent at high altitudes than at low levels, but they seem to have no connection with the position of the stations in or near the shadow-belt. While the reports of the various observers indicated a general agreement for the direction in which the bands lay, yet there was no uniformity in the direction of progression which seemed not to be related to the direction of the wind. In every case the speed of the bands was much less than that of the shadow itself, thus disproving the theory that the bands are diffraction fringes in the shadow of the moon." (R2) Note that the effects of altitude and wind on the shadow bands do not seem to be consistent from one case to another. (WRC)

X3. April 16, 1893. Chile. "During the eclipse of April 16, 1893, observed in Chile under the most favorable circumstances, the shadow-bands were very generally seen immediately after totality. They lay approximately northwest and southeast, and moved mostly towards the southwest at a speed variously estimated at from three to twenty miles an hour. The width of the bands appeared to vary from one-eighth of an inch to four inches, and their distance apart from one to ten inches. A significant fact was that, contrary to the observations in the previous eclipse, the bands were much less conspicuous on the mountain summit, occupied by the writer, than near sealevel, where they were also coarser, thus indicating the effect of increased thickness of atmosphere." (R2)

X4. June 8, 1918. Corona, Colorado. "Too cloudy to see shadow bands at beginning of totality; shadow bands about 10 cm wide and 20 cm apart. An entry two minutes later reads: Brisk wind from N. W. at end of totality, shadow bands 15 degrees S. of E. traveling 2.5 to 3 m. per sec. toward S. E. by S. Would this not imply the visibility of shadow bands at mid-totality? It is also interesting to note here that although a brisk wind was behind the shadow bands they were hardly moving as fast as the shadow bands that we observed at Beacon, N. Y., last year, although there the air was absolutely calm. If the shadow bands are due to surface air currents why should their motion not be speeded up appreciably by a stiff wind?

Observations of shadow bands during totality have a very important bearing upon the question of the cause of the shadow bands in general for if it can be proved beyond any question that shadow bands are seen---if only occasionally---during the total phase of the eclipse then the thin solar crescent immediately preceding and following totality must be eliminated as the source of light in the production of shadow bands and as the cause of their peculiar banded appearance. Eliminating the sun as the source of light in their production there is left only the general sky illumination, the corona or possibly light from conspicuous prominences. Also if the shadow bands are simply the shadows of surface air currents and can be seen occasionally during the darkness of total eclipse, one can think of many other occasions when they should be seen, for example daily around the time of sunrise and sunset when the glare of sunlight is sufficently reduced. Is it not also true that all of the so-called shadow bands produced by searchlights and other artificial means differ in some very important and essential details from the shadow bands that are associated with total solar eclipses?" (R7)

X5. September 22, 1922. Wallal, Australia. A good general impression of the shadow bands. "Suddenly, and before I had expected it, the shimmering, elusive, wave-like shadows began to sweep over me and the sheet. I grasped the rod and moved it back and forth until it was parallel to the crests of the waves as they moved forward. At the same time I began counting seconds ---one-and, two-and, three-and, etc.--- and I continued counting up to 150 before totality was announced. I think perhaps I counted a little too rapidly and, allowing for that, I judge that the bands began approximately 2 1/4 minutes before totality. They continued to move over me for only a short time, perhaps ten seconds, and then they were gone! They were faint, thrilling, ghost-like, but definite enough for me to be sure that I had seen the shadow bands..... I took my position at the second sheet to watch for the shadow bands again. They began at 15 seconds after totality and to my surprise they seemed to come from the opposite direction. They were even fainter than at the beginning and lasted perhaps five seconds. I moved the second rod until I judged it was parallel to the crest of the waves.

At the beginning of totality the bands came from west to east, moving in the same direction as did the moon's shadow as it swept across the earth; at the end they moved in the opposite direction, or so it seemed to me.

In a note concerning these observations Professor C. A. Chant concludes, with some uncertainty, that the direction of the line of crests of the shadow bands was at right angles to the direction of advance of the shadow cone and tangent to it. The bands were seen 2 1/4 minutes before totality, and as the cone moved forward with a speed of 25 miles per minute it would appear that the air producing the effect was about 56 miles in front of the cone. Professor Chant thinks it almost certain that the bands are caused by the light from the slender crescent of the sun shining through layers of air of continually changing density. From the fact that the bands lasted only a few seconds after totality he concludes that the atmosphere became quiescent at a shorter distance behind the shadow cone than it was in front of it." (R4)

X6. January 24, 1925. A total eclipse widely observed in the United States, with many accounts of shadow bands.

A general summary. "The reports differ considerably in detail, but in respect to the essentials they are much alike, and practically the same as those of the shadow bands seen at the times of various other eclipses. A composite of these reports may be summed up as follows: When the bands were seen: From totality to 2 or more minutes before and after, even 20 minutes being reported. Perhaps the extreme cases were owing to heated air from chimneys; one or more were proven at the time to be so caused. Length of bands: Broken, confused arcs, short splotches to 3 feet in length; generally 1 to 2 feet. Width of dark band: 1/4 inch to four inches, increasing with time before and after totality. Width of bright band: 1 1/4 inches to 3 feet, also increasing with time from totality. Velocity of travel: Stationary (shimmering) to very fast, estimated as possibly 30 to 40 miles per hour, at right angles to length of bands. Character: Narrowest, most distinct and most sharply defined immediately before and after totality, becoming broader, farther apart and more indistinct, unto disappearance, with increase in width of visible solar crescent. The bands were substantially the same after as before totality, and very much like the shadow bands cast upon a vertical screen by a distant, powerful arc light. Their description, therefore, appears to be practically complete in every detail. Their explanation, however, is not so satisfactory. Evidently, as Hastings pointed out long ago, the only conspicuous motion of the bands must be at right angles to their length. A shadow moving along itself, being everywhere alike, would not seem to be moving at all. Also, as he further explained, if these obscurations are owing to irregularities in the density of the atmosphere, they must lie mainly parallel to the crescent of the sun. This is because the ends of the crescent, more than its sides in proportion to their greater distance apart, would shine around every such disturbance, except those that lie parallel or nearly so to the crescent itself, and thereby restrict to the latter the deepest shadows. From these two generalizations, (1) that the shadow bands must parallel the solar crescent, and (2) move, apparently, at least, normally to their length, it follows that there can be no relation between the direction of their travel and the course of the wind at any level. " (R8) Such reasoning may be simplistic, as subsequent data will show. (WRC)

William Henry, in America, reports observing bright bands among the dark. (R5) This report is confusing because it does not specifically say that these bright bands were brighter than the background light level. (WRC)

Westerly, Rhode Island. "At Westerly, R.I., where I saw the shadow bands within the area of totality on January 24, 1925, there was one set of bands moving with the wind and others crossing them at angles so that at times a confused effect was produced." (R6) The rhomboids of X1 and other reports of short or splotchy bands could be the result of interfering sets of shadow bands. (WRC)

Newport, Rhode Island. "The only interesting thing which we observed which I have not seen reported in the papers, but which doubtless has been observed more satisfactorily by others, concerned the shadow bands. These were very marked indeed. They began a few minutes before the eclipse coming somewhat from the north of east; perhaps east-northeast. The bands were about 18 inches apart; according to some observers these were subdivided into much finer bands; but this we did not notice. There was no question about maxima with a distance apart about what I have given. Before the eclipse these shadows were almost vertical on a high board fence behind us and reached off in perfectly straight lines almost in the direction of the sun. It appeared then that there were vertical planes of darkness (if I may so qualify them) sweeping in more or less a southwesterly direction at right angles to the line drawn to the sun. On the white snow-covered field in front of us, the parallel lines pointing to the sun moved from left to right at a speed which has been estimated at 8 or 10 miles per hour. At first they were certainly not faster than this. The lines, converging on account of perspective, gave the effect of a vast wheel with fine spokes moving before us, the spokes pointing toward the sun. It was a very interesting and remarkable sight. After the eclipse the shadow bands came from the northwest moving directly toward the sun. They were far less distinct and separated, but looked much like a series of ripples on the surface of the snow, rather irregular, all of them scurrying as fast as possible, it seemed, toward the growing light. I should think their speed must have been twice that of the shadow bands before the eclipse. Unfortunately we did not notice the progress of these irregular shadows on the fence behind us, so I know nothing of the vertical component of their motion. There was very little surface wind at any time in any direction, I should think not enough where we were to explain the speed even in places entirely away from the fence. Of course, there may have been varying winds above. The ground was rather high on the ridge which runs northward from Newport about a mile from the center of the town. I have not yet seen any explanation of these phenomena which satisfies me, but I doubt if our very hurried observation (there were so many things to think about!) can be of much service." (R9) In this observation, the bands were radial to the sun, not parallel to the crescent. The directional and speed changes are also interesting. (WRC)

X7. March 7, 1970. Various locations. "....although the five teams organized by George M. Hurley of the Selbyville Central School in Delaware were all within a half mile of each other on Assateague Island in Maryland, their reports differ in many respects. One team did not see the bands before totality but did so after, while another experienced the opposite. Also, there was a wide difference in speed estimates and in the direction of band motion across the screens after totality. H. L. Bresnick, Rochester, New York, who was also on Assateague Island, reported that some people observed bands during totality. To Stu Block of Summit, New Jersey, the bands at Fort Monroe, Virginia, disappeared 10 seconds before totality, and reappeared seven seconds after. He also found that they flickered or shimmered, a property noted in about half of Mr. Paulson's reports. This effect added to the difficulty of judging the speeds of the bands. Peter Abramson and Jerry Mundell were at Nags Head, North Carolina, 12 miles outside the southern limit of totality. They saw bands on the beach two minutes before maximum partial eclipse, but none afterwards. Some reports mention two or more sets of bands. Observing from Chincoteague, Virginia, C. Robson, J. Brandick, and R. Whiteford saw two sets moving in opposite directions after totality. Mr. Brandick wrote, 'I got the impression that the shadow bands were recurring over themselves.' The effect resembled a turning wagon wheel on a movie-film, according to Mr. Robson. 'I saw color at the center of the board,' said Mr. Whiteford, 'perhaps red and a blue or blue-green.' (Could this have been merely the apparent color often associated with black-and-white flicker?) A similar 'overrunning' of two sets of bands was observed by the team of P. Gruntmeyer and G. Condon, near Chatham, Massachusetts, four miles outside the path of totality." ....From a team at Great Bridge, Virginia. "The bands were visually very distinct, especially right after totality. Shortly before second contact, very diffuse, almost imperceptible shadings were seen on the board. In about seven seconds they became much darker and conspicuous. Travelling from south-southwest to north-northeast, many of the bands were diffuse and had a netted appearance. We estimated their widths as half an inch to two inches, but uncertainly because of the speed at which the bands moved, 16 to 26 inches per second. Shortly after totality ended, the bands reappeared, moving in a similar but not identical direction. After two or three minutes the display ceased, but resumed 1 1/2 minutes later. During the following minute, the bands swept from north-northeast to south-southwest. Then they reversed direction twice again, and finally moved randomly for about 15 seconds before disappearing." (R12) All of these variations cannot be explained by the prevailing theories. (WRC)

Wallops Island, Virginia, using photometric equipment along with visual observation. "The orientation of the shadow bands remains tangent to the noneclipsed crescent of the sun. Little polarization of the shadow bands in evident. Shadow bands distinctly different in width and frequency are present simultaneously. Observers at Wallops Island Station reported two types of bands: (1) width, one foot; band separation, six feet; velocity, 50 feet per second. (2) width, 1.5 inches; band separation, three inches; velocity, 10 feet per second." (R13)

Wallops Island, Virginia. Summary of photoelectric and visual observations. "Most of the energy of the shadow bands ranges between 1 and 25 Hz, and appears not unlike plots of scintillation spectra. The onset, amplitude, and duration of the shadow bands

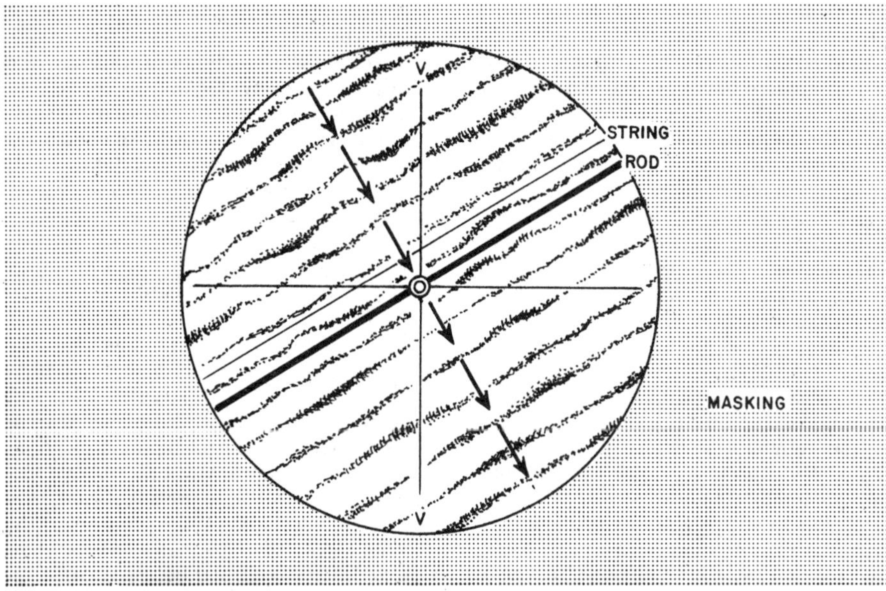

Typical experiment for observation of eclipse shadow bands. A white cloth on a table-like surface is commonly employed. The pivoted rod is set parallel to the bands.

seem to be spectrally related to the limb darkening of the Sun's photosphere. Other results of the experiment conclusively indicate that the shadow band phenomenon is a manifestation of atmospheric turbulence in the form of air packets of different density from that of their environment (density schlieren), made visible by the light from the crescent sun." (R14)

Dam Neck, Virginia. "Abstract. During the total eclipse of 7 March 1970 two types of measurements were made at a site in the center of the eclipse track at Dam Neck, Virginia. The OI forbidden 5577 A and 6300 A lines were measured looking vertically through the eclipse shadow. The intensities were approximately 1 kilorayleigh, essentially the values obtained in recent rocket measurements of normal airglow at altitudes above 100 km. This is consistent with the hypothesis that these lines originate primarily at altitudes well above the shadow from mechanisms which are not appreciably perturbed by the approaching shadow. Measurements were also made of the time variation of intensity of incident light during the shadow-band period before second contact. There are saw-tooth fluctuations with steep leading edges, as well as many high-frequency components of smaller amplitude. This time variation does not appear to be obviously related in a simple way to current theories of their production mechanisms, or to the visual appearance of shadow bands." (R15)

Neuse Park, North Carolina. A cadmium-sulfide photosensitive cell recorded the intensity of the light during the passage of the shadow bands. The signal was recorded on tape and played back as sound. "Following the eclipse, playback of the tape revealed that we had recorded two types of signals. One of these begins as a soft patter effect, like rain drops, which is first detectable above the low background noise level about 45 seconds before the beginning of totality or second contact. This appears to be caused by the true shadow bands and the sounds grow steadily in number and amplitude until, near totality, the effect is like that of a herd of horses running free. At totality the sound ceases abruptly. At third contact the shadow bands begin at maximum intensity and gently taper off in about 35 seconds. In addition there are several sharp spikes or flashes in the recording which we cannot explain although we feel it is unlikely that they were caused by electrical interference." (R16)

"Shadow bands have several fundamental characteristics: (1) The patterns consist of elongated, often wavy, bands or blotches of light and shadow from half an inch to several inches in width. Intervals of light and shadow may appear equal in extent, or gaps of several feet can appear between patterns of light or shadow. (2) The patterns are elongated tangent to the edge of the Moon's shadow on Earth. Therefore, near the center line of an eclipse the band orientation is the same before and after totality. Near the edges of the eclipse track the band orientation can shift as much as 90°. (3) While they sometimes seem stationary, bands may appear to move perpendicularly to their length at a speed of several feet per second. Individual blotches of light and shadow may appear, blend, and disappear with such rapidity that an individual pattern moves only a few feet before dissolving. (4) Bands are visible for several seconds to several minutes before and after totality---longer at partial and annular eclipses. The bands' contrast is strongest, and the spacing between patterns is least, when the uneclipsed solar crescent is smallest. (5) There is some suggestion that the contrast between light and shadow is greater at shorter wavelengths. Thus the pattern should be most visible in blue light and least noticeable in red." (R17) Some of these generalizations are inconsistent with some of the observations recorded above; viz., tangency of the bands to the edge of the moon's shadow. (WRC)

X8. General observations. A 1907 evaluation of the situation. "The first thing that struck me when reading observations of the shadow-bands was their often very indefinite character, their variableness, and their many apparent contradictions, which make them at first sight less useful than they might otherwise be. As instances, while one observer describes the bands as one inch wide and three inches apart, another at the same place estimated them at ten inches apart. Some describe their motion as being in the same direction before and after totality, others find the motion reversed after totality. Some describe the bands as of uniform width, others say they became narrower as the eclipse deepened; yet others that they became wider as the solar crescent decreased. In fact, there is hardly a detail of the data but what we apparently have contradictory testimony in relation to it." (R3)

Based on a lengthy study of many eclipses. "The critical reader is asked to keep in mind certain peculiarities of behavior of the bands which are demonstrated to exist. (1) It is only at points along the center line of the path of totality that the line of extension of the bands is the same after totality as before; an abrupt difference is always observed elsewhere, this difference approaching a right angle as the locations reporting are farther and farther out from the center line. (2) Although these abrupt changes in extension are noted between pre- and post-totality, an interval which seldom exceeds two or three minutes, nevertheless (and this is important in objecting to the 'air wave' refractionists) no abrupt changes during the pre-totality period, or during post-totality, ever seem to have been reliably reported. On the other hand, points near the border of the path of totality have frequently reported progressive changes in direction of extension. (3) The direction of progress of the bands, which is normal, or perpendicular, to their extension, is also normal to the shadow's edge, progressing either toward or away from it, with several reliable reports recorded of stationary bands. (4) The directions of progress with reference to the shadow may be any of the following:

Pre-totality	Post-totality	Example
away from	away from	Australia, 1922
away	towards	New York State, 1925
towards	away	New Haven area, 1925
towards	towards	Newport area, 1925
(with the following special cases)		
stationary	away	Canton I., Hellweg, 1937
stationary	stationary	Sumatra, Humphreys, 1901
stationary	stationary	Alcala de Chiver, 1905, Moye and Tramblay

(5) Any one mode of motion with respect to the lunar shadow will be maintained for a considerable distance along the eclipse track, but will change, apparently abruptly, into other modes as the shadow travels. (6) It is readily recognizable that the range of

speeds of the bands over the ground,---reported most often at from 5 to 10 miles per hour,---is but the hundredth part of the speed with which the shadow spot races across the face of the globe." (R10)

X9. No date given. Middletown, Connecticut. "It has long been supposed that the bands were due to motions of our atmosphere making irregular refraction in much the same way that a heated surface in summer makes objects appear to dance above it. The observations at Middletown showed that the bands were in some cases formed close to the ground and that they depended on swirls in the air caused by the wind. In one case they were observed chasing one another in all directions round a tree. In other cases they were quite poorly defined, just patches of light and shade, moving with all sorts of speeds and having different directions in different places. At various eclipses, they seem to have had more regularity, probably because the air currents were a little more regular." (R18) The Middletown bands were obviously highly disorganized. In several of the above reports, the bands are claimed to be independent of wind. (WRC)

References

R1. "Fringes of Light during Solar Eclipses," Eclectic Magazine, 60:131, 1863. (X1)
R2. Rotch, A. Lawrence; "Physical Observations during the Total Solar Eclipse," Science, 11:752, 1900. (X2, X3)
R3. Holmes, Edwin; "The Shadow-Bands," Popular Astronomy, 15:77, 1907. (X8)
R4. Chant, Jean L.; "The Shadow Bands at the Australian Eclipse," Popular Astronomy, 32:202, 1924. (X5)
R5. Stevens, Catharine O.; "Shadow Bands," English Mechanics, 121:190, 1925. (X6)
R6. Watts, Harvey M.; "Shadow Bands during the Period of Greatest Obscuration....," Popular Astronomy, 33:236, 1925. (X6)
R7. Lewis, Isabel M.; "The Shadow Bands during Totality," Popular Astronomy, 34:279, 1926. (X4)
R8. Humphreys, W.J.; "Shadow Bands," Popular Astronomy, 34:566, 1926. (X6)
R9. Feldman, Richard L.; "Shadow Bands," Popular Astronomy, 46:187, 1938. (X6)
R10. Feldman, Richard L.; "Shadow Bands, Part II," Popular Astronomy, 48:2, 1940. (X8)
R11. Paulton, Edgar; "Eclipse Shadow Band Motions---An Illusion?" Sky and Telescope, 25:328, 1963. (X6)
R12. Young, Andrew T.; "Shadow Bands and the March Solar Eclipse---I," Sky and Telescope, 40:176, 1970. (X7)
R13. Young, Andrew T.; "Shadow Bands and the March Solar Eclipse---II," Sky and Telescope, 40:242, 1970. (X7)
R14. Quann, J.J., and Daly, C.J.; "The Shadow Band Phenomenon," Journal of Atmospheric and Terrestrial Physics, 34:577, 1972. (X7)
R15. Kerr, Donald E., et al; "Brightness of Forbidden OI Lines and Properties of Shadow Bands....," Journal of Atmospheric and Terrestrial Physics, 34:585, 1972. (X7)
R16. Susel, Frank M.; "Recording the Eclipse Shadow Bands on Magnetic Tape," Royal Astronomical Society of Canada, Journal, 65:273, 1971. (X7)
R17. Marschall, Laurence A.; "Shadow Bands---Solar Eclipse Phantoms," Sky and Telescope, 67:116, 1984. (X8)
R18. Brown, Ernest William; "The Shadow Bands," Franklin Institute, Journal, 202:162, 1926. (X9)

GES2 Moving Shadow Bands in the Atmosphere

Description. Dark ripples or bands seen in motion in the air against a background of blue sky, cirrus clouds, or (from high vantage points) the ground itself. On occasion, the bands are observed directly in mist and fog. The bands may be straight or arcuate, single or in parallel formations.

Background. This phenomenon is probably identical with disturbances crossing halo displays (GEH11).

Data Evaluation. A surprising number of observations occurred during World Wars I and II in the vicinity of artillery and bomb impacts. The phenomenon seems rarer in peacetime but present nevertheless. Rating: 1.

Anomaly Evaluation. Most atmospheric ripples probably have simple explanations: they are simply visible manifestations of shock waves moving through the atmosphere at near the speed of sound. In fact, they are often termed "visible sound waves". The anomaly, which is only of marginal significance, is usually in the identification of the source of the disturbance. The long, straight ripples seen at high atltitudes, which cannot be correlated with any natural or artificial explosion, are the most curious. Rating: 3.

Possible Explanations. Shock waves or pressure waves in the atmosphere create bands of higher and lower refractive index, which manifest themselves visually as dark and light streaks, similar to shock waves made visible in aeronautical wind tunnels. The sources of the waves are natural and artificial detonations, supersonic aircraft, and possibly gravity waves.

Similar and Related Phenomena. Flashing volcanic arcs (GLD9); ripples crossing halo displays (GEH11); effects of thunder on rainbows (GEB9); atmospheric pressure waves (GSW).

Examples of Shadow Bands Moving in the Atmosphere

X1. July 31, 1850. Near Mont Blanc, France. 3:00 P.M.; altitude 1630 meters. "The same day, and at the same hour, I remarked much more curious shadows. On examining the strata of the air comprised between my eye and the bottom of the lake, towards the east, I saw four nearly parallel and equidistant bands, which, inclined toward the sun, appeared to rise from the surface of the water to a height of about 30 degrees. These sombre but indistinct bands had the same aspect as the obscure train produced by the cloud in a very different region of the sky. They were distinguished from it however by their size, which was much less; each extended only a degree in diameter. They were displaced in a perceptible manner toward the south-east, in proportion as the sun declined toward the horizon. They continued to be visible at least two hours, to my companions and me. No perceptible obstacle in the immense panorama which we commanded could produce shadows in their direction, and serve thus to explain their presence." (R1) Unlike most of the other examples in this section, no obvious cause can be found. (WRC)

X2. September 26, 1918. Meuse-Argonne region, France. "For several days the weather had been more or less rainy and wet, and the morning of September 26 found us covered by a very heavy band of fog which entirely excluded the sun. Soon after the attack opened, I had occasion to go to the top of one of the hills which flanked our position, and at a certain definite level above the battery a very considerable disturbance in the fog was noticeable after each discharge of the heavy rifles behind me. The visibility was such that the flash of the discharge could not be seen, but each time before the report reached us a band of greater density was clearly visible in the fog, moving with great rapidity up the valley toward us in the form of an arc. Its arrival was simultaneous with that of the sound of the discharge. This arc of greater fog density was perhaps six feet from its anterior to its posterior edge, and of about the same depth. It followed closely an altitude of some sixty or seventy feet above the floor of the valley and was clearly visible from both above and below that plane, but no similar phenomena were visible in any other plane." (R4)

X3. September 26, 1918. France. "The observer was looking at the battery in profile, facing the sun, which was on the other side of the gun. 'At the very moment of firing, upon the appearance of the flame,' states the officer, 'a shadow made its appearance and was propagated over the surface of the ground in the form of the arc of a circle, the center of the circle appearing to be located at the gun and the radius increasing with great rapidity. The width of this shadow appeared to be about 0.80 meters in the case of the 75's. This width was two or three times as great in the case of the 155's and furthermore the shadow appeared to be deeper so that the phenomenon was much more remarkable in the case of guns of the latter caliber than in the case of the 75's.'" (R3) Although the shadow in this instance is actually on the ground, it is so closely related to X2 that it is inserted here. (WRC)

X4. Early 1944. Anzio, Italy. During the battle at the beachhead, ripples from the concussions were observed crossing a virtually clear sky by an observer with the sun behind him. (R8)

X5. August 9, 1944. Near Southampton, England. "In the company of six other physicians, all considered dependable observers, on a chance glance at the sky I noted long narrow curved 'rolls' or 'bands' of what seemed to be areas of modified refractive index which passed rapidly overhead in a direction perpendicular to their lengths at a seemingly constant speed of about 3° to 5° per second. They occurred in single bands, occasionally in double ones, parallel and adjacent..... The bands were brief and evanescent; they appeared, moved rapidly at constant speed and vanished, seemed to travel both toward and away from the observers. Occasionally two bands met, travelling in opposite directions and passed without visible interference. The bands suggested a similarity to the bands seen in photographs of the 'sound wave' at the nose and tail of a bullet in flight." (R5) Although these shadow bands can probably be attributed to the battles across the Channel, it is difficult to account for bands moving in opposite directions. (WRC)

X6. 1944. Rouen, France. Expanding arcs of shadows passing across a layer of thin cloud. They radiated from the direction of battle. (R6)

X7. 1944. England. "It may be relevant to mention here another kind of optical phenomenon which has recently been described in letters to The Times and discussed in an article on September 3 under the heading of 'Flying Bomb Waves'. Such waves have been seen by observers of nearby explosions and have been described as 'a faint but distinct line in the form of a seemingly perfect arc centered on the spot where a bomb had disappeared'. These effects are connected with the explosion wave within a few hundred yards of the explosion itself, and it seems that the curved line represents the hemispherical wave of compressed air. In this case the optical effect might arise either from evaporation of water droplets in cloudy air or even as a refraction (mirage) effect from the relatively steep gradient of air density." (R7)

X8. March 1945. Bitche, France. "The air was very clear, and the sky was of a deep blue with cirrus clouds that were estimated to be at an altitude of 20,000 to 30,000 ft. During the preparation firing prior to breaking through the Siegfried Line many men saw dark shadows in the form of huge arcs that passed across the white clouds, and were seen only where there was a background of cloud." (R9)

X9. October 15, 1971. Lancaster, California. "It was about 5:45 p.m. Pacific daylight time on October 15, 1971, with the sun about 3° above distant hills and the sky quite cloudless. An aircraft perhaps 5,000 feet above Edwards Air Force Base passed almost overhead toward the setting sun. As it receded, two concentric well-defined dark semicircles appeared below and slightly behind the craft. These arcs rapidly expanded symmetrically to at least six times their original size, maintaining their spacing and sharpness. The total optical impression, which lasted only a fraction of a second, was of a double ripple in the sky that expanded away from the aircraft. It was immediately followed by the double concussion effect of a sonic boom." (R10)

X10. July 26, 1975. Woodford Green, England. "I witnessed a curious phenomenon from my home at Woodford Green, Essex, at 1830 GMT on Saturday, 26 July 1975. Much of the sky was covered with altocumulus developing into castellanus. The phenomenon was witnessed in a S.S.E. direc-

Shock wave from a supersonic plane visible over California (X9)

tion at an elevation of 45-60°. I saw a dark marking on the underside of an altocumulus which swept across it (an arc distance of about 5-10°) in a period on the order of a second. It moved in a roughly north-easterly direction, and vanished completely as it passed the edge of the altocumulus. The phenomenon was repeated at least once in much the same form, and in a minute or so after the first occurrence there was a repetition on a neighbouring altocumulus, but with the marking moving roughly the opposite way. The markings were not sharp or dark, but showed up as dull shadings on the white background of the surfaces of the altocumuli---there was perhaps a slight tinge of brown colouring in the markings. Their dimensions were on the order of one or two degrees, and there was some structure, mainly in the form of spots or cells of angular separation slightly less than 1/2°. The markings seemed somewhat elongated in a direction perpendicular to that of their motion, and there was possibly some banding perpendicular to the motion. The form changed somewhat during the observed lifetime of each marking. Another point noticed, in the same part of the sky, was a flock of soaring swifts (about a dozen, a fairly normal number). This was first noticed even before the dark markings, and was largely the reason why I was looking in that direction at all. Later there seemed to be gulls and other sea birds mingling with the swifts, but I could not be specific as to the species. While this was going on, the air became filled with flying ants, almost all of them the smaller males; I counted only three of the larger females among hundreds of males." (R12) The observer suspected that the bands might be clouds of insects. (WRC)

X11. July 27, 1976. Woodford Green, England. Same observer as X10 saw dull markings moving against a clear blue sky. Soon after it began to rain ants---so many that it was uncomfortable to remain outdoors. Observer now convinced that the dark shadows in the sky were in reality clouds of insects. (R13)

X12. February 3, 1977. Kitt Peak, Arizona. "On going outside to check the sky conditions I found a mixture of cloud types, mostly altocumulus and cirrocumulus. The latter had some rippled structure. My attention was caught by a patch of sky which displayed a wave-pattern moving apparently at high speed from west to east. The pattern contrast was very low: it was the motion that made it visible. My notes suggested that the pattern may not have been purely sinusoidal, but rather more of a square-wave pattern or that it at least contained some higher harmonics. The wavelength was on the order of one degree and the orientation of the troughs was north-south. I would not rule out that the velocity was that of sound. The phenomenon lasted for about three seconds, vanished, then reappeared for perhaps a couple of seconds more." (R14)

X13. No dates given. Europe. Summaries of three cases of moving shadows, arcuate in form, expanding away from artillery sites. See X2 and others for typical descriptions. (R2)

X14. No date or place given. Dark bands seen moving through cirrus clouds. (R11)

References

R1. Wartmann, E.; "On Atmospheric Shadows," Philosophical Magazine, 4:2:160, 1851. (X1)

R2. "Visible Sound Waves," Scientific American, 117:343, 1917. (X13)

R3. "Are Sound Waves Ever Visible?" Scientific American, 121:555, 1919. (X3)

R4. Saunders, Frederick A.; "Visible Sound Waves," Science, 52:442, 1920. (X2)

R5. Tabruh, Frank L.; "Can Anyone Explain?" Royal Astronomical Society of Canada, Journal, 38:351, 1944. (X5)

R6. Revell, D.G.; "Strange Phenomenon in the Sky," Royal Astronomical Society of Canada, Journal, 38:432, 1944. (X6)

R7. Goldie, A.H.R.; "Optical Phenomena in the Atmosphere," Nature, 154:738, 1944. (X7)

R8. Taylor, V.S.; "Optical Phenomena in the Atmosphere," Nature, 155:388, 1945. (X4)

R9. Jones, Arthur Taber; "Secondary Shock Waves and an Unusual Photograph," Amer-Journal of Physics, 15:57, 1947. (X8)

R10. "Seeing a Sonic Boom," Sky and Telescope, 44:205, 1972. (X9)

R11. White, R.; "Rapidly Moving Ripples in Cirrus," Weather, 30:205, 1975. (X14)

R12. White, R.; "Curious Phenomenon at Woodford Green," Journal of Meteorology, U.K., 1:163, 1976. (X10)

R13. White, R.; "Again---Curious Phenomenon at Woodford Green," Journal of Meteorology, U.K., 2:24, 1976. (X11)

R14. Livingston, W.C.; "Rapidly Moving Ripples in Cirrus Cloud," Weather, 33:38, 1978. (X12)

GES3 Colored Shadows

Description. Distinctly blue, purple, and green shadows cast by the sun. The effect intensifies as the sun sinks lower in the sky, with a definite green appearing at sunset. The presence of snow exaggerates the phenomenon.

Background. Perhaps the most sensitive description of the phenomenon is by the poet Goethe (X4), who, in addition to his literary accomplishments, conceived a theory of colors quite different from that of Newton. Of course, Newton won out.

Data Evaluation. A very common phenomenon that is mentioned only rarely in the literature surveyed so far. Rating: 1.

Anomaly Evaluation. Minnaert (R4) claims that the phenomenon is both physiological and psychological in origin. In other words the human eye-brain systems just happens to be built so that such illusions are "seen". It is curious that we are made this way but geophysically at least, there seems to be no anomaly. Rating: 4.

Possible Explanations. See above.

Similar and Related Phenomena. The green flash (GEL1); blue snow (GWP10).

Examples of Colored Shadows

X1. 1743. Europe. "Buffon had the merit of contributing powerfully to direct the attention of observers toward the coloured shadows that form in solar light. 'I observed,' says he, 'during the summer of the year 1743, more than thirty sun risings and as many settings. All the shadows that fell upon white, as upon a white wall, were sometimes green, but most commonly blue, and of a blue, as lively as the most beautiful azure. I shewed the phenomenon to several persons, who were as much surprised as myself. Difference of season has no effect upon it, for there are not eight hours (15th November 1743) since I have seen blue shadows; and whoever will give himself the trouble of looking to the shadow of his fingers at sunrise and sunset, upon a bit of white paper, will see like me this blue shadow,' &c." (R1)

X2. 1893. Ipswich, England. "The bright sunny weather we have had for so many weeks has given frequent opportunities of seeing the effect of the sun setting in a cloudless sky, which produces a blue shadow on a sheet of white paper, the hand, or any substance being held close in front of the paper. The shadow gradually becomes more blue after the sun is about five or six degrees above the horizon, attaining the deepest blue at the horizon." (R2) A letter is response stated that all shadows appear in the color complementary to that of the light from which they are cast. (R3)

X3. No date given. Germany. Quoted from Goethe's Theory of Colours. "During a journey in the Harz, in winter, I descended the Brocken at nightfall; there was snow on the white fields above and below me and the heath was covered with snow; all the sparse trees and projecting cliffs, all the groups of trees and masses of rock were completely covered with hoarfrost; the sun was just setting over the lakes of the Oder. By day, when the snow was tinged with yellow, the shadows were only a faint violet, but now, when the illuminated parts reflected an intenser yellow, they were decidedly bright blue. When, however, the sun was at last on the point of setting and its light, toned down by the atmosphere, coloured everything around me a most glorious purple, the tint of the shadows was seen to change into a green which for purity could compete with the green of the sea, for beauty with that of the emerald. The phenomenon became increasingly vivid, the surroundings changed into a fairy-world, for everything was covered with these two bright and beautifully matched colours, when finally, the sun once set, the gorgeous scenery changed to a grey twilight and afterwards to a clear night with moon and stars." (R4)

References

R1. Zschokke, and Treschsel; "On Coloured Shadoes," Edinburgh New Philosophical Journal, 2:32, 1826. (X1)

R2. Biddell, G. A.; "Blue Shadow," English Mechanic, 57:291, 1893. (X2)

R3. "Blue Shadow," English Mechanic, 57: 310, 1893. (X2)

R4. Minnaert, M.; "Coloured Shadows," The Nature of Light and Color in the Open Air, New York, 1954, p. 134. (X3)

GES4 Shadow Bands Seen through the Telescope

Description. Shadowy splotches and bands appearing in the telescopic extrafocal images of stars and other light sources.

Data Evaluation. A single paper by a well-known modern astronomer. Rating: 2.

Anomaly Evaluation. No mystery exists in the optical formation of the splotches and bands in the extrafocal image; rather it is the blobs themselves that excite curiosity. How are they formed? How do they maintain their stability so long? Are they arranged in any large-scale order? Aerial blobs represent a poorly explored and understood facet of geophysics. Rating: 2.

Possible Explanations. Aerial blobs, which are discrete volumes of air with differing densities, temperatures, etc., alter the optical path of starlight, thus forming streaks and bands on planes before and after the normal focal plane.

Similar and Related Phenomena. Eclipse shadow bands (GES1); the shadow bands seen in areas illuminated by searchlights (GES6).

Examples of Shadow Bands Seen through the Telescope

X1. General observations. "In this paper attention is called to some striking features of the stellar scintillations and excursions that are due to what I propose to call aerial blobs. Although many atmospheric disturbances refract, diffract, scatter, or absorb light from distant celestial and terrestrial sources in an irregular manner, aerial blobs are volumes of air of locally altered density, temperature, and water content that possess remarkable optical properties. Blobs in combination with the mirrors or lenses of a telescope often bodily displace the images of stars or focus them in points in front or behind the regular focal surface, as shown in Fig. 1.

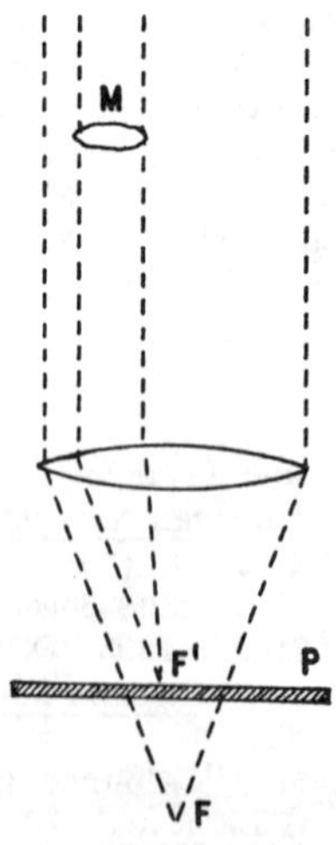

An aerial blob (M) displaces the focal point from F to F′ (Fig. 1, X1)

Figure 2 (top) depicts the structure of a typical extrafocal stellar image as obtained from an instantaneous exposure with the 200-in. telescope. The extrafocal image shown is composed of a network of dark and bright patches and lanes. (The inner circle is due to the obstruction offered by the observer's cage.) Notice the distinct points A, B, C; these are good stellar images resulting from the combined optical action of the telescope and various aerial blobs traveling through the cylindrical beam of light from a given star falling on the 200-in. mirror. If, instead of taking a still picture, the photographic plate is moved in a plane normal to the axis of the telescope, the points A, B, C will produce sharp striations. These will be inclined as shown if the blobs have any velocity component normal to the motion of the plate (Fig. 2, bottom).

Linear dimensions of aerial blobs have been observed ranging from millimeters to many meters. Blobs may be globular, lenticular, or cylindrical in shape, thus producing sharp pointlike or linelike extrafocal images of stars. Often hundreds of blobs are quite regularly spaced and drift with the winds at various altitudes up to 50 km or perhaps higher. Methods for the determination of the physical characteristics of aerial blobs, of their velocities, and of the altitudes at which they are found have been

Top: Possible structure of extra-focal stellar image of the 200-inch telescope with three sharply focused star images. Bottom: Drifted extra-focal image. (X1)

discussed in another place.

A most amazing feature of many aerial blobs is their durability and stability; some of them preserve their shapes for hours. The lifetimes of the blobs can best be observed through the partial or total condensation of their moisture content. Such condensation occurs, for instance, in the regions adjoining vapor trails caused by jet planes." (R1) Aerial blobs might be involved in the production of eclipse shadow bands (GES1) as well as those bands associated with artificial light (GES6). (WRC)

References

R1. Zwicky, F.; "Aerial Blobs," Science, 122:159, 1955. (X1)

GES5 Unusual Shadows Observed during Eclipses

Description. Crescents of light in the shadows of foliage, multipled shadows of posts, shadows with asymmetrical clarity, shadows with odd appendages, etc.

Data Evaluation. Crescents of light, like eclipse shadow bands, are commonly observed. The other shadow phenomena noted above have been reported rarely and only perfunctorily. Rating: 2.

Anomaly Evaluation. The light crescents are easily accounted for (see below). Shadow distortions and multiplication are more puzzling, but seem to be mainly consequences of the irregular, extended nature of the light source (solar crescent, Bailey's Beads, corona, etc.) No explanation has been found for the strange spurious appendages on some shadows. Rating: 3.

Possible Explanations. The bright crescents in the shadows of foliage are merely inverted images of the solar crescent formed by small gaps in the leaves, which act as pinhole cameras. The changing geometry of the light source naturally produces the multiple and confused shadows described in the other examples.

Similar and Related Phenomena. Eclipse shadow bands (GES1); the strange rays and image distortions of Brocken Specters (GEI1).

Examples of Unusual Shadows Observed during Eclipses

X1. April 17, 1912. Southsea, England. "Shadows.---This was a very interesting line of study for one not engrossed with telescopic views. I saw in the dappled shadow cast by bushes just bursting into spring greenery numberless small crescents of light, giving a quaint and pleasing look on the ground. In cycling along, the shadow of the different parts of the cycle seemed to be fringed with points, while interstices were turned into bright little crescents. My own head, as projected on the ground, had such a curious effect that I made a couple of rough sketches. That made at 11h. 58m. seemed to show whiskers on each side of the head, although I

Eclipse shadows of observer's head during different stages of solar eclipse (X1)

am quite free from such appendages. At this time these dusky fringes showed on both sides, and had their points or curvature, turned upwards. The second sketch, made at 12h. 31m., or 33 minutes later, only showed fringes on the right-hand side of the head, and the points were now turned downwards. These changes were doubtless due to the varying position of the bright crescent of sunlight." (R1)

X2. June 8, 1918. Illinois. During a partial eclipse of the sun the shadows of foliage contained numerous small crescents of bright light. As would be expected with a pin-hole-camera effect, the crescents were inverted. (R2)

X3. March 7, 1970. Virginia Beach, Virginia, and Mexico. "A visual observation that is only indirectly related to shadow bands was noted at Virginia Beach, Virginia, by Dorothy Beetle and also by my wife and me in Mexico. Before totality the shadows of observers were sharp on the east side and fuzzy on the west; after totality the reverse occurred. This effect is due to the asymmetry in shape and brightness of the solar crescent." (R5)

X4. General observations. "Some Curious Effects. Once the first contact takes place, the eclipse proceeds until the Sun's disc is almost completely covered. A curious effect, is, however, frequently observed under trees during the first stages of an eclipse. While the Sun's disc is clear, one can, if one takes a stand under a spreading tree, often see on the ground a number of overlapping images of the Sun, caused, of course, by the interstices between the leaves acting in some respects as lenses. As the eclipse proceeds these circular images change to crescents, getting ever narrower and narrower. The size of these sun-crescents depends on the height of the tree, so that, speaking generally, the higher the tree the bigger the images. Other curious effects may also be observed; for instance, vertical posts may cast double, triple, quadruple, or even quintuple shadows." (R3, R4)

References

R1. Markwick, E. E.; "The Solar Eclipse, 1912, April 17," English Mechanic, 95: 295, 1912. (X1)

R2. Moulton, Robert H.; "Shadows Cast during a Partial Eclipse," Scientific American, 119:31, 1918. (X2)

R3. Lloyd, J. A.; "The Coming Solar Eclipse," English Mechanics, 2:126, 1927. (X4)

R4. Arakawa, H.; "Crescent-Shaped Shadows during a Partial Eclipse of the Sun," Weather, 16:254, 1961. (X4)

R5. Young, Andrew T.; "Shadow Bands and the March Solar Eclipse---I," Sky and Telescope, 40:176, 1970. (X3)

GES6 Non-Eclipse Shadow Bands

Description. The projection of moving shadow bands on surfaces by artificial lights and the uneclipsed sun. These bands closely resemble eclipse shadow bands (GES1), and possess the same irregularity and unpredictability.

Data Evaluation. The data are sparse and testimonial in nature. Observations are doubtless often written off as due to "atmospheric turbulence" and hence not appreciated. Next to nothing is known about the dimensions, velocity, and origins of these bands. Rating: 3.

Anomaly Evaluation. The non-eclipse shadow bands involving searchlights and window glass may originate in the wavy structure of some drawn glass, but shadow motion still remains a problem here. Other bands are thought to be caused by moving inhomogeneities in the atmosphere---like those producing eclipse shadow bands. Here, the anomaly is the unexpected existence of uniform, persistent, parallel "aerial blobs". Composite rating: 2.

Possible Explanations. Abnormal refraction due to nonuniform glass structure and/or aerial blobs.

Similar and Related Phenomena. Eclipse shadow bands (GES1); low-sun shadow bands (GEL-10).

Examples of Non-Eclipse Shadow Bands

X1. Circa 1915. Near Bournemouth, England. A searchlight beam from the Isle of Wight projected upon a wall on the shore of the mainland resulted in a constant progression of shadow bands moving from right to left at the speed of a man running at the double. (R3)

X2. No date given. Quiberon Bay, France. "The observations were made in Quiberon Bay, during the stay of the North Squadron. During the night exercises the powerful searchlights from the warships repeatedly rested on the white vertical walls of the buildings on shore. The dark bands, very clear, with an undulating motion, were from 10-15 cm. wide, and had a motion of 4-5 metres per second. The wall forming the screen was oriented to the East, and was placed perpendicularly to the rays from the projectors, of which the distance was about 3 1/2 kilometres. Under those conditions the bands were displaced from the North to the South. They were best seen at about 10 metres distance. Similar fringes were seen on another wall oriented North-east to South-west, thus being very oblique to the line of rays, and in this case, as before, the shadows moved from right to left." (R1)

X3. General observations. "M. Raymond has communicated to the Astronomical Society of France a method of obtaining ocular evidence of the existence of waves and currents in the atmosphere by projecting on a wall a magnified image of the sun. The first observation of this kind was made by Ventura. The focussing can be adjusted as to bring out sharply the vibrations of the air in the form of markings which cross the sun's image. A remarkable regularity and parallelism of the markings, and consequently of the stream lines of the atmosphere, is frequently observed. Puiseux has called attention to the analogy between these phenomena and the fringes and shadows which are so often seen moving over the ground and along walls during total solar eclipses, especially just before and after totality, when the visible disc of the sun is reduced to a narrow crescent. These fringes have given rise to numerous discussions. Their explanation is undoubtedly to be found in the refraction of the sun's rays by ripples in the upper atmosphere, running parallel to the general direction of the thin solar crescent. When the entire disc of the sun is visible, the shadows cast by its various parts overlap and become confused and hence invisible. A parallelism between the course of the fringes and that of very high cirrus clouds crossing the face of the sun has also been observed." (R2)

X4. General observations. "I wonder whether one of your readers could explain a phenomenon that has been puzzling me on those rare occasions when the sun shines through my bow window and traces the shadows of the frames on the opposite wall. Between the frames, where the light passes through only clear glass, there are fairly regular dark stripes, mostly vertical but for some panes horizontal. They disappear when the window is opened so that they depend on the glass, presumably some polarization effect resulting from the manufacture of the plate. More puzzling is the fact that these dark lines move, even when the window remains stationary. Sometimes a continuous movement can be traced to the edge of a cloud passing over the sun. But frequently there is considerable slow to-and-fro movement, looking rather like a Moire fringe, when there is no cloud in the sky and no visible movement anywhere! The dark lines are not consistent in appearance---sometimes they are even diagonal, but at any moment of time they are very well defined, remarkably parallel and equidistant." (R4) A letter in response blames the bands on ripples in the glass created by the drawing mode of manufacture. (R5) The variable appearance of the bands and irregular motion does not seem to be explained entirely by manufacturing artifacts. Could the patterns seen in searchlight beams be a consequence of glass in the searchlight? (WRC)

References

R1. Butler, Charles P.; "Terrestrial Shadow Bands," Knowledge, 3:431, 1906. (X2)

R2. "Visible Vibrations of the Atmosphere," English Mechanic, 89:364, 1909. (X3)

R3. Barlow, E. W.; "Optical Phenomena," English Mechanic, 102:74, 1915. (X1)

R4. Semler, E. G.; "Moving Shadows on the Wall," New Scientist, 14:362, 1962. (X4)

R5. Campbell, Ian; "Moving Shadows on the Wall," New Scientist, 14:482, 1962. (X4)

GES7 Persistent or "Living" Shadows

Description. Shadows that persist for several seconds after the object casting them has been moved.

Data Evaluation. A single report, although very convincing. Rating: 2.

Anomaly Evaluation. Persistent shadows must be classed as only curiosities, but they are so delightful that they must not be omitted. Rating: 4.

Possible Explanations. The shadows are temperature-sensitive rather than simple obscurations, as described in X1.

Similar and Related Phenomena. Lightning shadowgraphs (GLL22).

Examples of Living Shadows

X1. General observations. "The accompanying photograph (not reproduced) furnishes a view of a 'sump-hole' at the oil-field located in Kern County, Southern California. The field promises to become one of the great petroleum-producing regions of the State. One very remarkable feature about the reflections at this hole is the fact that after the object casting the shadow moves from the identical spot, or is moved away, the shadow remains for some seconds in the same place, without following the substance, or changing its position, in the least. Usually the shadows remain all the way from five to ten seconds after the substances have changed position, fading away, and finally disappearing altogether. The longer the shadow is cast in one particular spot, the longer it remains there. Very careful investigations of this remarkable peculiarity have been made, and the explanation given is that when the crude oil is first pumped up into the 'sump-holes,' or temporary earth reservoirs, it is very heavily charged with gas. After remaining in the reservoir for a short time it becomes quite warm from the sun's rays, and the gas is rapidly discharged from the heavy body of petroleum, and rises to the surface, and passes off into the atmosphere. Millions of minute bubbles are constantly rising to the surface of the crude oil, especially when the sun pours down its fervid rays. This gas is extremely volatile and sensitive to the condition of temperature. So sensitive is it that whenever a shadow is cast on the surface, and to a certain extent lowers the temperature on that particular spot, it lessens the number of bubbles that rise. This effect serves to emphasise the shadow cast. When a person stands for a few minutes in one spot, and then suddenly shifts his position his reflection will still be indicated in the same place." (R1)

References

R1. "Shadows That Live," English Mechanic, 91:538, 1910. (X1)

GES8 Curious Mountain Shadows

Description. Remarkable movements and distortions of the shadows of mountains. For example, the Adam's Peak shadow (X1) stretched out for scores of miles on the plains before the observer seems to rise up vertically and then fall. Further, mountain shadows almost always display sharp, triangular shapes regardless of their true shapes.

Background. Dwellers along the flanks of mountain ranges occasionally observe the shadows of the major peaks cast upon the sky when the sun is low. While spectacular to the uninitiated, such shadows are not unusual enough to catalog.

Data Evaluation. The rise and fall of Adam's Peak, in Sri Lanka, has long been a traveller's attraction. The "unnatural" sharpness of mountain shadows, too, has been recognized for many years. Scientific reports on these phenomena, though, are scarce. Rating: 2.

Anomaly Evaluation. All of the phenomena reported below seem to have prosaic explanations, except perhaps for X2. Rating: 3.

Possible Explanations. The Adam's Peak phenomenon is most likely a Brocken Specter. Mountain shadow sharpness and triangularity can be attributed to perspective; i.e., a progressive "stretching out" of the shadow of the higher portions of the peak on a low horizontal plain.

Similar and Related Phenomena. Brocken Specters (GEI1), especially the projecting rays.

Examples of Curious Mountain Shadows

X1. Frequent occurrence. Adams Peak, Sri Lanka. "One of the most startling and extraordinary of all atmospheric phenomena may be seen during fine weather from the summit of Adam's Peak in Ceylon. This mountain rises extremely abruptly from the low country, and by its elevation of 7200 feet above the sea commands a most extensive and map-like view of all the low lands to the south-west and north-west for a distance of 50 miles or more up to the sea coast. The upper part of the mountain is an acute cone of solid rock, some 2000 feet or more in height, that rises in perfect isolation above the range of which it forms a part, and also above all the neighbouring mountains to the east and north. The phenomenon, which is described by all who have witnessed it as of the most striking character, is seen at sunrise, and consists apparently of an enormous elongated shadow of the mountain projected to the westward, not only over the land but over the sea, to a distance of 70 or 80 miles. As the sun rises higher it rapidly approaches the mountain, and appears at the same time to rise before the spectator in the form of a gigantic pyramid of shadow. Distant objects, a hill or a river (or even Columbo itself, at a distance of 45 miles), may be distinctly seen through it, lighted up by the sunlight, diffused most probably by the surrounding illuminated atmosphere; so that the shadow is not really a shadow on the land, but a veil of darkness suspended between the observer and the low country. All this time it is rapidly rising and approaching, and each instant becoming more distinct, until suddenly it seems to fall back on the spectator, like a ladder that has been reared beyond the vertical; and the next instant it is gone." This author considered it to be a mirage phenomenon. (R1, R2)

Another opinion. "A great peculiarity has been noticed by many travellers about the shadow of Adam's Peak at sunrise. The shadow, instead of lying flat on the ground, appears to rise up like a veil in front of the spectator, and then suddenly to fall down to its proper level. Various theories have been propounded to account for this, and it has usually been supposed to be due to a sort of mirage. The author, in the course of a meteorological tour around the world, spent a night on the top of the peak, 7,352 feet above the sea, and obtained unmistakable evidence that the appearance is due to light wreaths of thin morning mist being driven past the western side of the mountain by the prevailing northeast monsoon up a neighbouring gorge. The shadow is caught by the mist at a higher level than the earth, and then falls to its own plane on the ground as the condensed vapour moves on. The appearance is peculiar to Adam's Peak; for the proper combination of a high isolated pyramid, a prevailing wind, and a valley to direct suitable mist at a proper height on the western side of a mountain, is only rarely met with. Any idea that the appearance could be caused by mirage is completely disproved by the author's thermometric observations.

Another curious but totally distinct shadow effect is sometimes seen from Adam's Peak just before and at the moment of sunrise. The shadow of the base of the peak stretches along the land to the horizon, and then the shadow of the summit appears to rise up and stand up against the sky. The first part seems to be the natural shadow lying on the ground, and the sky part to be simply the ordinary earth shadow of twilight, so clearly projected against the sky as to show mountainous irregularities of the earth's surface. As the sun rises the shadow of the summit against the sky gradually sinks to the horizon, and then the ordinary shadow steadily grows shorter as the sun gets higher in the usual manner.

The author found a similar effect only at sunrise on Pike's Peak, Colorado. Towards sunset the shadow creeps along the prairie to the horizon, and then begins to rise in the sky till the sun has just gone down, and the anticrepuscular shadow rises too high to catch the outline of the peak." (R3, R4) The theory proposed in the first paragraph is simply a type of Brocken Specter. (WRC)

X2. November 11, 1947. Red Sea, Indian Ocean. "1440 G. M. T. (sunset). At about 1450 the sunset glow broke up, leaving blue sky from horizon to zenith in the sector between W x S and SW. On the W x S bearing, the division between bright orange glow and blue sky was at right angles to the horizon, whereas on the SW bearing the orange glow

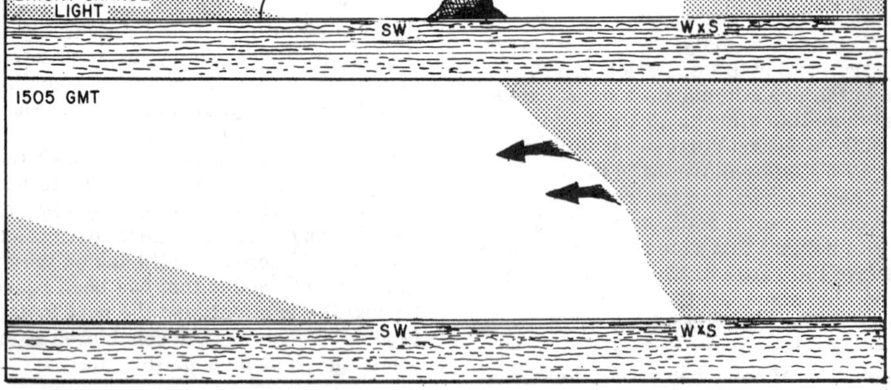

Peculiar sky coloration following sunset over the Red Sea. The blue area may have been a mountain shadow. (X2)

was at an angle of 20° to the horizon. This condition prevailed for about 15 minutes, when the orange glow in the W x S direction gradually covered the blue sky and joined the orange light in the SW, which had remained stationary. When the phenomenon was at its best, the summit of Jebel Aduali could be seen between the two sections of orange light." The editor of the Marine Observer ventured that the blue sky might have been the shadow of some inland mountain. (R5)

X3. General observations. "For an observer standing at the summit, any mountain---regardless of its profile---casts a sharp triangular shadow when the Sun is low. That the shape of a mountain's shadow should apparently bear little or no resemblance to the mountain casting it has puzzled many observers. Now two American scientists have turned to the laws of geometrical optics to find a simple explanation.

William Livingstone of the Kitt Peak National Observatory in Arizona and David Lynch of California Institute of Technology in Pasadena, California, have found that any summit features in a mountain's shadow are cast on points on the ground or in atmospheric haze many tens of kilometers distant, while the shadow of the slopes is cast nearby. As a result, any mountain has a triangular shadow, the apex angle of which is linked directly to the ratio of the mountain's height to its breadth. The quite remarkable sharpness of the summit portions of the shadow, in spite of the half degree diameter of the Sun, appears to be due to the contrast between bright sunlight and dark shadow being reinforced for a summit observer through many kilometres of hazy atmosphere. The shadow becomes less distinct away from the summit." (R6)

References

R1. Abbay, R.; "Remarkable Atmospheric Phenomena in Ceylon," Philosophical Magazine, 5:2:58, 1876. (X1)

R2. "Remarkable Atmospheric Phenomena in Ceylon," Scientific American Supple-2:500, 1876. (X1)

R3. Abercromby, Ralph; "The Peculiar Sunrise-Shadows of Adam's Peak in Ceylon," Report of the British Association, 1886, p. 528. (X1)

R4. Abercromby, Ralph; "The Peculiar Sunrise-Shadows of Adam's Peak in Ceylon," Philosophical Magazine, 5:23:29, 1887. (X1)

R5. Rees, R.G.; "Sunset Phenomenon," Marine Observer, 18:193, 1948. (X2)

R6. "The Point of Mountain Shadows," New Scientist, 81:186, 1979. (X3)

GEZ ANOMALOUS GEOMAGNETIC DISTURBANCES

Key to Phenomena

GEZ0 Introduction

The magnetic anomalies described here are transitory---sudden deviations of the compass needle and jostlings of magnetometers. Permanent magnetic anomalies caused by local distortions of the earth's magnetic field (some of which are quite remarkable) are to be found in the Catalog volumes covering the field of geology (E).

Transitory magnetic anomalies are of two types: (1) Those where no source of the disturbance can be identified; and (2) Those that are correlated with other phenomena but where it is difficult to understand how the magnetic effects are produced.

Careful observers of the compass needle noted long ago that it quivered during sun-induced magnetic storms. Possibly some of the disturbances noted below are of this type, but generally they are much smaller and of shorter duration than those due to magnetic storms. The origins of many remain mysterious.

Auroras and thunderstorms are manifestly electrical in nature, and it is conceivable that electrical currents set in motion during these phenomena may perturb the geomagnetic field. Earthquakes, too, though not obviously electrical, may actually generate considerable electricity via the piezoelectric effect. Even meteors leave strongly ionized clouds in their wakes. The gist here is that many natural phenomena have the potential for disturbing the geomagnetic field. A geomagnetic disturbance has many potential causes.

GEZ1 Unexplained Magnetic Disturbances

Description. Sudden, large-amplitude changes in the geomagnetic field that cannot be assigned to any specific cause. The directional components and/or amplitude of the field may be affected. The changes may occur over a period of a few seconds or extend for several months.

Background. Geomagnetic storms excited by solar activity are, as far as possible, excluded here. Auroral activity is symptomatic of geomagnetic storms.

Data Evaluation. A small handful of good quality examples are available, but they are weakened by the possibility of unappreciated local, artificial magnetic disturbances. Rating: 2.

Anomaly Evaluation. If all artificial sources and geomagnetic storms can be eliminated, we are left mainly with the geophysical phenomena delineated below. None of these potential sources of magnetic disturbance is well-understood. Rating: 2.

Possible Explanations. Magnetic perturbations caused by earthquakes and volcanos may contribute to this phenomenon, although seismo- and volcano-magnetic effects seem to be very weak (GQM1). Strong earth currents, possibly created by nearby lightning, perhaps even tornados (GWT1), may also have roles. Secular changes in the (hypothetical) core dynamo might account for long-period changes (X7).

Similar and Related Phenomena. The other entries in this section (GEZ); the magnetic effects of earthquakes and volcanos (GQM1); tornado electricity (GWT1); auroras (GLA), which are often associated with geomagnetic effects.

Examples of Unexplained Magnetic Disturbances

X1. November 24, 1894. Arctic Ocean. During Nansen's Expedition. "One day---it was November 24th---he came in to supper a little after 6 o'clock, quite alarmed, and said, 'There has just been a singular inclination of the needle to 24°, and, remarkably enough, its northern extremity pointed to the east. I cannot remember ever having heard of such an inclination.' He also had several others of about 15°. At the same time, through the opening into his observatory, he noticed that it was unusually light out of doors, and that not only the ship but the ice in the distance was as plainly visible as if it had been full moonlight. No aurora, however, could be discerned through the thick clouds that covered the sky." Nansen still thought the disturbance to be of auroral origin. (R1)

X2. December 28-30, 1908. Northern Hemisphere. "An examination of the magnetograms of the Coast and Geodetic Survey magnetic observatory at Cheltenham (MD) showed no disturbance either magnetic or mechanical at the time of the Messina earthquake. It called attention, however, to several magnetic disturbances of unusual character occurring during the three days following the earthquake. From an inspection of the magnetograms of the Baldwin (KS) and Porto Rico observatories, it was found that similar disturbnaces were recorded simultaneously at the three observatories....The peculiar nature of the disturbances is that they were of short duration and occurred at times when the magnetic conditions were otherwise unusually calm. They might be compared to short thunder showers occurring on a bright sunny day, were it not for the fact that they were evidently world-wide in extent rather than local.....The most marked of these disturbances was recorded on December 29th, and the Cheltenham record of it is reproduced herewith (not reproduced). There was a steady decrease in horizontal intensity amounting to about 30 gammas (0.00030 C.G.S.), followed by a corresponding increase, and an increase of west declination amounting to about 3' followed by a corresponding decrease. The variation in vertical intensity was much less marked, and on account of the large scale value would hardly be perceptible on a reduced copy of the curve, and is accordingly omitted in the figure. There was a decrease of about 6 gammas (0.00006 C.G.S.), followed by a corresponding increase. The other disturbances were similar to this one but of less than half the amplitude. It may be simply coincidence that these peculiar disturbances occurred on the days immediately following the Messina earthquake, but there is at least a possibility that they were caused by some after effects of that convulsion of nature." (R2)

Requests for magnetic data from other observatories revealed that the disturbance extended at least as far as Honolulu. (R3)

X3. March 1, 1912. No place given. "A terrestrial magnetic disturbance occurred on March 1. The magnetic needle was disturbed before noon and at 2:05 p.m. moved to 6° west and stood there for four minutes. Later it moved to 7° west and remained there for some time; at 3:38 fell to 0°; 3:43 moved to 20° east and, quieting down, stood at 20° east for a minute; 3:45 swung to 14° east and came to rest at 9° east; 3:50 fell to 0°; 5:03 moved to 10° east; 5:10 5° east; 6 p.m. 7° east, increasing to 9° east; 7 p.m. 7° east. From 7 to 10 p.m. the needle moved slowly four times between 7° and 9° east, requiring about 45 minutes to each move." (R4) Such activity of the compass needle is rather typical of strong magnetic storms. All auroral records would have to be checked before admitting this as a real magnetic anomaly. (WRC)

X4. December 23, 1928. South Atlantic Ocean. "Steering by Standard Compass S. 17° E. (Error 21° W. = S. 38° E. True), Steering Compass S. 22° E., both compasses were suddenly deflected 90°, North Point swinging to East, and then swung freely back to normal. Period of disturbances about three minutes." (R5)

X5. March 30, 1937. South Pacific Ocean. "..... between the islands of Hiva Oa and Ua Pow (Marquesas), a magnetic disturbance was experienced. The compass suddenly became very erratic, at one time pointing as much as 20° to the east of magnetic north. Within two minutes the compass-card settled down, and an azimuth showed the error to be normal again." (R6)

X6. September 30, 1952. North Atlantic Ocean. "1932-2052 G.M.T. Abnormal magnetic variation was experienced during this period. At 1932 the deviation was 3°E, at 2024 it was 1°W. The deviation then increased to 4 1/2°W by 2040 and decreased to the normal deviation of 3°E at 2052. Numerous bearings of celestial bodies were obtained having both E'ly and W'ly hour angles and the gyro compass remained correct throughout. No abnormal static was experienced." A check of magnetic observatories revealed no magnetic storms at that time. (R7)

X7. Circa 1969. Worldwide phenomenon. "It now seems that around 1969 a spectacular change took place in the geomagnetic field. The change was almost synchronous over the whole of the Earth's surface, took place in less than two years, and is now known to have consisted of a 'jerk': a step change in secular acceleration of the magnetic field that has it origin inside the Earth. (R9) This 'jerk' is obvious in records of secular variation, particularly those from European observatories. For over 20 years the secular variation was about 6 arcminutes per year and was drifting slowly downward. In 1969, the curve began to rise with a much steeper slope. (R8) Note that this type of magnetic disturbance is on a much longer timescale and global in scope---quite different from most of the others above. (WRC)

References

R1. Nansen, Fridtjof; "We Prepare for the Sledge Expedition," Farthest North, 2:11, London, 1897. (X1)

R2. Hazard, D. L.; "Peculiar Magnetic Disturbances in December, 1908," Terrestrial Magnetism and Atmospheric Electricity, 14:37, 1909. (X2)

R3. Faris, R. L.; "The Peculiar Magnetic Disturbances of December 28-31, 1908," Terrestrial Magnetism and Atmospheric Electricity, 16:13, 1911. (X2)

R4. Brush, Judson; "Magnetic Disturbance March 1," Popular Astronomy, 20:254, 1912. (X3)

R5. Gordon, A. L.; "Magnetic Disturbance," Marine Observer, 6:261, 1929. (X4)

R6. Wilde, H. J.; "Magnetic Disturbance," Marine Observer, 15:6, 1936. (X5)

R7. Sell, R.; "Abnormal Magnetic Variation," Marine Observer, 23:148, 1953. (X6)

R8. Malin, S. R. C., and Hodder, B. M.; "Was the 1970 Geomagnetic Jerk of Internal or External Origin?" Nature, 296: 726, 1982. (X7)

R9. Whaler, K. A.; "Geomagnetic Impulses and Deep Mantle Conductivity," Nature, 306:117, 1983. (X7)

GEZ2 The Effect of the Moon on the Geomagnetic Field

Description. The variation of the geomagnetic field with lunar position relative to the earth, as measured by changes in the intensity and/or direction of the terrestrial field or the general level of geomagnetic activity.

Data Evaluation. Although a large quantity of geomagnetic data has been amassed from all over the world, its statistical analysis indicates only a weak lunar effect. Even this slight correlation is disputed by some. Rating: 3.

Anomaly Evaluation. Since the moon periodically penetrates the earth's magnetic tail, small perturbations of the geomagnetic field at the earth's surface are not surprising, although the precise physical mechanisms are not completely clear. Rating: 3.

Possible Explanations. Disturbances caused by the moon's passage through the earth's magnetic tail may propagate up the tail and be detected by surface instruments.

Similar and Related Phenomena. Lunar effects on meteor flux (AY); the moon and the weather (GWS1).

Examples of the Effect of the Moon on the Geomagnetic Field

X1. Statistical analysis. Compass deflections, 1840-1845, in Philadelphia. "The effect is very small, scarcely much beyond the probable error; but the table indicates that the north end of the magnet is deflected to the westward 0'.1 at the full and as much to the eastward at the day of new moon, the range between full and new moon being 0'.2." (R1) Studies a century or so later have been scarcely more definitive! (WRC)

X2. Statistical analysis. Relation of magnetic storms to lunar phase, 1882-1884, England. "For some time I have known how closely magnetic storms seemed to approximate to lunar phases, but have refused to admit the evidence because of the distance between sun and earth. Since the 30th ultimo I have placed the storms of 1882-83-84 under further scrutiny, and the facts as to coincidence are absolute. The storms of those years numbered 40, of which 18 were at new and 14 at full moon. The doubling or duplication of spots is well known, and if similar doubling be permitted in these storms, the alliances reach 38. The question at issue is a remarkable one, because the solar synodic period and our moon's lunation period are each nearly 27 days. All the years show a similar feature, and in spite of prejudice one seems compelled to admit some sort of effect by earth and moon upon the sun...." (R2) Of course, it is the sun which influences the earth. The coincidence between lunar and solar periods also confused more recent studies reported below. (WRC)

X3. Statistical analysis. A study, similar to that of X2, in which the great magnetic storms from 1880-1909 are shown to be related to lunar phase. (R3)

X4. Statistical analysis. Level of geomagnetic activity in Finland. "Abstract. The geomagnetic activity in Sodanklya during some 10 years shows a fairly systematic double-waved variation in a period corresponding to the synodic revolution of the moon, the chief activity minimum occurring during the new moon. A decrease in the activity also appears at the times of the inferior conjunctions of the planets Mercury and Venus. This fact suggests that all these celestial bodies are surrounded by a magnetic field." (R4) See also GEZ4.

X5. Statistical analysis. Geomagnetic disturbances, Australia. "Abstract. It is shown that occurrences of geomagnetic disturbances of various intensities are not uniformly distributed in lunar phase. There is a tendency for storms to occur preferentially near first and third quarters and to avoid dates corresponding to new moon." (R5)

"Abstract. A decrease in frequency of magnetic storms at new moon which was previously demonstrated has been further analyzed and shown to apply only to the 'gradual-commencement' type of storm, whereas sudden-commencement storms show an increased frequency at that time. Suggestions that Venus and Mercury influence magnetic disturbances near their inferior conjunctions are verified and extended, and it is shown that the position of the moon at these times is also important. The possibility that the 'blue clearings' of Mars are related phenomena is also considered." (R6)

X6. Statistical analysis. K_p index of geomagnetic activity. "Abstract. The behavior of the daily K_p index of magnetic activity is investigated as a function of lunar phase by the method of superposed epochs. The analysis indicates a slight but statistically significant enhancement of geomagnetic disturbance during several days following full moon and, with marginal significance, a slight diminution of disturbance during several days preceding full moon. A small dip in K_p, found at the phase of new moon, is shown to be without statistical significance." (R7)

X7. Statistical analysis. "Abstract. The geomagnetic daily planetary indices from 1932 to 1961, and also the starting days of great magnetic storms from 1840 to 1954, are examined for possible dependence on lunar phase. It is concluded, in agreement with Bartels but contrary to the findings of Bigg, that there is no evidence of any such dependence." (R8)

X8. Statistical analysis. Another study reaching the same negative conclusions as those in X7. (R9)

X9. General observations. "Abstract. Difficulties in testing the statistical significance of lunar effects on the frequency of severe magnetic disturbances are discussed. It is shown that a valid case has not yet been made that the effects are due entirely to chance. The theme that statistical methods should be used as an aid to physical understanding and not as a substitute for it is developed." (R10) This statistical impasse resembles that encountered in the attempts to demonstrate a solar influence upon the weather (GWS). (WRC)

X10. Statistical analysis. Abstract. An analysis of 31 years of K_p data suggests a variation of geomagnetic disturbance with lunar phase. A general increase in geomagnetic activity of about 4% begins after full moon and lasts for seven days. A general decrease in geomagnetic activity of about 4% is found for the seven days preceding full moon. A study of randomized data indicates that the probability that these variations should have occurred by chance is less than 5%. The effect is found to be associated with the K_p data derived from periods of geomagnetic quiet conditions; it is not evident in the data from disturbed periods." (R11) This analysis was extended to show that the variation requires that lunar latitudes at full moon be within about 4° of the ecliptic plane. (R12)

X11. Statistical analysis. Abstract. Evidence is presented indicating that two factors play a part in the lunar modulation of geomagnetic activity: (1) Phase of the moon: the average level of geomagnetic activity is diminished for several days before full moon and enhanced for several days after full moon. (2) Celestial latitude of the moon: activity is diminished when the moon is more than 4° from the plane of the ecliptic and enhanced when the moon is closer to the plane. Approximately from first quarter to last quarter, when these two factors work together the greatest amplitude of modulation occurs; when they work against each other no significant modulation of geomagnetic activity is detected. There is no evidence of any significant lunar influence on geomagnetic activity over the new-moon half of the orbit. These results are discussed in terms of recent observations of the earth's magnetic tail." (R13) About this time, satellite investigations were revealing possible physical mechanisms for lunar influence, taking the problem out of the realm of astrology. (WRC)

Other researchers maintained that the periodicity of geomagnetic activity attributed to the moon, as claimed above, was actually due to solar activity, which has a similar period. (R14) The authors of R13 replied, defending their results and conclusions.(R15)

X12. Statistical analysis. Anomalously large variations in the geomagnetic field appeared to be correlated with daily positions of the sun and moon for some Australian stations. (R16)

X13. Statistical analysis. Abstract. "Micropulsation records obtained at California stations during a 6-year period, from August 1, 1960, to July 30, 1966, have been analyzed for a possible full-moon effect on hydromagnetic emission activity..... In all cases there was an indication of an enhancement of hydromagnetic emission activity commencing the day before the full moon and, at least in the case of storm days, extending into the full moon day. A full-moon effect on the K_p index has been previously suggested but, as in the present case, there were insufficient statistics to establish the effect beyond doubt. This new result for hydromagnetic emissions provides additional evidence for a small lunar disturbance of the magnetosphere around the time of the full moon." (R17)

References

R1. Bache, A. D.; "Abstract of a Discussion of the Influence of the Moon on the Declination of the Magnetic Needle.....," American Journal of Science, 2:31:98, 1861. (X1)

R2. Adams, Alex. J. S.; "Magnetic Storms and the Moon," British Astronomical Association, Journal, 15:106, 1904. (X2)

R3. Henry, John R.; "Large Magnetic Storms in Relation to Certain Longitudes of the Moon," English Mechanic, 103:90, 1916. (X3)

R4. Sucksdorff, E.; "The Influence of the Moon and the Inner Planets on the Geomagnetic Activity," Geophysica, 5:95, 1956. (X4)

R5. Bigg, E. K.; "The Influence of the Moon on Geomagnetic Disturbances," Journal of Geophysical Research, 68:1409, 1963. (X5)

R6. Bigg, E. K.; "Lunar and Planetary Influences on Geomagnetic Disturbances," Journal of Geophysical Research, 68:4099, 1963. (X5)

R7. Bell, B., and Defouw, R. J.; "Concerning Lunar Modulation of Geomagnetic Activity," Journal of Geophysical Research, 69:3169, 1964. (X6)

R8. Davidson, T. W., and Martyn, D. F.; "A Supposed Dependence of Geomagnetic Storminess on Lunar Phase," Journal of Geophysical Research, 69:3973, 1964. (X7)

R9. Michel, F. C., et al; A Search for Correlation between K_p and the Lunar Phase," Journal of Geophysical Research, 69:4177, 1964. (X8)

R10. Bigg, E. K.; "Lunar Influences on the Frequency of Magnetic Storms," Journal of Geophysical Research, 69:4971, 1964. (X9)

R11. Stolov, Harold L., and Cameron, A. G. W.; "Variations of Geomagnetic Activity with Lunar Phase," Journal of Geophysical Research, 69:4975, 1964. (X10)

R12. Stolov, Harold L.; "Further Investigations of a Variation of Geomagnetic Activity with Lunar Phase," Journal of Geophysical Research, 70:4921, 1965. (X10)

R13. Bell, Barbara, and Defouw, Richard J.; "Dependence of the Lunar Modulation of Geomagnetic Activity on the Celestial Latitude of the Moon," Journal of Geophysical Research, 71:951, 1966. (X11)

R14. Rassbach, M. E., and Dessler, A. J.; "The Lunar Period, the Solar Period, and K_p," Journal of Geophysical Research, 71:4141, 1966. (X11)

R15. Bell, Barbara, and Defouw, Richard J.; "Discussion of Paper by M. E. Rassbach, A. J. Dessler, and A. G. W. Cameron,..." Journal of Geophysical Research, 71:5770, 1966. (X11)

R16. Malin, S. R. C., and Winch, D. E.; "Anomalous Luni-Solar Daily Variation in the Geomagnetic Field at Watherloo, Western Australia," Nature, 218:941, 1968. (X12)

R17. Fraser-Smith, A. C.; "A Possible Full-Moon Enhancement of Hydromagnetic Emission Activity," Journal of Geophysical Research, 74:2987, 1969. (X13)

GEZ3 Effects of Solar Eclipses on Geomagnetism

Description. The disturbance of the terrestrial magnetic field by solar eclipses.

Data Evaluation. Specific experiments and statistical analyses are practically nonexistent. One rather old experiment reports a positive effect, but with no detail. A more recent statistical study also confirms the reality of the effect. Rating: 3.

Anomaly Evaluation. Since the moon's shadow on the earth's upper atmosphere will affect ionization levels, a slight magnetic perturbation is not as startling as it first might appear. Rating: 3.

Possible Explanations. The sudden change in solar illumination changes ionization levels in the upper atmosphere and current flow. A magnetic perturbation is a possible consequence. Also---though less likely---temperature changes at the earth's surface might af- earth currents, which would then modify the magnetic field.

Similar and Related Phenomena. The lunar influence on the geomagnetic field (GEZ2); the effect of solar eclipses on torsion pendulums (AS); the so-called eclipse wind.

Examples of the Effects of Solar Eclipses on Geomagnetism

X1. February 1934. Pacific Ocean. "'Commander Akiyosi's apparatus,' Dr. Johnson relates, 'in principle at least, consisted of a small mirror attached to the needle of a compass. A narrow beam of light was directed upon this mirror, from which it was reflected to a sensitized paper slowly moving over a revolving drum. Any changes in position of the needle were easily detectable. During the eclipse the needle rotated slightly, the north end coming to the east by about one minute of arc, then going back to its original position at the end of totality. It seems strange that an eclipse should affect the compass but such was actually the case.'" (R1)

X2. Statistical analysis. 164 solar eclipses between 1906 and 1976. "The large, persistent conditional probabilities, however, require some sort of geomagnetic reverberation effect between the Earth and Moon." (R2) I. e., a large effect was noted.

References

R1. "Compass Needle Swings Aside during Eclipse," Science Newsletter, 26:167, 1934. (X1)

R2. Meisel, David D., and Blazejewski, Anthony F.; "Search for Worldwide Solar Eclipse Perturbations of Geomagnetic Activity 1906-1976....," Journal of Geophysical Research, 84A:2539, 1979. (X2)

GEZ4 The Effects of the Planets on the Geomagnetic Field

Description. Changes in the strength and/or direction of the geomagnetic field or a general change in the level of geomagnetic activity correlated with the positions of one or more of the major planets relative to the earth. Mercury, Venus, and Jupiter are the planets that supposedly influence the geomagnetic field.

Data Evaluation. Data exist in profusion; but their statistical analysis seems to lead to conflicting results. The reality of this phenomenon, therefore, cannot be said to be assured. Rating: 3.

Anomaly Evaluation. Planetary influence upon the geomagnetic field does not suggest astrology at all. Interior planets with strong magnetic fields might well shield the earth from some of the solar wind at inferior conjunctions. Exterior planets, of course, could not do this. It turns out that neither Mercury nor Venus seem to have sufficiently strong magnetic fields to disturb the geomagnetic field at conjunctions. Therefore, if this phenomenon is a real one; it is not readily explained. Rating: 2.

Possible Explanations. Shielding of the earth by the interior planets may be more effective than anticipated. Also, on another track, the activity of the geomagnetic field is strongly influenced by solar activity, which, in turn, is purported to be affected by the positions of the planets. If such is true, planetary effects on geomagnetic activity would be merely secondary.

Similar and Related Phenomena. The possible influence of planetary positions on solar activity (AS); the correlation of terrestrial weather with planetary positions (GWS8).

Examples of the Effects of the Planets upon the Geomagnetic Field

X1. Statistical analysis. "Venus has a magnetic field five times as strong as the earth's at a distance of about eight and a half million miles from both planets. Dr. J. Houtgast of Utrecht, The Netherlands, has estimated. Although scientists have speculated for years about the possibility of a magnetic field for earth's sister planet, Dr. Houtgast is the first to suggest its numerical value. His estimate is based on the marked drop in magnetic activity 'from seven days before to one day after' Venus comes between the sun and earth. He studied records of changes in the earth's magnetic activity covering the period 1884-1953, during which there were 44 times when Venus was lined up between the earth and sun. At this time, the planet acts as a shield, deflecting some of the solar particles responsible for magnetic activity." (R1) Venus' magnetic field is no where near this strong, but the observations of geomagnetic disturbances should be valid. (WRC)

X2. Statistical analysis. Finnish geomagnetic records show a decrease in activity at the times of inferior conjunctions of the planets Mercury and Venus. (R2)

X3. Statistical analysis. "Suggestions that Venus and Mercury influence magnetic disturbances near their inferior conjunctions with the earth and verified and extended, and it is shown that the position of the moon at these times is also important." (R3)

X4. Statistical analysis. "Statistical evidence indicates that the positions of the Moon, Mercury, and Venus affect magnetic activity frequency observed at the Earth, and the position of the Earth affects the frequency of blue clearings on Mars. This study shows that these effects may be explained as a result of the action of shock and bow waves formed by these bodies in the supersonically streaming interplanetary plasma. The attenuation of large kinetic energy variations in the streaming plasma behind such bodies is shown to be equal to the square of the ratio of the Mach number upstream to the Mach number downstream. For typical solar induced activity, this implies an attenuation coefficient of approximately 1/2-1/3. It is also shown that an activity increase is expected in the bow wave. The observational data fit a model with bow waves of Mach numbers 2.5 and 15 corresponding to the two waves predicted by the theory. The Moon's effect varies from that of the planets in a manner that can be explained by its closeness to the magnetosphere." (R4)

X5. Statistical analysis. "Our study has shown that we experience a maximum in geomagnetic disturbance when Venus is at inferior conjunction during times of ashen light occurrence. This result is in direct contrast to previous studies which have shown a general decrease in geomagnetic activity at inferior conjunction for times of non-ashen light occurrence (since the ashen light is observed about a dozen times a year its frequency of occurrence is negligible when compared with its frequency of non-occurrence." (R5) The ashen light of Venus is a soft luminance of the entire planet. See AV. (WRC)

X6. General observations. Comments on the methods used in correlating terrestrial magnetic activity with planetary positions. No specific results presented. (R7)

References

R1. "Venus' Magnetic Pull Five Times Earth's," Science News Letter, 67:281, 1955. (X1)

R2. Sucksdorff, E.; "The Influence of the Moon and the Inner Planets on the Geomagnetic Activity," Geophysica, 5:95, 1956. (X2)

R3. Bigg, E.K.; "Lunar and Planetary Influences on Geomagnetic Disturbances," Journal of Geophysical Research, 68: 4099, 1963. (X3)

R4. Atkinson, Gerald; "Planetary Effects on Magnetic Activity," American Geophysical Union, Transactions, 45:630, 1964. (X4)

R5. Levine, Joel S.; "On the Possibility of Auroral Activity on Venus," Strolling Astronomer, 19:149, 1965. (X5)

R6. Nelson, John H.; "Certain Planetary Angles Related to Magnetic Storms," Cycles, 30:129, 1979. (X6)

R7. Maxey, E. Stanton; "An Analysis of the Terrestrial Magnetic Correlates of Venus/Jupiter Critical Angles, 1868-1980," Cycles, 34:159, 1983. (X6)

GEZ5 Meteor Activity Correlated with Geomagnetic Activity

Description. Disturbance of the geomagnetic field by single meteors or changes in the general level of geomagnetic activity correlated with specific meteor showers.

Data Evaluation. Only two meteors have been reported to influence the geomagnetic field. Several statistical analyses of the effect of meteor showers on the geomagnetic activity are available, but overall they seem inconclusive. Rating: 3.

Anomaly Evaluation. Large, low-level meteors leave long, highly ionized trails behind them. Meteor showers, too, must change ionization levels in the upper atmosphere. Thus, correlations of meteor activity and geomagnetic activity, if ultimately proven, would not be particularly anomalous because a reasonable physical mechanism exists. Rating: 3.

Possible Explanations. Electrical current flows in meteor-created ionized regions perturb the geomagnetic field.

Similar and Related Phenomena. Earth currents correlated with large meteors (GEZ6). earth currents caused by auroral activity (GLA20).

Examples of Meteor Activity Correlated with Geomagnetic Activity

X1. July 30, 1868. Observatory of Itataya, Brazil. An immense detonating meteor. "This aerolite so disturbed the magnetic instruments that the declinometer turned its pole from the north toward the west and stuck itself in the box where it found resistance; the horizontal magnetometer turned toward the west eight divisions of the scale; the vertical magnetometer fell in its center of gravity, and finally, the compass oscillated 15° from north to west.....It is, therefore demonstrated for physics that an aerolite has an intense action upon the north pole of magnets, powerfully attracting them." (R1) The 1908 Tunguska Event produced similar, but lesser, effects (X6).

X2. Statistical analysis. From the analysis of geomagnetic activity data from several U.S. stations. "A striking feature of the activity levels thus far available is the occurrence of occasional large increases in the frequency band centered at 1.5 cps. In particular, the data from the Denver, Colorado, recording station show an increase in this band of about 5 or 6 times the usual level on the night of December 12-13, 1959. Since this is also approximately the expected time of the maximum of the Geminid meteor shower, a connection between the two events was suggested." (R2)

X3. Statistical analysis. "Abstract. Increased activity of magnetic micropulsations with periods of 5 to 30 seconds and magnetic flux densities of 20 to 320 mγ was found to accompany the η Aquarid, δ Aquarid, and Perseid meteor showers in 1958. Conflicting reports are discussed." (R3)

X4. Review of reports of this phenomenon. "Experimental work already carried out

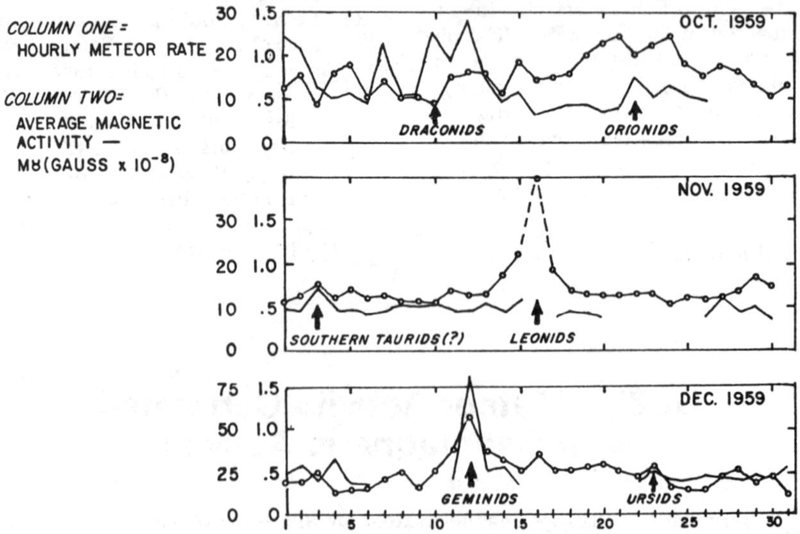

Meteor frequency (circles) compared with geomagnetic activity (X2)

thus gives no evidence of a magnetic effect due to a single meteor, with the possible exception of the first worker, Kalashnikov. The correlation to date with showers also lacks conviction...." (R4) The magnetic perturbations caused by the Tunguska Event must have been overlooked (X6). (WRC)

X5. Statistical analysis. "Abstract. Magnetic effects due to the passage of the Geminid meteors through the ionosphere were computed using an equation derived by Chapman and Ashour. The observed micropulsation activity at Jicamarca, Peru, and Kingston, Rhode Island, was compared with the predicted value. No striking similarity was found either in magnitude or phase." (R6)

X6. June 30, 1908. Irkutsk Observatory, Russia. Following the Tunguska Event. "Magnetic disturbances similar to those subsequently recorded after atmospheric nuclear explosions were recorded at the Irkutsk Observatory." (R7)

X7. General observations. Experiments and statistical analyses prior to 1965 were not well-adapted to detect the effect of meteors on geomagnetic activity. A discussion follows of how future studies should be carried out. (R5)

References

R1. "An Aerolite," Scientific American, 19: 276, 1868. (X1)

R2. Jenkins, Alvin W., Jr., et al; "Observed Magnetic Effects from Meteors," Journal of Geophysical Research, 65:1617, 1960. (X2)

R3. Campbell, Wallace H.; "Magnetic Micro-Pulsations Accompanying Meteor Activity," Journal of Geophysical Research, 65:2241, 1960. (X3)

R4. Ellyett, C.D., and Fraser, B.J.; "On the Correlation of Meteors with Micropulsations," Journal of Geophysical Research, 68:5937, 1963. (X4)

R5. Chapman, Sydney, and Ashour, Attia A.; "Meteor Geomagnetic Effects," Smithsonian Contributions to Astrophysics, 8:181, 1965. (X7)

R6. Green, Richard G.; "An Analysis of the Magnetic Effect from Shower Meteors," Journal of Geophysical Research, 72:2309, 1967. (X5)

R7. Hughes, David W.; "Tunguska Revisited," Nature, 259:626, 1976. (X6)

GEZ6 Terrestrial Electrical Effects Correlated with Meteors

Description. Abnormal electrical currents in conductors, humans reporting electrical shocks, St. Elmo's Fire, and other evidence of unusual atmospheric and terrestrial electricity correlated with the observation of meteors. Usually, this phenomenon involves a single, large, low-level meteor, which more often than not impacts nearby.

Background. As with many other startling phenomena reported in the Catalog of Anomalies, the surprise and shock experienced by the observer may engender psychological reactions erroneously reported as "electricity".

Data Evaluation. The data are soft: mainly testimonial and collected under trying conditions. Furthermore, St. Elmo's Fire and other manifestations of strong electrostatic fields are rather common and may be merely coincidental with the meteors. Rating: 3.

Anomaly Evaluation. A large, low-level meteor leaves behind an impressive trail of ionized gases, in which strong electrical currents may flow. Such currents could induce electrical phenomena at the surface. So, a reasonable physical mechanism can be found for this phenomenon. Rating: 3.

Possible Explanations. See above.

Similar and Related Phenomena. Geomagnetic effects of meteors (GEZ5); meteors correlated with auroral displays (GLA4).

Examples of Electrical Effects Correlated with Meteors

X1. November 13, 1833. Poland, Ohio, and Fredonia, New York. Near the time of the great meteor shower. "Dr. Kirkland, at Poland, Ohio, 'on retiring to rest, a little after 10 o'clock, discovered brilliant electrical sparks emitted from his clothes on any slight motion.' While riding in the town of Fredonia, (says a correspondent of the New York Daily Advertiser,) in the evening of the 14th inst. between 6 and 7 o'clock, the night after the meteoric display, the tips of the ears of my horse, for a half an inch in length, became luminous, and similar in appearance to phosphorescent bodies. It remained for some minutes." (R1) Such electrostatic phenomena are, of course, quite common during cold, dry weather and could well be just coincidental. (WRC)

X2. October 13, 1839. Cape of Good Hope, South Africa. A great fall of stones with much noise. "Many felt a curious sensation, especially about the knees, as if they had been electrified. At the time of the occurrence I was on the very skirts of its influence, on the edge of the Karoo, in company with the Hon. Mr. Justice Menzies. At the moment of the explosion I witnessed a volume of the electric fluid forcing its way from the west in the form of a Congreve rocket; it exploded almost immediately over my head, into apparent globules of fire, or transparent glass." (R2)

X3. January 1844. Argentina. During the fall of an iron meteorite. "At the same time I and my companions all agreed that we had experienced a violent electric shock; but probably this sensation may have been but the effect on our drowsy senses of the indescribably intense light and noise." (R3)

X4. July 7, 1883. Lachine, Quebec, Canada. A remarkable aerolite. "This brilliant body had a solid nucleus that appeared to the eye about four feet square, and a strange indescribable noise was caused by its flight through the air. Simultaneously, as it seemed to Mrs. Popham, she received a paralysing shock that affected her from head to foot, as if the entire contents of a highly-charged battery had been discharged into her body at once.... Mr. Horace Baby also saw the glare caused by the flight of the meteor, although he did not actually see the body itself. He said that he felt a tremendous shock, and that he could feel the electricity oozing out of his finger-ends for some time after." (R4, R5) Such subjective reactions may not actually be due to electricity. Psychological reactions to such a startling event could be very similar. (WRC)

X5. February 24, 1885. North Pacific Ocean. On the bark Innerwich. "All at once a large mass of fire appeared over the vessel, completely blinding the spectators; and, as it fell into the sea some fifty yards to the leeward, it caused a hissing sound, which was heard above the blast, and made the vessel quiver from stem to stern. Hardly had this disappeared, when a towering mass of white foam was seen rapidly approaching the vessel. The noise from the advancing volume of water is described as deafening. The bark was struck flat aback; but, before there was time to touch a brace, the sails had filled again, and the roaring white sea had passed ahead. To increase the horror of the situation, another vast sheet of flame ran down the mizzen-mast, and 'poured in myriads of sparks' from the rigging." (R6)

X6. November 26, 1919. Lake Michigan. "Witnesses saw the great ball of fire, white and orange-colored, which illuminated the whole country, pass over the lake south of Grand Haven. It fell into the waters, creating a fog of steam clouds. Telegraph operators of the New York Central lines reported wire trouble as far north as Grand Rapids, Mich. The power plant of the Calumet Electric Light Company at Kalamazoo was put out of commission, and another plant near Berrien Springs was disabled." (R7)

X7. April 30, 1927. Gulf of Mexico. An exceptionally bright meteor seemed to fall between the ship and the horizon. Wireless operators reported unusually severe static. (R8)

X8. March 24, 1933. Adrian, Texas. "The passing of the meteor created an electrical effect, which appeared in the radio set of the pilot, resembling frictional static sometimes caused by dust in the air. This effect lasted for a little while after the meteor had passed." (R9)

X9. No date given. Near Vladiovostok, Russia. "During the fall of one of the largest bolides, near Sikhote-Alin, near Vladiovostok (USSR), an electrician on a telephone pole received a strong electric shock from disconnected wires at the instant the bolide became visible. The shock may have been due to other causes, but the possibility of strong electromagnetic effects is not ruled out." (R10)

References

R1. Olmsted, Denison; "Observations on the Meteors of November 13th, 1833," American Journal of Science, 1:25:363, 1834. (X1)

R2. Thompson, G.; "Fall of Meteorites at the Cape of Good Hope," Philosophical Magazine, 3:14:391, 1839. (X2)

R3. Greg, R. P.; "Fall of a Large Mass of Meteoric Iron at Corrientes in South America," Philosophical Magazine, 4:10:12, 1855. (X3)

R4. "An Extraordinary Meteor," Symons's Monthly Meteorological Magazine, 18:120, 1883. (X4)

R5. Claypole, E. W.; "The Lachine Aerolite," Nature, 28:319, 1883. (X4)

R6. Science, 5:242, 1885. (X5)

R7. "A Celestial Short-Circuit," Literary Digest, 64:27, February 28, 1920. (X6)

R8. Hague, J. W.; "Meteors," Marine Observer, 5:61, 1928. (X7)

R9. Ellis, R. W.; "Observation of a New Mexico Meteor from the Air," Science, 78:58, 1933. (X8)

R10. Condon, Edward U.; Scientific Study of Unidentified Flying Objects, New York, 1969, p. 746. (X9)

GEZ7 Geomagnetic Disturbances Correlated with Stellar Activity

Description. Perturbations of the terrestrial magnetic field correlated with stellar activity, such as flares.

Data Evaluation. A single observation which is weakened by the nearly simultaneous appearance of a solar flare. Rating: 3.

Anomaly Evaluation. While solar-flare X-ray fluxes are strong enough to ionize the earth's upper atmosphere and, in turn, perturb the geomagnetic field; it would be surprising to find this effect significant at stellar distances. Rating: 2.

Possible Explanations. X-rays from nearby stars (including the sun) increase ionization levels and currents in the upper atmosphere, which then disturb the geomagnetic field.

Similar and Related Phenomena. None.

Examples of Geomagnetic Disturbances Correlated with Stellar Activity

X1. October 15, 1967. Kodaikanal, India. "We have examined the possibility that an X-ray flare from Sco X-1, detected in the course of a balloon flight on October 15, 1967 at Mildura (34° 12'S; 142° 6'E), had a daytime geomagnetic effect. We observed a conspicuous perturbation in the form of an increase in the horizontal component of the magnetic field observed at the equatorial station at Kodaikanal (10° 14'N, 77° 28'E). On October 15, 1967, Sco X-1 flared up at about 0700 UT, reaching a maximum in less than 10 min and then decreasing during the next 20 min. The perturbations in the horizontal component of the magnetic field started around 0715 UT, and rose to a maximum at 0725 UT (an increase in the H component of about 14γ), about 16 min after the flare had reached a maximum at around 0710 UT. Unfortunately, a small optical solar flare occurred about 15 min before the Sco X-1 flare, during the period 0647-0710 UT, with a maximum at 0650 UT." Further study of the data led the researchers to the conclusion that the Sco X-1 flare was more likely to be the cause of the geomagnetic perturbation. (R1) Obviously, the solar flare muddied the waters considerably. (WRC)

References

R1. Sastri, J. Hunamath, and Murthy, B. Suryanarayana; "Geomagnetic Effect Associated with an X-Ray Flare from Sco X-1," Nature, 257:35, 1975. (X1)

GEZ8 Gravity Waves Correlated with Geomagnetic Storms

Description. The positive correlation of signals from gravitational wave detectors with geomagnetic disturbances.

Background. The reality of this phenomenon would not necessarily imply a causal connection between gravitational waves and geomagnetism; rather it would suggest that gravitational wave detectors do not really detect the hypothetical gravitational waves at all but are actually acting as magnetometers. Although theory predicts gravitational waves, no one has shown conclusively that they exist. The true nature of the signals from gravitational wave detectors has been the subject of much heated controversy.

Data Evaluation. Several statistical analyses show strong correlations between signals from gravitational wave detectors and geomagnetic storms. Rating: 2.

Anomaly Evaluation. No one believes that the data cited below proves that gravitational waves and magnetism are physically related---such a discovery would be a first-class anomaly. Instead, the reality of this correlation almost certainly means that gravitational wave detectors are also magnetometers, although not designed as such. In this interpretation, no anomaly exists. Rating: 4.

Possible Explanations. The metal supports or other features of gravitational wave detectors are sensitive to magnetic fields.

Similar and Related Phenomena. None.

Examples of Supposed Gravity Waves Correlated with Geomagnetic Storms

X1. General observations. "Exactly 4 years ago the first report of gravitational radiation was published, and a new field of research was established. Since then many observers have built devices for detecting gravity waves, many have sought to explain what astronomical bodies could produce such waves, and a few have asked what the signals could be, if not gravitational waves from space. The original report was made by Joseph Weber at the University of Maryland, College Park, and he is still detecting coincident signals between detectors at College Park and Argonne National Laboratory outside Chicago.

About 1 year ago a team of Russian researchers found a correlation between some of the events reported by Weber and the planetary geomagnetic activity index K_p. This index is the mean of measurements of fluctuations of the earth's magnetic field at 12 locations and is primarily an indicator of auroral activity. The Russian paper suggested the possibility that pulsations of the earth's magnetic field could have caused the signals reported as gravitational radiation, but the analysis was based on a very small sample of data, namely, the 17 events originally reported by the group at Maryland."

A sample of 262 gravitational radiation events was next analyzed by a group at Bell Laboratories. The geomagnetic correlation did not disappear, increasing the likelihood that the gravitational wave detectors were somehow detecting changing magnetic fields rather than gravitational waves. (R1)

References

R1. Metz, William D.; "Gravity Waves: Correlation with Geomagnetic Storms," Science, 180:1161, 1973. (X1)

TIME-OF-EVENT INDEX

PLACE-OF-EVENT INDEX

FIRST-AUTHOR INDEX

SOURCE INDEX

SUBJECT INDEX